## Themenhefte

SCHWERPUNKTPROGRAMM **UMWELT**
SCHWEIZ. NATIONALFONDS ZUR FÖRDERUNG DER WISSENSCHAFTLICHEN FORSCHUNG
PROGRAMME PRIORITAIRE **ENVIRONNEMENT**
FONDS NATIONAL SUISSE DE LA RECHERCHE SCIENTIFIQUE
PRIORITY PROGRAMME **ENVIRONMENT**
SWISS NATIONAL SCIENCE FOUNDATION

# Environmental Policy Between Regulation and Market

C. Jeanrenaud (Ed.)

Birkhäuser Verlag
Basel · Boston · Berlin

Editor

Prof. Dr. Claude Jeanrenaud
Université de Neuchâtel
Institut de Recherches Economiques et Régionales (IRER)
Pierre-à-Mazel 7
CH-2000 Neuchâtel

A CIP catalogue record for this book is available from the Library of Congress, Washington D. C., USA

Die Deutsche Bibliothek - CIP-Einheitsaufnahme

**Environmental policy between regulation and market/**
C. Jeanrenaud (ed.). - Basel ; Boston ; Berlin : Birkhäuser, 1997
(Themenhefte SPP Umwelt)
ISBN-13:978-3-7643-5319-3 e-ISBN-13:978-3-0348-9012-0
DOI:10.1007/978-3-0348-9012-0

NE: Jeanrenaud, Claude [Hrsg.]



Camera-ready copy prepared by the editor
Cover design: Markus Etterich, Basel
Printed on acid-free paper produced from chlorine-free pulp. TCF ∞
ISBN-13:978-3-7643-5319-3
ISBN-13: 978-0-8176-5319-4

9 8 7 6 5 4 3 2 1

# Table of Contents

# Foreword

Economics certainly has an important role to play in the design of environmental policies. Programmes that rely on market-based incentives generally offer better efficiency and effectiveness. Market-based instruments are more flexible and give the agents directly concerned greater freedom to choose means, thus making it possible to reduce abatement costs and achieve environmental targets faster and more reliably. Yet the obstacles encountered during the implementation phase must not be underestimated. Therefore, achieving cost-savings and environmental benefits requires carefully designed programmes.

The Priority Programme «Environment» of the Swiss National Science Foundation has devoted part of its funds to analysing environmental protection and sustainable development from an economic point of view. Four projects focused more specifically on the role of market-based instruments in environmental policy. It is possible to discourage behaviour which represents a threat to the environment by means of prohibitions – the traditional approach – or by market-based incentives. Each project dealt with a particular instrument – charges, tradeable permits, covenants and ecological fiscal reform – to be evaluated from the standpoint of its ability to reduce abatement costs, its capacity to achieve environmental targets and its acceptability.

A team from the University of Geneva, headed by *Beat Bürgenmeier*, looked into incentive taxes. In this connection, it compared the effectiveness of a fuel tax with that of a tax on new cars. It further conducted a survey in order to determine how taxes are perceived and accepted by economic agents. The study by *Klaus A. Vallender*, from the University of St. Gallen, covered ecological tax reform. It was aimed at determining whether the tax system should be supplemented, overhauled or even reformed more radically to ensure that individuals and industry adopt environmentally friendly behaviour. The Institute for Economic and Regional Research of the University of Neuchâtel, headed by *Claude Jeanrenaud*, focused on tradeable permits, instruments which have been little used to date in European countries. In the main body of the research project, the authors simulated the functioning of a marketable permit system for $NO_x$ in a region offering favourable conditions for implementation and estimated the potential cost-savings for companies in the area if emission trading was authorized. The work of the fourth team, headed by *Ernst A. Brugger* and *Barbara Gorsler* (Beratungen Hanser and Partner, Zurich), pertained to voluntary agreements between industry and regulators, as well as the advantages of this type of instrument compared with other strategies.

In June 1995, on the occasion of a conference held at the University of Neuchâtel, representatives of the different teams had an opportunity to present their research findings. Foreign keynote speakers were invited to comment on the work done within the framework of the Priority Programme «Environment». They also had a brief to take stock of experience gathered in their countries with market-based instruments. These contributions are also included in this volume.

*Alberto Majocchi*, from the University of Pavia, presents the conclusions of the Italian White Paper on Fiscal Reform, whose recommendations run along the lines of an ecological tax reform. *Albert L. Nichols*, of National Economic Research Associates, Cambridge, Mass., a well-known expert in the field of environmental policy, analyses experiments with tradeable permit systems designed to reduce $NO_x$ emissions in the northeastern part of the US. He demonstrates that careful design of trading programmes is vital to the success of the operation. The article by *Peter Bohm* (University of Stockholm) shows that tradeable permits have an important role to play in international agreements aimed at averting global threats. The Basle region was the first in Europe to introduce a permit market for atmospheric pollutants. *Elke Staehelin-Witt* (B,S,S. Economic Consulting, Basle) describes this experiment and explains why the results failed to live up too the expectations. An important albeit rather controversial aspect of ecological fiscal reform – that of the double dividend – is taken up by *Stefan Proost* and *Denise Van Regemorter*, of the Catholic University of Leuven. The Netherlands have acted as pioneers in the implementation of covenants, as the first agreements were signed in that country over ten years ago. *Wim Hafkamp* (Erasmus University, Rotterdam), draws the lessons from the some thirty voluntary agreements concluded between the Dutch government and industry.

Two contributions pose the theoretical framework: one, by *Claude Jeanrenaud*, concerns market-based instruments as a whole, while the other, authored by *Philippe Thalmann* of the Swiss Federal Institute of Technology, deals exclusively with incentive taxes. Experience shows that economic instruments are often superimposed on regulatory instruments rather than replacing them. From this perspective, it is easier to understand how important it is to find the right policy mix. In his contribution, *Hans Vos*, the co-author of the recent OECD study on economic instruments, examines the relationship between direct regulation and incentive-based regulation. By way of conclusion, *Anyl Markandya* gives us his thoughts on the lessons which can be drawn from theoretical work on market-based instruments and from their utilization in the environmental policies of the industrialized countries. He reminds us that implementation-related difficulties must not be underestimated and that it is often necessary to choose between efficiency and administrative simplicity.

Our thanks go to the Swiss National Science Foundation, which funded the four projects dealing with economic instruments. We are particularly grateful to Dr. Rudolf Haeberli, Director of the Priority Programme «Environment», for his constant encouragement throughout our work and for his support with regard to the writing of this book. Our sincere gratitude goes to the main research collaborators, *Reto Jacobs* (University of St. Gallen), *Marc-Alain Stritt* (University of Neuchâtel) and *Nicolas Wallart* (University of Geneva), all of whom spared no effort to ensure that this project was brought to a successful completion.

Neuchâtel, August 1996

Claude Jeanrenaud

# *1. Introduction*

Environmental Policy Between Regulation and Market
ed. by C. Jeanrenaud


# Economic Instruments for Environmental Policy

**Claude Jeanrenaud**
**University of Neuchâtel, Neuchâtel, Switzerland**

## 1. Environmental policy - trying to catch its second wind

From the mid-70s up until the beginning of the present decade, environmental policy benefited from a particularly favourable climate. New programmes enjoyed practically unlimited popular support, and moved easily through the different stages up to implementation. Even the tightening of standards at close intervals did not lead to protests from the economic circles consulted -- at least not openly. Politicians found it difficult to resist the adoption of more stringent measures, even when they might have had some doubts as to the usefulness of the programmes. The cost of the regulations for the economy was not a subject of discussion; moreover, there was no information available in that connection. The only concern was reaching environmental targets, whatever the cost. All these measures led to a significant improvement in environmental quality, but their limits have become clear today.

Since the early 90s, there has been a change in the way in which the different groups of actors perceive environmental problems. With the recession and persistently high unemployment rates, creating jobs and maintaining living standards are viewed as priorities whereas threats to the environment appear more remote.

Over the past two decades, significant progress has been made with regard to air and water protection. Attention has shifted towards the danger of a change in the overall climate. However, the steps needed to stabilize emissions are encountering stiffer resistance. They are very costly, the dangers appear remote and uncertain, and in any event action taken within a small country only represents a symbolic contribution to solving a world problem. Lastly, there is a real threat to the competitiveness of energy-intensive industries.

Accordingly, the attitude of industry has changed. Any new measures which impose additional costs on manufacturers come in for close scrutiny indeed and are no longer accepted without resistance. Moreover, steps which are not limited to making companies bear abatement costs meet with stiff opposition. Proof of this is provided by the tax on heavy lorries, the $CO_2$ tax or the fertilizer tax.[1] It is also open to question whether economic circles' interest in voluntary agreements might not also reflect their desire to play a more active role in the shaping of environmental policy and to place limits on the power of the administration... The financial difficulties of public bodies further hinder them from becoming involved in environmental protection programmes.

Do economic instruments offer a means of breaking the current environmental policy deadlock, thereby enabling such policy to catch its second wind? Isn't there a need for greater boldness and imagination when it comes to designing programmes? Lastly, if the principle of subsidiarity were applied better, giving the cantons greater leeway, wouldn't this make it easier to reach targets? These are the questions which we shall endeavour to answer in the pages which follow.

---

[1] It will take larger than initially planned to introduce these taxes. The bill on a $CO_2$ tax and a tax on fertilizers were rejected by the Swiss Federal Parliament. A new $CO_2$ draft bill will shortly be submitted to the bodies concerned.

## 2. Design of environmental policies

*When the invisible hand fails*

For the economist, damage to the environment constitutes a negative externality, i.e. costs imposed on others without compensation. In this case, the market is no longer able to fulfill its role, which consists of informing the party responsible for the damage of the costs caused by his behaviour and persuading him to take this into account in his decisions.

Natural resources also possess the characteristics of public goods, i.e. goods which we enjoy in common, with no possibility of rationing utilization through pricing. The natural tendency is to behave as a free-rider, using natural resources without worrying overly much about their deterioration. Both externalities and public goods create situations that make it impossible for the market to function properly. When this happens, there are several possible interpretations which determine the choice of corrective policies:

- The fact that individuals do not take proper care of environmental resources reflects an educational deficiency. It is the State's duty to use information campaigns and social pressure to persuade individuals to modify their behaviour.
- The iron hand of the State must replace the defective market.
- Individuals must feel that they have to dip into their own pockets to cover the costs of the damage for which they are responsible; accordingly, they will take better care of nature.
- Natural resources lack the protection which would be provided by a private owner.

Each interpretation leads to a different environmental policy approach.

*Governmental response to market failure*

When it comes to preventing or remedying damage to the environment, governments may choose between several strategies and have access to a whole set of instruments, ranging from police action to steps which are better suited to the logic of the market. Three types of approaches may be differentiated:

- The first consists of allocating budgetary resources to public anti-pollution investments. The construction of plants for the treatment of waste water or noise-reducing walls in built-up areas along communication routes are two examples of steps which come under this first strategy.

- The second is characterized by the application of measures to modify the behaviour causing the damage. As natural resources are public goods, the aim is to reduce free-riding incentives by making agents aware of the negative impact of their acts on the environment. For example, having to pay a refuse tax encourages households to rely more on recyclable products and to try and sort refuse.
- The last approach consists to leaving it up to the actors concerned -- those responsible for the damage and the victims -- to negotiate a reduction in the damage on a voluntary basis. Here, the role of the State is limited to providing a clear definition of property rights. If these rights are granted to the victims, they may, if need be, assert them in court.

Based on the three above-mentioned approaches, it is possible to establish a typology of strategies and draw up a list of corresponding instruments.

Table 1. Design of environmental policy.

| Approaches | Strategies | Instruments |
|---|---|---|
| I Direct State intervention | | Investments in public projects<br>Research |
| II Behaviour modification | • Information and moral suasion | Information campaigns, education |
| | • Direct regulation (command-and-control) | Standards, permits, specifications, bans, zoning |
| | • Incentive-based regulation | Charges, tradeable permits, deposit-refund systems, joint implementation |
| | • Privatisation | Private property of environmental resources |
| III Liability and property-rights approaches (« Laisser-faire ») | | Allocation of property rights<br>Legal liability for damages |

*I Direct State intervention*

The State budget has been one of the primary sources of financing for environmental protection programmes. For obvious reasons, when emissions come from a great many low-intensity sources -- in particular households -- it is more effective for authorities to handle prevention or abatement action. Direct State intervention does not rule out the simultaneous application of economic instruments, such as refuse taxes or deposit-refund systems. Public spending on environmental protection accounted for 4.0% of total expenditure by the Swiss authorities (*Finances publiques en Suisse, 1995*). However, financial statistics underestimate the scope of such charges because expenditure to protect the environment appears in practically all areas of governmental policy without being counted as such. This is particularly true in the field of transport. According to certain estimates, half of all spending on the Swiss national road network is due to constraints linked to the protection of the landscape and the well-being of the residents affected. Now that they are facing difficult budgetary situations, most governments tend to opt for anti-pollution strategies that place the burden on private actors. However, since the cost of regulations is less strictly controlled than public investments, there is a greater risk that inefficient measures will be adopted (The Economist, 1996).

*II Behaviour modification*

We shall concentrate on strategies to change behaviour, as they are the ones which play the most important role in environmental policies. As we have seen, the root of the problem is the fact that environmental damage constitutes external costs which do not appear in the utility functions of those who cause such damage. Accordingly, there is a need to internalize these costs in order to persuade those responsible to give more consideration to the impact of their acts on the environment. There are several strategies to achieve this goal.

The first consists of *educating and informing* individuals, to make them aware of the consequences of their behaviour on the well-being of the population or on future generations. For example, energy-saving campaigns are to be viewed from this perspective. The moral pressure of the group reinforces the effects of the information policy. Thus, the idea is to persuade polluters to exercise self-discipline. If stronger moral pressure is desired, publishing the names of those who fail to comply with prevailing legislation is apparently a very effective and inexpensive means (cf. in this connection the list of fines published regularly by the Los Angeles Times). In 1986, the US Congress amended the Superfund law to require manufacturers to report the quantities of some 300 toxic chemicals they release each year. "Companies just want to get off the lists of top polluters" (Business Week, 1993). Moreover, it was this obligation to publish its emissions of toxic products for the first time in 1987 which led the company Monsanto to make an enormous effort, on a voluntary basis, to reduce its emissions of pollutants into the atmosphere. The fact that companies publish their environmental balance sheets on a voluntary basis is undoubtedly one of the results of this policy. The same applies to conventions freely concluded between companies and regulatory agencies.

The second strategy, *direct regulation*, also known as "command-and-control" (CAC), is the most common. It consists of influencing behaviour directly and continues to play a decisive role, *inter alia* in countries which have applied new, incentive-based strategies. As noted by the authors of the assessment on the use of economic instruments drawn up by OECD, in no country is there a tendency to replace the general approach based on constraint by a purely incentive approach (OECD, 1994). There is a wide range of available instruments: standards, permits, bans, zoning, etc. However, the two main means of action are emission standards, which limit the quantities or concentrations of pollutants which may be emitted, and restrictions concerning the use of specific equipment or procedures. Prescribing a technology offers the advantage of avoiding actual monitoring of emissions because it is enough to ensure that it has indeed been used. The limits to this approach appear when the emphasis is placed on the cost-effectiveness of policies. Simplicity is obtained at the cost of loss of flexibility, since the company is no longer in a position to choose the abatement technology it feels is most appropriate and least expensive. Yet it is usually more familiar with abatement technologies than the regulatory agencies. Moreover, in view of the fact that such technologies are evolving, banning a facility or a procedure rapidly becomes a source of rigidity.

The third strategy is based on the *introduction of incentives*. There are many such measures capable of modifying behaviour: charges, tradeable permits, deposit-refund systems, etc. All these instruments have one characteristic in common: they make the polluter aware of the opportunity cost of environmental resources and lead to the internalization of damage. The two main instruments are charges and tradeable permits.

A charge is said to be an incentive-based instrument when its sole aim is to modify behaviour and it is neutral from the point of view of the State budget. In this case, all revenue raised is refunded to the population in an appropriate form, in which connection we may speak of ecological tax reform (cf. in this connection Vallender and Jacobs in this volume). Granted, ecotaxes often play a fiscal role in addition to acting as an incentive, in which case they are not pure incentive charges. In its analysis of economic instruments, the OECD (1989) notes that out of 150 instruments identified, the proceeds from more than half ended up in the State's budget.

Tradeable permits are the second major category of incentive-based instrument. Rather than acting directly on prices, the government fixes the maximum quantity of emissions by giving companies a corresponding amount of permits. Each permit entitles the holder to emit a set quantity of pollutants, for example one ton. Companies have the option of buying or selling certificates (trading) or even holding on to them for future use (banking). Price is determined by supply and demand on the market. As long as the price of the permit is lower than the marginal abatement cost, the polluter prefers to use such rights rather than take technological measures to combat pollution.

It may be noted that there are certain analogies between the *joint implementation* used in the framework of international agreements and a system of tradeable permits. The Framework Convention on Climate Change concluded in 1992 offers such a possibility. It provides that a signatory State may achieve the target set for it by financing the reduction of greenhouse gases in other countries (costs in developing countries are often only a fraction of costs in industrialized countries). Joint implementation is one form of the tradeable permit system in the broad sense of the term (OECD, 1994) and represents a first step towards a system of tradeable quotas at the international level (cf. Bohm, 1994).

Seeking to protect a natural resource by placing it under the protection of private ownership is another means of persuading individuals to do a better job of conserving environmental resources. If the joint ownership of environmental goods is the cause of their degradation, privatization offers a solution to this problem. As Baumol and Oates note (1979), this strategy is only effective in cases where damage comes from a limited number of clearly identifiable sources, on which the owner can act. Consequently, it is not much help in the case of air pollution.

*III Liability and property rights approaches*

The strategy of "laisser-faire" is based on the following premise: the reason why the market fails to protect environmental resources properly is the lack of ownership rights over such resources. However, if the polluter's responsibility is clearly defined, the victim can seek to obtain compensation and (or) a reduction in the damage. Even in cases where the law does not protect the victim, negotiation is possible, but compensation must be paid to the victim. It has been shown that, from the viewpoint of economic efficiency, the result is the same whether the law favours the victim or the polluter. This bargaining mechanism is known in literature as Coase's theorem. The conditions for it to operate and lead to an efficient outcome are strict and hardly ever met. Transaction costs must be nul or very low, a prerequisite which limits the possible application of this bargaining mechanism to situations featuring a very small number of polluters and victims. Consequently, Coase's theorem has a very limited practical scope, except perhaps in negotiations between States aimed at solving a problem involving transboundary pollution (Mäler, 1990). In cases where the risk of environmental damage is low but costs are high, strict liability for the polluter does however represent a possible means of internalization. In this case, the risk of having to compensate the victim, or the insurance premium, plays a comparable role to that of a Pigovian tax (Cropper and Oates, 1992), as the polluter is led to take preventive measures. Thus, this instrument has points in common with market-based instruments, without however offering the same properties of efficiency as a charge or a tradeable permit system. The fact that compensation is paid to the victim can lead the latter to avoid taking all possible preventive measures. Moreover, the possibility that abatement costs may exceed the cost of the damage to the victim cannot be ruled out.

*Delimitation of economic instruments*

We shall now concentrate exclusively on economic instruments. Even though the notion of economic instrument is widely used in literature, a clear definition is rarely provided (cf.

however OECD (1994), Opschoor and Turner (1994)). Without wishing to attach undue importance to a precise definition, we would like to list the primary characteristics:

- Economic instruments serve to modify the behaviour of polluters by persuading them to take account, in their choices, of the damage for which they are responsible.
- Behaviour modification is based on incentives rather than constraints.
- The polluter is informed of the cost of utilizing environmental resources either directly (charge) or indirectly (market).
- The polluter is free to choose the most suitable means of reducing his emissions.

In principle, the term "economic instrument" is reserved for measures based on market incentives. This definition would rule out voluntary agreements concluded between regulators and industry. However, owing to the diversity of such covenants, they deserve a closer look (cf. Brugger and Gorsler, and Hafkampf in this volume). These are contracts signed voluntarily between the administrative authority responsible for environmental policy and industry (individual company, corporate group or branch of industry). Another possibility is contracts between a company and a group of persons affected by pollution. The contracts are signed voluntarily, with the signatory companies undertaking to achieve an environmental target. No formal sanction is foreseen if the target is not met (Potier, 1994). In practice, covenants include the following elements: target to be reached, timetable for implementation, modalities for the information transmitted to the population and, in particular, publication of results (cf. for example the convention signed between the Swiss Confederation and importers aimed at reducing consumption of petrol for automobiles, ORCA, 1995).

The key to understanding the nature of these conventions is the reasons that push companies to sign them. They are often tied to another form of intervention, and thus make it possible to avoid or delay the introduction of more stringent direct regulations. They may also be associated with a charge which is not levied if the targets adopted in the covenant are met. If this is the case, they have an incentive effect, even though this effect is obtained indirectly. In this form, covenants have several points in common with economic instruments. The proposal to introduce a dual instrument designed to reduce VOC emissions runs along these lines (Juen *et al,* 1992).

Lastly, covenants could constitute a means of introducing a system of tradeable permits without any drastic changes in current practices.

Table 2. Delimitation of economic instruments.

<table>
<tr><td></td><td colspan="3">Approach : Behaviour modification</td></tr>
<tr><td>Characteristics</td><td>Direct regulation</td><td>Incentive-based regulation</td><td>Information, moral suasion</td></tr>
<tr><td>Rigidity</td><td>Strict CAC*</td><td rowspan="2"></td><td rowspan="2"></td></tr>
<tr><td rowspan="3">Flexibility</td><td>Soft CAC**</td></tr>
<tr><td>Smart CAC***</td><td><u>Economic instruments</u><br>charges, tradeable permits</td><td>Self-regulation</td></tr>
<tr><td></td><td>convention linked to an incentive-based instrument<br>agreement between countries (« joint implementation »)</td><td>Covenant</td></tr>
</table>

* technology-based regulation ** emission-based regulation
*** cost-effective regulation

## 3. A cost-effective environmental policy

Before going to the heart of the matter, an important distinction should be made between two types of incentive charges: optimal charges (or "true" Pigovian-style charges) and cost-effective charges. An optimal charge obliges the polluter to bear the marginal social cost of his activity, thereby making it possible to internalize damage. The cost of collecting the information needed to calculate an optimal charge is enormous, owing to the great many agents involved and the difficulty of assigning a monetary value to environmental damage. Moreover, the very idea that an optimal level of pollution can be defined on the basis of a comparison of marginal costs and advantages is far from being generally accepted. It may be noted that the same comments apply to tradeable permits.

The economic instruments used in environmental policies do not possess these optimality properties. They are aimed at reducing emissions enough to reach an environmental target deemed acceptable. This type of instrument is designated by the term of "cost-effective charge" or "cost-effective tradeable permit".

We shall show that, under certain conditions, the three main instruments used, namely charges, tradeable permits and direct control, have the same properties as far as cost-effectiveness is concerned. Thus, they make it possible to reach the emission reduction target with a minimum of resources.

To simplify matters, let us consider a uniformly mixed pollutant. This hypothesis makes it possible to focus solely on emissions of pollutants without taking account of the localization of sources and possible hot spots.

*Cost-effective charges*

Let us suppose that the companies concerned are seeking to minimize their costs. They are subject to a unit charge on emissions of "x francs per ton of pollutant", and have a choice between reducing their emissions or paying the charge. As long as the charge exceeds the marginal abatement cost, it is more advantageous to reduce emissions. The company that reduces its emissions to the point where the marginal abatement cost equals the unit charge minimizes its costs. As all companies are subject to the same unit charge and apply the same rule to decisions, the charge leads to an automatic equalization of marginal abatement costs. In equilibrium, this means that the cost-effectiveness condition is met; efficiency cannot be further increased by shifting the abatement effort from one company to another. It should be noted that the regulator does not need to know the marginal abatement costs to achieve emission reductions at the lowest cost.

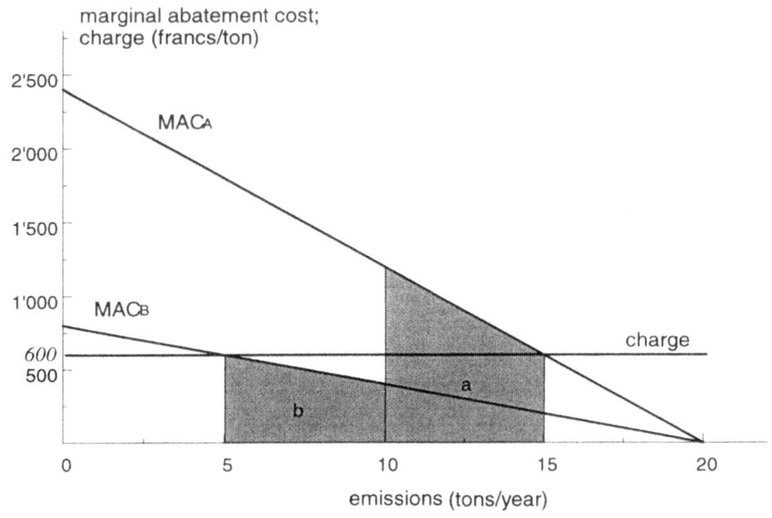


Figure 1. Cost-effective charge.

In the example shown in Figure 1, companies A and B each emit 20 tons of a given pollutant. The regulatory agency has set the target of halving overall emissions and accordingly levies a unit charge of CHF 600 per ton. As the marginal abatement cost for A ($MAC_A$) is higher than the corresponding cost for B, company A will not put as much effort into abatement and will therefore pay the charge on a greater amount of residual emissions. In equilibrium, the marginal abatement cost is the same for A and B; it corresponds to the amount of the unit charge. In comparison with an across-the-board reduction, the same abatement target is reached at a significantly lower economic cost, the gain (CHF 2,000) being represented by the differences between the two surfaces (*a* - *b*). In comparing costs, it is necessary to take the difference between expenditure which represents a true economic cost (abatement) and expenditure which merely constitutes a transfer (payment of the charge). Efficiency corresponds to the minimization of the economic cost alone. The total cost for B comes to CHF 7,500, i.e. CHF 4,500 of abatement expenditure and CHF 3,000 of charges on residual pollution. If the regulatory agency had limited itself to imposing a 10-ton reduction on each company, B's total (CHF 2,000) would have been much lower.

The instrument shown in this example is a cost-effective charge, because its introduction leads to the hoped-for reduction in emissions. Thus, the regulatory agency required first-rate information on participations' reactions. This hypothesis is unrealistic. In the absence of such information, the regulatory agency must proceed by trial and error and adjust the rate of the charge if it appears that it was initially set too low or too high. The fact that the State controls price but not quantity is without any doubt one drawback with incentive charges as compared with tradeable permits. Changes in close succession could well make companies apprehensive and lead to non-rational investment. Companies should be informed of incentives sufficiently well in advance, and such measures should remain unchanged during a relatively long period corresponding to at least the time it takes to write off the cost of the anti-pollution facilities (Juen *et al,* 1992).

The use of incentives poses a second problem. It costs companies a good deal more in abatement-related charges and expenditure to achieve the same pollution-reduction target than it does with regulations and tradeable permits distributed free of charge. As there is a risk that exporting companies will become less competitive, arrangements are usually made for refunds to be made at the border. The system becomes more complex, administrative costs rise and the question may well be asked whether it would not be simpler to use an instrument which does not cover residual emissions. In this connection, the dual instrument proposed by Juen *et al* (1992) aims precisely to avoid the taxation of residual emissions.

*Cost-effective permit system*

There are several stages to setting up a permit market. The regulator begins by adopting an environmental quality goal and determines on this basis the maximum amount of emissions that can be tolerated. It distributes to the companies concerned, according to arrangements to be defined (auction or free distribution, cf. Nichols and Stritt in this book), a quantity of emission permits that is compatible with the goal adopted. Each permit entitles the company holding it to emit a corresponding quantity of pollutants (e.g. 1 ton). It can also make additional abatement efforts and sell the permits which have become available, if this is to its advantage. The decision-making rule is very simple: the company will take steps to reduce its emissions as long as the marginal abatement cost is lower than the price of the permit. If the reverse holds true, it will purchase the emission rights it needs. The market-clearing price shows participants the opportunity cost of their emissions. If companies strive to minimize their production costs, if the market is competitive and if companies are properly informed, the system will be cost-effective (Baumol and Oates, 1988). Without any intervention by the regulatory agency, the permit market will automatically spread the abatement effort out among the participants in a way which minimizes total abatement costs. In equilibrium, all gains from trade have been exploited. As with a charge, marginal abatement costs are the same for all sources (the cost-effectiveness condition). Consequently, the tradeable permit system features the same effectiveness characteristics as an emission tax. Lastly, the regulatory agency does not need to calculate the participants' marginal abatement costs.

In the case shown in Figure 2, the regulatory agency wishes to place a cap of 24 tons on emissions. The quantity of permits placed in circulation is represented by the vertical straight line S. The marginal abatement costs for the two companies, $MAC_A$ and $MAC_B$, also represent demand for permits. The equilibrium price works out at CHF 650, at the intersection of supply, S, and demand for permits, D = $MAC_{A+B}$. Suppose that the two companies have received 12 permits each. Company A, which has higher marginal costs, will buy 5 additional permits on the market and will make less abatement efforts. Company B will sell 5 permits and will reduce its emissions by 13 tons.

The financial effort made by the companies is significantly less than in the case of a charge, because there is no payment on residual pollution. It is interesting to note that the incentive is not linked to a financial burden borne by the company. The permits received free of charge constitute an asset for the company. The possibility of selling these assets on the market is an incentive to make abatement efforts. Moreover, free distribution avoids the administrative costs of a refund at the border.

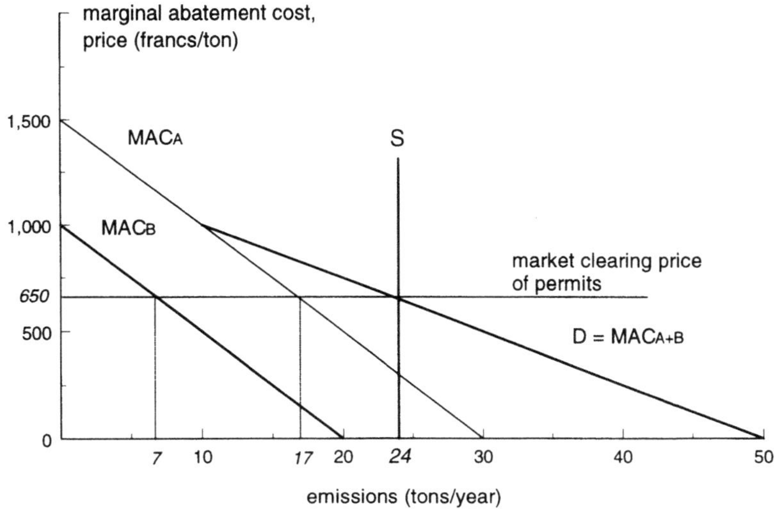


Figure 2. Cost-effective permit system.

*Direct control ("smart" CAC)*

Under certain conditions, direct regulation also features the same cost-effective characteristics as charges or permits, in which case we speak of "smart" CAC (Burtraw, 1994). Suppose that the regulatory agency has full information concerning emissions and abatement costs. This enables it to calculate the emission-reduction effort which each participant would have to make to achieve the overall abatement target while respecting the condition of cost-minimization (identity of marginal abatement costs for all). Here, it should be noted that the marginal cost at equilibrium corresponds to the amount of a unit charge making it possible to obtain the same overall reduction in emissions. The abatement effort will no longer be uniform, but will differ depending on the specific situation of each participant (Common, 1989). The abatement effort made by each company will be the same as with a cost-effective tax or a cost-effective permit system.

To reach the conclusion that the three instruments -- command-and-control, emission charges and tradeable permits -- are equivalent, we had to assume that the regulator had access to information enabling it to calculate the abatement effort to be made by each participant as well as the marginal cost at equilibrium. This is the *sine qua non* for direct regulation to be as cost-effective as economic instruments. Yet this hypothesis is not realistic, because the government has no means of obtaining such data (Tietenberg, 1990). As Hahn and Hester (1989) note, "only in rare instances will authorities possess the information to identify standards and technologies that minimize the costs of achieving a particular environmental target". Moreover, the costs of dealing with each emitter differently depending on the level of his abatement costs would no doubt be associated with unequal treatment and be deemed unlawful. Economic instruments require much less information -- this being supplied free of charge by the market -- which is their main advantage (Cropper and Oates, 1992, Common, 1989).

*Long-run benefits*

Short-term benefits stem from the ability of economic instruments to spread the abatement effort out among companies so as to keep the cost of achieving the environmental target as low as possible. These are short-run benefits for a given state of technology.

Likewise, there are long-term advantages to economic instruments as compared with direct regulation: they stimulate the development of more efficient and less costly abatement technologies (Tietenberg, 1985). With direct regulation, once the polluter has complied with the emission standard $e_1$, he no longer has much incentive to spend money on R&D to discover a more effective abatement technology (shift from $MAC_1$ from $MAC_2$ in Figure 3). The gain which he hopes to achieve corresponds to the surface $a$. Moreover, bringing out a new technology is a potentially risky step because the regulator could take advantage of this information to introduce more stringent standards.

Now suppose that there is a unit charge of $t$ per ton. With the initial technology, the company reduces its emissions up to $e_1$ and bears an overall burden equal to $(a+b)$ in the form of abatement expenditure, to which must be added the payment of the charge on residual emissions $(c+d+e)$. If the company manages to reduce its marginal abatement costs to $MAC_2$, it will achieve a cost-saving corresponding to the surface $(c+a)$. Thus, there is a greater incentive to adopt a more efficient technology with charges. An asymmetry exists between charges and permits: with the latter instrument, only companies which sell permits have an incentive to adopt new abatement technologies. In this connection, selling permits off at auction is preferable to distributing them free of charge (Maleug, 1989 and Milliman and Prince, 1989, quoted by Goodstein, 1995).

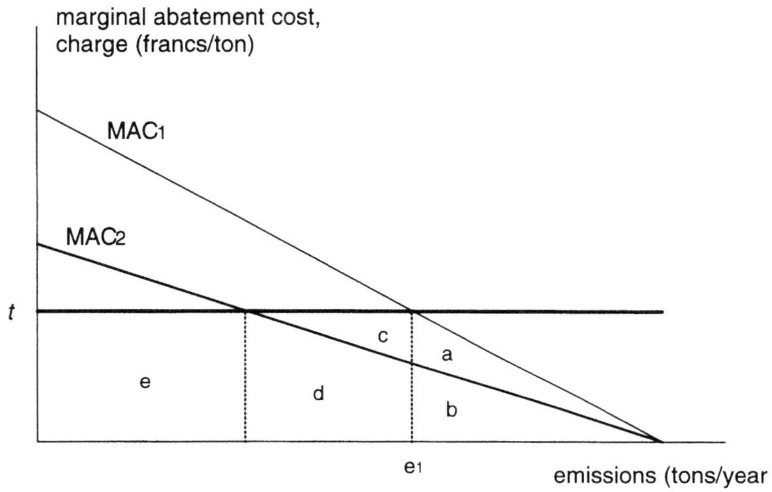


Figure 3. Long-run benefits: direct control vs charge.

## 4. Charges vs tradeable permits

The fact that charges and tradeable permits feature the same cost-effective properties does not however mean that the two instruments are identical. There are relatively major differences between them.

The first criterion to be considered is the reliability with which the emission reduction target is achieved ("dependability" criterion, Baumol and Oates, 1979). Here, it is appropriate to recall the main difference between the two instruments. A system of tradeable permits enables the government to control the quantities emitted, and the market establishes the charge. The reverse applies when a charge is used: the government fixes the price, while the market determines the quantity by which emissions will diminish. From this point of view, a permit system has an indubitable advantage over a charge, especially when pollution-caused damage is characterized by a threshold effect (OECD, 1994, cf. also Weitzman, 1974). Granted, the rate of the charge may be modified if it appears that the incentive effect is insufficient, but proceeding by trial and error would give companies the wrong signals which are likely to lead to inefficient investment decisions. Moreover, a permit system offers an advantage in times of inflation and economic growth. In such a situation, a charge maintained at a constant level would gradually become less effective unless the rate was adjusted regularly, something which is difficult to imagine. The disadvantage with tradeable permit systems is that they penalize new companies when the initial rights allocation is done on the basis of past emissions (grandfathering). However, this problem can be avoided simply by choosing another mode of allocation (cf. Stritt and Staehehin-Witt in this volume).

Adopted by the OECD in 1982, the *"polluter pays* principle (PPP) is a basic principle of environmental policies, even if there is a certain amount of vagueness as to its exact scope. Initially, the PPP simply meant that polluters had to bear themselves the cost of the abatement measures adopted by the authorities, without benefiting from subsidies (OECD, 1994). This recommandation was primarily designed to avoid distortions in international trade and, secondarily, to enhance effectiveness. If we accept this definition of the PPP which leads simply to excluding subsidies from abatement-related expenditure, then charges, tradeable permits (even allocated free of charge) and direct regulation all satisfy the "polluter pays" principle. However, one sometimes comes across a broader interpretation of the PPP ("enlarged PPP") which originated in the welfare economy. Compliance with this enlarged version requires the polluter to bear the marginal social cost of his activity and necessitates the full internalization of external environmental costs. From this perspective, compensation would be calculated on the basis of damage done. None of the above instruments -- direct regulation, cost-effective tax or cost-effective permit system -- satisfies this principle. Cropper and Oates (1992) base themselves on a criterion of equity and on the "polluter pays" principle to justify a certain preference for charges as against permits allocated free of charge. In fact, only a true Pigovian tax would satisfy the enlarged PPP.

The fact that charges generate revenue is sometimes put forward as an argument in their favour. A pure incentive tax should not bring in additional revenue for the State budget (principle of budgetary neutrality). In order to ensure this, it is enough to arrange for the yield to be refunded to the population according to arrangements to be defined. One possible means consists of paying a set amount to each inhabitant. However, there is a more advantageous way of proceeding with this refund. The taxes levied by the State lead to distortions in choices and give rise to an excess burden which is added to the normal tax load. By swapping environmental taxes for distortionary taxes (Pearce, 1991), it is possible to improve the well-being of the population. Baumol and Oates (1988) speak in this connection of a negative excess burden on the economy. Here, the incentive tax produces a double advantage: it reduces not only damage to the environment but also the distortionary costs of the tax system (Goulder, 1995; Proost and Regemorter in this book). The thesis of the double dividend is however debatable (Barde and Ownes, 1996).

In a permit system, participants must be able to come into contact with each other in order to trade emission rights. This trading gives rise to *transaction costs* which are sometimes quite sizeable: in one of the markets operating in southern California, the costs linked to the finding a buyer, negotiating and obtaining authorization from the regulator reportedly range from $15,000 to $30,000 per transaction. The expenses borne by the seller are equivalent. In the final analysis, the transaction cost appears to represent from 10 to 30 per cent of the value of a transaction (OECD, 1992). This is why the structure of the emission sources is one criterion in the choice of an instrument: when there are many low-intensity sources (e.g. households), a tax on emissions on products is more suitable than a permit system for emissions (cf. Albrecht in this volume).

Another advantage of emission trading as against a tax is that it is closer to the instruments with which the regulators are used to working ("familiarity" criterion, Baumol and Oates, 1988). There are many similarities between emission trading and direct regulation of emissions. When defined in this manner, a standard represents a right granted to an emitter to discharge a given quantity of pollutant into the atmosphere over a set period of time. It makes no difference whether the standard is set in the form of a concentration or a quantity. An emission permit also represents a right to emit a certain quantity of pollutant. The only difference -- albeit a rather important one -- is due to the fact that in a permit market, companies which make significant abatement efforts have the option of selling their excess rights. On the other hand, the tax represents a totally new system of emission control which brings about more fundamental changes in practices (Cropper and Oates, 1992).

## 5. Real world complexity

The pollutants considered up until now, called uniformly mixed pollutants, spread out evenly into the atmosphere. This means that the localization of the emission sources does not matter and economic instruments -- either charges or tradeable permits -- can concentrate on the quantities emitted. However, as soon as nonuniformly mixed pollutants are considered, instrument design must be re-examined. Even though air quality standards may be respected on the whole, they may be exceeded locally. The problem of "hot spots", or high concentrations of a pollutant in a particular zone, is often evoked. Moreover, it is one of the arguments frequently put forward against the use of economic instruments. The possibility of being able to pay to avoid abatement measures and cause a risk to the health or well-being of the population may legitimately come as a shock. However, a comment is in order here: the risk of exceeding air quality standards is not limited to economic instruments. Already today, air quality standards are exceeded in many zones and it has become clear that it is very difficult to resolve the problem using the traditional regulatory approach.

Do economic instruments offer better prospects? If we consider charges, the answer is clear: there is a need to differentiate rates depending on the localization of sources and the destination of the pollutants. This type of tax is known in literature under the name of "ambient charge" (Tietenberg, 1990). There is no gainsaying the fact that such an instrument has scant hopes of a warm reception by political decision-makers.

This idea has already been applied to tradeable permits for some time now (cf. Montgomery, 1972) and has been the subject of a considerable amount of research (cf. in particular Krupnick, Oates and Van De Verg, 1983; Atkinson and Tietenberg 1982). One possible option consists of designing a permit system which takes account of air quality in the zone towards which the pollutants flow and hence the damage done (ambient permit system). The trading rule must be adjusted to take into account the impacts of individual sources. If source A causes twice as much damage as source B owing to its localization and the wind direction, A will have to buy two permits from B to obtain a single right (trading ratio 1:2). Although the principle behind the functioning of such a market is relatively simple, implementation-related difficulties make it less appealing (high cost of information for participants and regulators). One variant consists of only authorizing trading within relatively narrow zones (Field, 1994).

In a world where players only have access to *imperfect information*, charges and permits no longer lead automatically to equivalent results. When uncertainty exists as to abatement costs, it is necessary to take a look at the marginal cost-benefit curves. Here, the rule for making a choice, which is known under the name of Weitzman's theorem, is the following: when marginal abatement costs vary relatively little but benefits increase rapidly as environmental quality is improved, it is preferable to rely on an instrument which concentrates on quantities (tradeable permits). Conversely, if abatement costs vary considerably while benefits remain more or less constant, it is better to use a charge (Weitzman 1974; Cropper and Oates 1992).

The cost-effectiveness of a permit system is based on the hypothesis of *perfect competition*. Yet such a situation is highly unlikely in the real world. The hypothesis of a market dominated by a few large emitters appears more realistic. Exactly how effective is a tradeable permit system under these circumstances? Hahn and Hester (1989), who have studied the functioning of tradeable permit systems in the US, feel that although the presence of dominant companies can modify the distribution of benefits, it does not call into question the cost-effectiveness of the instrument. Cost-savings with regard to abatement remain significant (cf. also Oates and Strassmann, 1984). However, a very real problem arises when a dominant firm uses permits to reduce competition. For example, imagine what could happen if a tradeable permit system were adopted in Swiss airports with a view to reducing noise pollution. Companies with low-noise fleets might be tempted to hold on to unused rights in order to force certain competitors off the market.

Are high *transaction costs* really an obstacle to the proper functioning of a permit system? The trading system for sulfur dioxide ($SO_2$) in the US, which was established under the Clean Air Act Amendment of 1990, provides interesting information in this respect. The programme's aim is to reduce overall emissions by 10 million tons between 1995 and the year 2000 with regard to the reference year of 1980. In order to achieve this, a programme of tradeable permits called "allowances" was introduced. It has been estimated that using a market instrument should lead to cost-savings of between $1 billion and $2.3 billion as far as antipollution costs are concerned (cf. Nichols in this book). However, observers were very surprised to note that permit prices ($131 at the first annual auction) stabilized at a particularly low level. This price is considerably lower than all the forecasts ($300 to $1500 according to the sources, Burtraw, 1994; New York Times, 1995; Ellerman *et al,* 1996). Several explanations have been put forward to explain this difference: an inappropriate auction mechanism which leads companies to announce values lower than their marginal abatement costs (Cason, 1995), or high transaction costs (Stavins, 1995). If this latter explanation were to prove true, one could indeed wonder as to the effectiveness of the programme. Two MIT researchers (Ellerman and Montero, 1996) note that the equilibrium price does not say anything about the functioning of the market other than a low price provides a strong support for the market-revealing properties of the system. The low price levels observed at present are said to be due to an external cause: the fact that electricity producers can procure low-sulphur fuel following a drop in transport costs.

## 6. Acceptability

Alongside ecological effectiveness (the guarantee that the goals will be met) and economic efficiency (minimization of abatement costs), acceptability is the third major criterion in the choice of instruments. Programmes must obtain the support of the main groups concerned by environmental policy: authorities, industry and environmentalists (OECD, 1994).

The first question which may legitimately be asked is the following: why do the recommendations of economists carry so little weight when politicians decide on the broad lines of environmental policy? Whereas the overwhelming majority of work highlights the many advantages of market-based instruments, regulations remain the dominant approach in most if not all actual environmental policies (Common, 1989). Apparently, the possibility of reducing the burden for companies is not the primary concern, while the assurance of being able to reach the target set as well as distributive aspects appear to be the determining factors.

The preferences and interests of three categories of actors -- authorities, industry and environmentalists -- determine the acceptability of the different types of instruments. One could expect companies to accept tradeable permit systems or even direct regulation ("soft CAC") relatively well and be dead set against charges and permits sold at auction. Yet they seem to accept regulatory instruments well and feel more reservations about economic instruments, something which can seem surprising at first sight. Dwyer (1992) shows that the "political market", which makes it possible to obtain exemptions or extensions at no cost, functions much too well for companies to be interested in the creation of a market for pollution rights. The explanation provided by Baumol and Oates dovetails with this: "Taxes on environment damage, when once enacted, are not easily escaped or repealed; in short they work. Enforcement or direct controls, on the other hand, typically allow a certain amount of leeway; the polluter may be able to negotiate with the regulator agency or take its case to the courts, where for a variety of reasons, it may find an easy escape" (Baumol and Oates, 1979). Thus, the effectiveness of economic instruments appears to be one of the reasons for their lack of popularity.

Hahn and Hester (1989) have concentrated on the attitude of interest groups -- industry, environmentalists and regulator -- in experiments involving the introduction of tradeable permit systems in the US. Industry has generally supported tradeable permit programmes, albeit with no great show of enthusiasm. Environmentalists have been strongly opposed to emission trading, using as an argument the fact that these programmes allow polluters to increase their emissions, thus leading to a deterioration in the quality of the environment. Yet there is a more fundamental objection to market mechanisms: benefitting from an intact natural environment represents a fundamental right of the population, which cannot be called into question solely by the desire to reduce companies' costs (cf. also Manfrini, 1991, who defends a similar thesis). The environmentalists' arguments have been heard, because law-makers have placed significant restrictions on the trading of rights, thereby reducing potential cost-savings on abatement expenditure. These restrictions have also favoured internal trading -- within the same company -- to the detriment of external trading. However, environmental protection groups have not systematically opposed the sale of tradeable emission permits: they have primarily done so when there has been a threat to environmental quality goals. The lead trading programme and the pollution rights trading programme in the Dillon Reservoir have been relatively well accepted.

Environmental regulators show a clear preference for regulatory instruments. Several authors have sought to determine why. Here are a few explanations that have been put forward:

- They note that the regulatory instruments used up until now have yielded good results from the point of view of effectiveness.
- They are more aware of implementation-related difficulties with market-based instruments. When the law prescribes a given technology or procedure, it is sufficient to make periodic inspections to ensure that it has been put in use. On the other hand, an emission charge or a permit system requires virtually continuous monitoring.
- They are averse to risks and hesitate to experiment with new instruments with which they are not familiar. Bureaucratic inertia (or inertia of the *status quo*) has also been suggested as a possible explanation for the persistently dominant role of binding measures (Tietenberg, 1986).

Lastly, it should be noted that the caricatural and sometimes dogmatic way in which incentive instruments are presented in literature, together with economists' lack of interest in practical implementation-related problems, have not served the cause of an environmental policy which relies on market-based mechanisms (cf. Kete, 1994).

## 7. Policy mix

Having access to instruments is not enough -- it is necessary to choose the right instruments suited to the situation. In this respect, it may be noted that the choice of instruments must be made pragmatically, on a case-by-case basis. This is why it is not advisable to formulate general recommendations. As Helm and Pearce (1990) note, "there is no first-best solution. The pursuit of single-instrument solutions is naive and possibly even dangerous". In particular, there are no grounds for claiming that economic instruments are always preferable to direct regulation. Lastly, it may be noted that design is at least as important as the choice of the instrument itself (cf. Bohm in this volume). Yet it is surprising to note to what extent programmes, even very costly ones, feature major design flaws. Sometimes, a few simple adjustments would have sufficed to enhance their economic and environmental effectiveness decisively.

The authors of the balance sheet on the use of economic instruments drawn up by the OECD come to the conclusion that market-based instruments are often more effective when they are properly associated with or complement direct regulation (OECD, 1994). Moreover, they are generally applied in this form, i.e. they are superimposed on existing regulation rather than replacing it. However, nowhere in literature is there a convincing demonstration of this proposition. On the other hand, one may wonder whether associating rigid rules with economic instruments might not sacrifice a large share of the efficiency gains. The problems encountered in the first experiments with permits in the US seem to confirm this hypothesis (Atkinson, 1994). Today, there is a tendency towards purer forms of use of economic instruments, like emission right trading of $SO_2$ in the US within the framework of the acid rain control programme (Klaassen, 1994).

When concentrations of pollutants in the atmosphere reach periodic peaks -- e.g. summer smog episodes -- the use of a single instrument, whether it be a charge or direct regulation, is not optimal. It is preferable to combine two instruments: a market-based instrument making it possible to achieve sufficient emission reductions to ensure that air quality standards are respected in normal periods, and regulatory measures (e.g. restrictions on automobile circulation) in order to cope with periods of crisis. Neither charges nor permits lend themselves to short-term manipulations when the state of the environment suddenly worsens, requiring urgent measures (Baumol and Oates, 1988).

One interesting means of combining the different instruments optimally consists of leaving companies free to choose between several options. The idea is to simultaneously introduce two or more different instruments while leaving it up to the polluter to choose the one which best suits him. This approach has been proposed in Switzerland under the name of a dual instrument with a view to reducing VOC emissions. Each VOC emitter would be free to sign a covenant with the (federal or cantonal) regulatory authority or pay a charge on its emissions. The covenant would define the objectives to be reached while taking account of available technology and abatement efforts already made. In the event of failure to respect this covenant, the company would have to pay the charge on all of its emissions. The charge would play its role fully, encouraging the company to respect the objective of emission reduction and the timeframe laid down in the contract (Juen *et al,* 1992). The advantage with such a system is that it would make it possible to achieve the goal without imposing an excessively heavy burden on the company. No tax would be levied on residual pollution if the terms of the contract were respected. It may be noted that a comparable result would be achieved with a tradeable quota system.

Roberts and Spence (1976) have proposed combining a tradeable permit system with a charge -- with a view to setting a ceiling on the permit price -- and a subsidy. The latter instrument would be designed to encourage participants to reduce their emissions beyond the admissible amount authorized by the quotas. In the same perspective, two authors (Chichilnisky and Heal, 1995) suggest associating a charge and tradeable quotas in a policy of prevention of climate change. $CO_2$ emissions would be limited by a tradeable quota system, but firms would be authorized to emit more than their quota on the condition that they paid a charge on excess emissions. A company whose rights were insufficient would have three options: reduce its emissions, buy quotas on the market or pay the charge. The advantage with the charge is that it fixes a ceiling on the quota price, thereby avoiding one of the drawbacks with a tradeable permit system -- uncertainty with regard to costs. On the other hand, the quantities discharged would no longer be directly monitored. Several Swiss communes apply a system quite close to the one described above in order to encourage households to sort their refuse. At the beginning of the period, each household receives a given number of "official" bags which represents its quota of unsorted refuse to which it is entitled. Unused bags can be given away, sold (trading) or kept for the next period (banking). Households have the option of buying additional bags from the communes, but these are sold at a relatively high price. This price, which is the equivalent of a bag tax, sets a ceiling on the price of bags traded between households.

## 8. Market-based instruments: how do the results measure up?

The question of concern to those responsible for governmental policy is determining whether economic instruments make it possible to achieve environmental targets. The answer is without any doubt an affirmative one: economic instruments work. As Nichols notes, another advantage of a well-designed permit trading system is that it makes it possible to reach emission reduction targets more rapidly. Moreover, it is difficult to imagine that companies would agree to spend large sums of money to purchase permits if they had doubts as to the quality of controls in the firms which sold them the permits. In addition to the control conducted by the authorities, there is the reciprocal monitoring of the companies which participate in the market.

There are many examples of market-based instruments which have led to an important reduction in damage to the environment. The Swedish tax on $NO_x$, which made it possible to reduce emissions by 30 to 40%, the sulfur tax in that same country, lead trading in the US, the tax on non-biodegradable plastic bags in Italy and the household waste tax levied in several Swiss cantons are but a few examples of instruments which generate a broad incentive effect.

Have the economists' promises with regard to cost reductions been borne out? Literature contains contradictory information in this connection. Are cost-savings disappointing and promises irrealistic, as some claim (cf. in particular Kete, 1994)? Or on the contrary, have economic instruments enabled participants to achieve substantial cost-savings? Most authors conclude in favour of the second hypothesis (EPA, 1985; Hahn and Hester, 1989; Burtraw, 1994; Tietenberg, 1986 and 1990; Nichols and Stritt in this volume).

Some comments are in order with regard to the way in which cost-efficiency gains are measured. Some authors compare the costs of direct regulation with those of an ideal situation called "least cost" (Tietenberg, 1990). In this case, the reduction of abatement costs stemming from the application of economic instruments is a result of two different mechanisms: on the one hand, the shift in abatement efforts towards establishments where marginal costs are lower, and on the other hand, the increase in emissions in zones where limits have not yet been reached, the goal being merely to respect ceilings within the area as a whole. When estimated in this way, the CAC cost/least cost ratio ranges from 1.07 to 22.0 for air pollution and from 1.12 to 3.13 for water pollution. Abatement costs are up to 22 times higher when a

CAC-type strategy is adopted. For example, a figure of 2.0 means that the potential saving represents 50% of the abatement costs of the traditional command-and-control approach. However, one may wonder as to the relevancy of this type of estimate. It is a well-known fact that the ideal instrument which would make it possible to achieve the lowest cost allocation -- the ambient permit system or ambient charge -- poses practically insurmountable implementation problems. Moreover, can we really compare the two instruments if environmental goals differ? The scale which Nichols gives -- a 1:2 ratio -- is indeed less spectacular, but without doubt closer to reality. The gains estimated by Stritt (in this volume) are situated in the same scale of things.

To date, little has been done to take advantage of the possibilities offered by the utilization of economic instruments in environmental policies. Potential applications are very important when it comes to facing environmental problems within national borders. Here, we are thinking in particular of the field of transport: utilization of market-based mechanisms to reduce congestion (road pricing), charges making it possible to internalize external environmental costs, tradeable quota systems designed to encourage manufacturers to develop their own fuel-efficient cars, etc. Moreover, there have been hardly any efforts to explore mixed strategies combining direct regulation and market-based instruments, where the company is able to choose the instrument which suits it best. Institutional changes, and in particular broader application of the subsidiarity principle, could favour the emergence of bolder environmental policies. But it is doubtless policies designed to combat global threats which offer the most promising prospects for the application of market-based instruments.

*Acknowledgements*
*This article is part of a research project on market-based instruments funded by the Swiss National Science Foundation within the Priority Programme « Environment » (Project No. 5001-035281). For helpful comments and discussions of an earlier draft I would like to thank Nathalie Schwab Christe.*

## References


**Atkinson, S. E. and Tietenberg, T. H.** (1982) The Empirical Properties of two Classes of Designs for Transferable Discharge Market. *Journal of Environmental Economics and Management,* 9 : 101-121.

**Atkinson, S. E.** (1994) Tradeable Discharge Permits : Restrictions on Least-Cost Solutions. *In:* Economic Instruments for Air Pollution Control, Ger Klaassen and Finn R. Forsund (eds), Kluwer Academic Publishers, Dordrecht, 3-21.

**Barde, J. P. and Owens, J.** (1996) L'évolution des écotaxes. *L'observateur de l'OCDE,* 198 : 11-16.

**Baumol, W. J. and Oates, W. E.** (1979) Economics Environmental Policy and The Quality of Life. Prentice-Hall, Inc., New York.

**Baumol, W. J. and Oates, W. E.** (1988) The Theory of Environmental Policy. Cambridge University Press, New York.

**Bohm, P.** (1994) Making Carbon Emission Quota Agreements More Efficient : Joint Implementation Versus Quota Tradeability. *In:* Economic Instruments for Air Pollution Control, G. Klaassen and F. R. Forsund (eds), Kluwer Academic bublishers, Dordrecht, 187-208.

**Burtraw, D.** (1994) Agency in International Pollution Permit Trading. *In:* Economic Instruments for Air Pollution Control, G. Klaassen and F. R. Forsund (eds), Kluwer Academic Publishers, Dordrecht, 249-273.

**Cason, T. N.** (1995) An Experimental Investigation of the Seller Incentives in the EPA's Emission Trading Auction. *The American Economic Review,* 85 (4) : 905-922.

**Chichilnisky, G. and Heal, G.** (1995) Markets for Tradeable $CO_2$ Emission Quotas : Principles and Practice. Working Paper n° 153, OCDE, Paris.

**Common, M. S.** (1989) The Choice of Pollution Control Instruments : Why is so Little Notice Taken of Economist's Recommendations ? *Environment and planning,* A (21) : 1297-1314.

**Cropper, M. L. and Oates, W. E.** (1992) Environmental Economics : A Survey. *Journal of Economic Literature,* 30 : 675-740.

**Département fédéral des finances** (1995) Finances publiques en Suisse. Berne.

**Dwyer, J.-P.** (1992) La politique de la Californie en matière de permis négociables d'émission et ses applications dans le cadre de la lutte contre les gaz à effet de serre, *in :* Le changement climatique, OCDE (éd.), Paris, 41-80.

**Ellerman, A. D. and Montero, J. P.** (1996) Why Are Allowance Prices so Low ? An Analysis on The $SO_2$ Emissions Trading Program. CEEPR Working paper n° 96-001, Massachusetts Institute of Technology (MIT), Center of Energy and Environmental Policy Research, Cambridge, New York.

**Emerson, P. M. and Gaines, S. E.** (1995) Why Not Trade Pollution, Too ? *The New York Times,* 22.

**Field, B. C.** (1994) Environmental Economics : An introduction. McGraw-Hill, Inc., New York.

**Goodstein, E. S.** (1995) Economics and the Environment. Prentice Hall, Inc., New Jersey.

**Goulder, L. H.** (1995) Environmental Taxation and the Double Dividend : A Reader's Guide. *International Tax and Public Finance,* 2 : 157-183.

**Hahn, R. W. and Hester, G. L.** (1989) Marketable Permits : Lessons for Theory and Practice. *Ecology Law Quartely,* 16 (36) : 361-406.

**Helm, D. and Pearce, D.** (1990) Assessment : Economic Policy Towards the Environment. *Oxford Review of Economic Policy,* 6 (1) : 1-16.

**Juen, C. ; Moser, B. ; Taormina, A. M. and Zürcher, M.** (1992) Politique de l'environnement conforme à la logique de marché. SDES, Genève.

**Kete, N.** (1994) Air Pollution Control in the United States : A Mixed Portfolio Approach. *In:* Economic Instruments for Air Pollution Control, Ger Klaassen and Finn R. Forsund (eds), Kluwer Academic Publishers, Dordrecht, 122-144.

**Klaassen, G. and Forsund, F. R.** (1994) Economic Instruments for Air Pollution Control. Kluwer Academic Publishers, Drodrecht.

**Krupnick, A. J. ; Oates, W. E. and Van De Verg, E.** (1983) On Marketable Air Pollution Permits : The Case for a System of Pollution Offsets. *Journal of Environmental Economics and Management,* 10 : 233-247.

**Maleug, D. A.** (1989) Emission Trading and the Incentive to Adopt New Pollution Abatement Technology. *Journal of Environmental Economics and Management,* 16 : 52-57.

**Manfrini, P. L.** (1991) Fixation des normes d'émission : approche technique ou économique. Conférence, Université de Neuchâtel, Switzerland.

**Mäler, K. G.** (1990) International Environmental Problems. *Oxford Review of Economic Policy*, 6 (1) : 80-108.

**Milliman, S. R. and Prince, R.** (1989) Firm Incentives to Promote Technological Change in Pollution Control. *Journal of Environmental Economics and Mangement,* 17 (3) : 247-265.

**Montgomery, W. D.** (1972) Markets in Licenses and Efficient Pollution Control Programs. *Journal of Economic Theory,* 5 : 395-418.

**Oates, W. E. and Strassman L., D. L.** (1984) Effluent Fees and Market Structure. *Journal of Public Economy,* 24 (1) : 233-247.

**OCDE** (1992) Le changement climatique : concevoir un système de permis négociables. OCDE, Paris.

**OCDE** (1994) Gérer l'environnement : Le rôle des instruments économiques. OCDE Paris.

**Opschoor, H. and Turner, K.** (1994) Environmental Economics and Environmental Policy Instruments : Introduction and Overview. *In:* Economic Incentives and Environmental Policies : Principles and Practice, H. Opschoor and K. Turner (eds), Kluwer Academic Publishers, Dordrecht, 1-38.

**ORCA** (1995) Projet d'ordonnance sur la réduction de la consommation spécifique de carburant des automobiles, DFTCE, Berne.

**Pearce, D. W.** (1991) The Role of Carbon Taxes in Adjusting to Global Warming. *Economic Journal,* 101 : 938-948.

**Potier, M.** (1994) Les accords volontaires sur l'environnement. *L'observateur de l'OCDE,* 189 : 8-11.

**Roberts, M. J. and Spence, M.** (1976) Effluent Charges and Licenses Under Uncertainty. *Journal of Public Economy,* 5 (3 and 4) : 193-208.

**Stavins, R.** (1995) Transaction Costs and Tradeable Permits. *Journal of Environmental Economics and Management, 29* : 133-148.

**Tietenberg, T. H.** (1985) Emissions Trading : an exercise in reforming pollution policy. Resources For the Future, Inc., Washington.

**Tietenberg, T. H.** (1986) Uncommon Sense : The Program to Reform Pollution Control Policy. *In:* Regulatory Reform : What Actually Happened, L. W. Weiss and M. W. Klass (eds), Little, Brown and Company, Boston, 269-303.

**Tietenberg, T. H.** (1990) Economic Instruments for Environmental Regulation. *Oxford Review of Economic Policy,* 6 (1) : 17-33.

**Tietenberg, T. H.** (1994) Economics and Environmental Policy. Edward Elgar Publishing Limited, Hants.

**Walter, F. and Suter, S.** (1993) Umweltabgaben in Europa. *Schriftenreihe Umwelt,* 198, Bundesamt für Umwelt, Wald and Landschaft (ed.), Ökonomie, Berne.

**Weitzman, Martin L.** (1974) Prices vs. Quantities. *The Review of Economic Studies*, XLI : 477-490.

# *2. Environmental Charges*

Environmental Policy Between Regulation and Market
ed. by C. Jeanrenaud


# 2.1 Environmental Taxes : Analytical Framework

**Philippe Thalmann**
**Swiss Federal Institute of Technology, Lausanne, Switzerland**

This overview of environmental taxes addresses the following questions:

- What are environmental taxes?
- How should revenues from environmental taxes be used?
- Can an environmental tax be used to raise revenues?
- Do legal issues matter for the last two questions?

## 1. Incentive taxes

A small lake is much appreciated by leisure fishers for its quietness and fish population. Unfortunately, rain washes fertilizer spread on neighboring fields into the lake. The fish population must compete for oxygen with the expanding seaweed and the fishers catch less. Accordingly, the local authorities charge the farmers for the fertilizer they spread on their fields. This is a classical incentive or "Pigouvian" tax: a levy that forces the farmers to take account of the damage they inflict on other users of the lake, who cannot charge a price for the "costs" inflicted on them. This levy does not serve to cover any actual costs or expenses incurred by anyone.

Yet suppose that the threat to the lake comes from a factory that draws water from the lake. It discharges the water into the atmosphere as vapor, thereby lowering the level of the lake and reducing the fish population. The local authorities charge the factory for the water pumped out of the lake. This price does not serve to cover costs to the authorities: there are none. If fishers were charged for fishing in the lake, the authorities would lose revenues when fishing is less productive. There is no such charge here, so that the price for water does not compensate the authorities for revenues lost. The price is set equal to estimated losses incurred by the fishers from a lower water level. It is called a price or a user fee, but it should rather be called an environmental tax. Indeed, the factory pays for the use of a natural resource and for the losses it inflicts on other people who cannot charge it for that. The authorities levy that tax by virtue of their power to raise taxes, not by virtue of ownership of the lake.

The example of the charge for water drawn from the lake shows that there can be an environmental tax without any damaging emissions. Even the charge on fertilizer is not directly a tax on the emissions into the lake. Environmental taxes can take many forms:

- Taxes on natural resources: primary goods, land, fossil energy. Some of these goods are free due to a lack of property rights.
- Taxes on intermediate goods (fuels, electricity) or final goods (consumer goods, packaging). These are environmental taxes if their scope and rate depend on the use of natural resources in the production, utilization or elimination of these goods.
- Taxes on emissions: materials discharged into the environment, garbage, noise, etc.

In principle, the rate of the environmental tax should reflect the damage inflicted on everyone else. Returning to the example, the price of water should reflect the damages inflicted on fishers. How can we determine average or marginal damage? Suppose that the first cubic meters ($m^3$) drawn from the lake do not matter for fishing, but that beyond a certain point the threat to the fish population increases exponentially as the level of the lake decreases. On the other hand, the usefulness of the water to the factory decreases as it pumps more of it out. Consequently, the factory's demand for water is downward sloping. Figure 1 illustrates these assumptions.

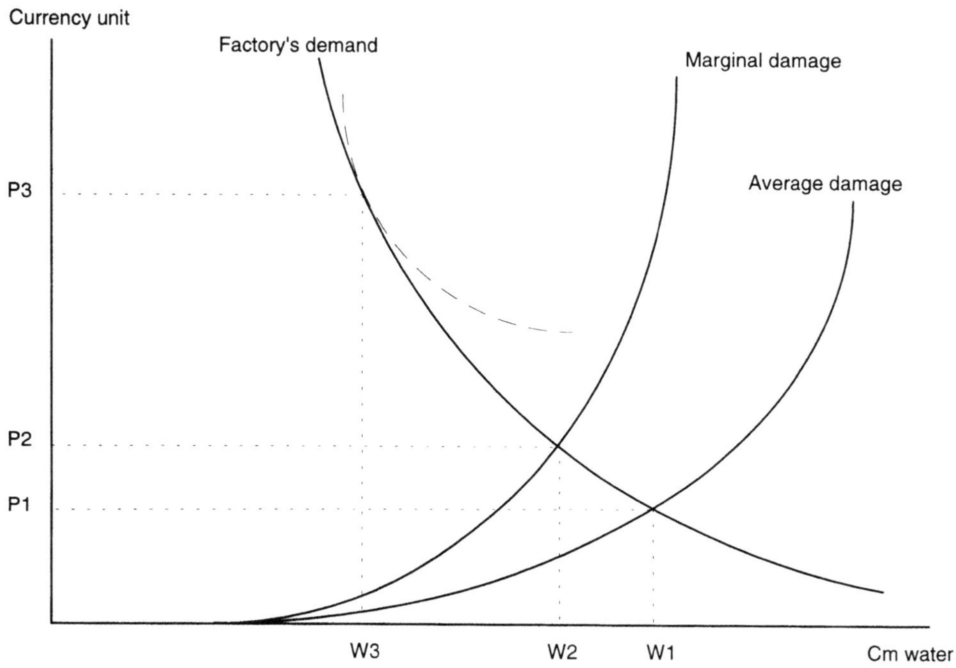


Figure 1. Equitable, efficient, and revenue-maximizing price.

The damage from the last $m^3$ pumped is higher than the average damage. The authorities could also charge the average damage P1, so that the total amount paid is equal to the total damage inflicted on fishers: P1*W1. Price P1 shall be called the "equitable" tax. It is important to note that revenues from the tax are not necessarily used to compensate the fishers.[1]

The equitable tax does not give the factory the right incentives for determining its consumption of water. Indeed, all the water drawn from the lake beyond W1 has a lesser value for the firm than its "cost" for the fishers. The authorities should set a uniform price of water

---

1 This is not exactly the "polluter pays" principle. This principle requires the polluter to pay for the expenses incurred to alleviate the damage he causes to the environment.

at P2, in order to induce the firm to pump only so much water as has a greater value to itself (and presumably to society) than its cost for other users of the lake. With this price, the revenues from the tax P2*W2 are superior to the damages inflicted. Price P2 shall be called the "efficient" tax. Price P3 will be defined below.

## 2. The unwanted revenues of environmental taxes

In Figure 1, the efficient tax induces the factory to pump an amount W2 of water, which corresponds to the greatest net value for society. The value of a quantity of water W for the factory is given by the surface under its demand curve between 0 and W. This is the amount the factory would be willing to pay for the right to pump W. The cost of this water to the fishers is equal to the surface under the marginal cost curve between 0 and W. This is the amount the fishers would be willing to pay the factory for not pumping W out of the lake. The net value to society of water used by the factory is equal to the value to the factory minus the cost to the fishers. This net value is maximized for W = W2.

In fact, the firm does not benefit from the total value of this water, as it must pay a tax amounting to P2*W2. However, the net value to society includes these tax revenues, just as if the authorities had handed this money back to the firm or to the fishers. The analysis of incentive taxes does not concern itself further with the use of revenues.

Ignoring the use of revenues from the incentive tax is fine when they are just a drop in the great ocean of public revenue and expenditure. This may be the case for the typical textbook example of incentive taxes, such as the ones used above. Presumably the value to society of the last dollar of public spending is equal to its value to any citizen, i.e. one dollar; if it were higher, the government would increase global welfare by increasing its revenues and spending; whereas if it were less, the government would return the dollar to the taxpayers who could put it to a more valuable use.

Revenues from the incentive tax become interesting when the above assumptions are not valid:

- The revenues are not negligible.
- The authorities cannot collect or spend the marginal dollar without distortions.
- The authorities do not pursue the greatest social welfare.

Incentive taxes considered recently to protect the environment would potentially produce very large revenues because they apply to broad and rather inelastic tax bases: use of carbon (production of $CO_2$), use of energy. This is why the question of the use of revenues has come to enrich the theory of incentive taxes. Economists realized that environmental taxes could yield additional efficiency gains, by using revenues to lower some other distortionary tax. They called this the "double dividend" of environmental taxes.

Unfortunately, the question of how to use the revenues from incentive taxes is threatening to delay if not prevent the implementation of environmental taxes on any significant scale. Many actors in the political arena claim that environmental taxes must be pure from any fiscal pursuits. They call it hypocritical if the government claims environmental concern and fills its coffers with the proceeds from its interventions. Implicit in these views is the assumption of a Leviathan government, independent and egoistic. Actors on one side of the political spectrum call for neutral and fair distribution of the unwanted proceeds from environmental taxes, while actors on the other side demand that the revenues of the tax be used to further promote environmental protection.

One problem with the first view is the impossibility of passing money to economic actors in a "neutral" and fair way, as it is impossible to collect revenues in such a way. As the goal is to avoid environmental taxes representing an additional tax burden, their proceeds should be returned to taxpayers. However, the more closely this money is refunded to those who paid it, the smaller the incentive effect is.[2]

---

2 If the money is returned to the factory that pumps water from the lake, the incentive effect of the tax is lost. Redistribution does not destroy the incentive effect if the beneficiaries do not see a relationship between the tax they pay and the refund they earn. This is possible when the number of polluters is large and when each is refunded an amount independent of his contribution, say the same amount for all.

Any distribution is basically equivalent to public spending. Transfers and subsidies already account for the bulk part of public spending. The only difference with the redistribution of the revenues from environmental taxes lies in the tying of these revenues: the authorities can only use them for a particular form of transfer to designated beneficiaries.

The debate about the form of redistribution of the unwanted proceeds from incentive taxes should be avoided altogether. Simply add the revenues of any pure incentive tax to the general budget. If a majority considers this budget to be too large, a separate discussion can address ways and means of reducing its size.

When authorities raise environmental taxes with the explicit aim of obtaining revenues or when the revenues are sizable, the question of their use cannot be separated from the decision to introduce such a tax. This is the topic of the next section.

## 3. Revenues as a primary objective of environmental taxes

The authorities may set the tax at a level different from both the equitable and the efficient rate. For instance, the authorities could not or would not measure the damages, or the authorities deliberately set the tax at a level that yields the amount of revenues they wish to collect from the firm, more or less than either P1*W1 or P2*W2 in Figure 1 above. If the authorities wish to maximize revenue from the factory, given its demand for water, they should set the price at P3. This price is defined, in Figure 1, by the tangency of the highest unit hyperbola with the demand curve. Price P3 shall be called the revenue-maximizing tax.

What should we think of environmental taxes as a deliberate means of raising revenue? One way to answer is to compare these taxes with ordinary revenue sources because they are just one means of collecting money from the taxpayers. The main difference between all these sources of public funds is the pattern of distribution of the burden among the citizens (and possibly foreign contributors). Historically, the following arguments have come to dominate as criteria for determining a pattern of burden:

- Efficiency (convenience or, more recently, "optimal taxation"): raise revenues on tax bases that are difficult to hide.
- Fairness (compensation or benefit principle): each citizen pays for his use of public services.

- Equity: citizens with a greater ability to pay contribute more.

Optimal taxation aims at minimizing "dead-weight losses" or "excess burden", i.e. the losses that occur for everyone when taxpayers try to avoid a tax. Ideally, taxation should be neutral with respect to economic choices. As this is not feasible since it is not possible to charge the same rate for all alternatives, a second-best solution that minimizes distortionary losses is sought. This solution is associated with the names of Ramsey and Boiteux. However, the weakness of this approach is that it does not really care about who bears the tax and who bears the "distortions". Yet, the selection of "optimal taxes" can include such considerations. The strength of the optimal taxation approach is to recognize that there is a gap between desired and actual patterns of contributions.

Although the compensation principle is not widely used, its popularity is on the rise again. However, it cannot serve as an important benchmark for financing public spending, because (a) benefits are difficult to measure, (b) charging for public services would discourage their use, and (c) subsidies and transfers increasingly outweigh public services as such. Authorities invoke the compensation principle for allocating the tax bill across generations: borrowed money finances durable public investments, and taxes reimburse the debt over the lifetime of the asset, presumably in accordance with the services from the assets.

Ability to contribute is the dominant tax criterion today. It buttresses income taxes as well as consumption taxes. The problem with this criterion is that ability to pay is not exogenous. Therefore, taxes on ability to pay are to a large extent taxes on effort, sacrifice and entrepreneurship (labor, savings and risk-taking). Hence the growing popularity of consumption taxes: people should not contribute on the basis of what they give to society but on the basis of what they take from society.

In the same logic, a new criterion has been proposed, which I shall call the environmental criterion: tax people on the basis of what they take from the environment. If environmental use is considered to be analogous to the use of public services, the environmental criterion is fair in the sense of the compensation principle.

Environmental taxes must also confront the question of efficiency. They can address two forms of inefficiency: the excessive use of a natural resource (Pigou), and distortions in economic choices brought about by any public levy (Ramsey, Boiteux). The incentive dimension of environmental taxes is obviously perfect for the first inefficiency. However, such taxes typically fare much worse with regard to the second form of inefficiency. Their tax base is narrow, so they impose a sizable excess burden on taxpayers. In order to address both forms of inefficiency, the optimal tax rate is a weighted average of the optimal Pigouvian tax and the optimal Ramsey-Boiteux tax (Sandmo, 1975). Formal models also show that the tax rates on goods and activities related to the polluting good or activity (complements or substitutes) should be set at the Ramsey-Boiteux level: there is no need to give indirect incentives for reducing the polluting activity when direct taxation is available.

In sum, the environmental principle of taxation, which might serve as an ethical basis for environmental taxes as revenue sources, is compatible with efficiency and fairness. It is more difficult to make it dovetail with equity. In several instances, people with a low ability to pay inflict greater damage on their environment. Simulations of the kind of straight carbon and energy taxes that are envisioned in Europe and the U.S. have shown that households in the first quintile of income would pay about twice as much tax, compared to their total spending, as households in the other quintiles (Poterba, 1991, Smith, 1993). These simulations do not even account for the fact that poorer people often bear the brunt of taxes aimed at wealthier people, who can pass it down to them (for instance through lower wages for the unqualified and reduced employment).

Some environmental taxes would impose a larger burden on the rich. Returning to the lake of my earlier example, imagine that speedboats cruise it. Their noise is a nuisance to fishers and causes stress in the fish population. A tax on speedboats provides proper incentives (or "disincentives"). It is also a tax which targets the wealthy, as speedboats are typically reserved for the happy few. Say the authorities spend the revenues of the tax for the benefit of the poor. This tax clearly flattens the income distribution, though only weakly. Several countries apply similar taxes to large or foreign cars. This example shows that environmental taxes can be used as instruments for income redistribution.

Yet it is unlikely that large scale environmental taxes could be raised without imposing a disproportionately large burden on low-income citizens. One can imagine amendments that alleviate the burden for them, akin to the lower VAT rates on necessary goods. However, any such solution is a move away from environmental protection. It is preferable to compensate excessive burden directly with instruments targeted at the needy. In particular, part of the revenues could be spent in favour of the poor.

Finally, it must be ascertained that environmental taxes qualify as revenue sources for the needs of the government. Can environmental taxes provide sizable revenues that are stable both through the business cycle and over a sufficient time horizon? This all depends on the tax base. Energy and water use, waste and land occupation are four tax bases that are stable both over the business cycle and in the long run (Von Weizsäcker, 1989). Note that taxes raised precisely on such stable bases are not incentive taxes.

## 4. Legal aspects

Does a tax, introduced with a view to obtaining revenues, qualify as an environmental tax just because it is levied on an activity that hurts the environment? In the preceding section, I called such a tax an environmental tax, but the question is more delicate because legal definitions are involved.

In a legal constitution, the authorities cannot exact any tax from their subjects without a legal basis. In most democracies, the government can tailor a law to introduce any type of tax provided the parliament approves the law. In Switzerland, a constitutional amendment is necessary, subject to a referendum. For most environmental taxes, like those of the introductory examples, such a procedure is too costly and time-consuming. The authorities would rather introduce the law at a lower level, maybe an ordinance, based on their general power to act on behalf of environmental protection. For this reason, it is important to decide that a tax is introduced with a view to protecting the environment. This forces the authorities to define the base and rate in accordance with damages inflicted. The above speedboat tax cannot be implemented under environmental regulations if there are other sources of noise on the lake (fishingboats, passenger ships).

## 5. Conclusions

The authorities pursue many goals, ranging from stabilizing business activity to correcting market failures, from helping the distressed to promoting the arts. The diversity of instruments does not match the diversity of goals. The authorities can hire people to work, build infrastructures, pay out transfers and subsidies, and regulate. Such intervention is generally very costly, so funds must be found. The authorities are entrusted with so many tasks precisely due to their ability to raise the necessary funds through taxes.

Environmental taxes are an original and interesting type of instrument because they make it possible to pursue an important goal, namely protection of the environment, without any cost for the authorities other than collecting the tax. Better still, they provide revenues for other tasks. They must seem very attractive to the authorities.

Obviously, there are no free lunches. Instruments that please the finance minister are not popular with taxpayers. Hence the great skepticism encountered by any proposal to introduce such taxes on a significant scale. To make it easier for us taxpayers to swallow the pill and to learn to love it, lawmakers are proposing small environmental taxes which will not significantly affect public revenues, private balances or even emissions. In general one would say that is fine: at least little damage is done from a bad implementation of the tax. They cost little to put in place and they serve to learn about the effects of environmental taxes.

However, the assumption that small taxes cost little to implement may prove wrong. It is true that implementing environmental taxes could simply amount to raising the regulated prices of utilities or increasing some existing taxes or duties. Thus the administrative costs as such can indeed be held down. What is costly is the political debate, the scientific research and the legal paperwork necessary today to implement even the smallest environmental tax. Since those costs are the same for a larger tax, one may be well advised to go all the way to an effective tax at once rather than in steps. The authorities could still introduce the tax gradually, but would take the decision to raise this tax once and for all.

## 6. The chapters in this section

Car driving generates a wide array of external costs: different types of dangerous exhausts, noise, risk for other users, congestion, etc. In addition, it uses an exhaustible resource that may be under-priced. To address all of these externalities, the authorities would need as many regulatory or fiscal instruments. They face, however, the allergy of people in general and enthusiastic car drivers in particular to the sheer number of impediments to driving. As a result, the authorities may try to impose only few taxes or restrictions. In his contribution, Nicolas Wallart shows what tax base is the most appropriate to internalize the external costs of automobiles. He does not consider the combined effects of several instruments. Instead, he insists aptly on the practical difficulties in enforcing ideal taxes in the face of the unlimited heterogeneity of cars.

In December 1994, the Italian Finance Minister presented a White Paper on tax reform. Even though this text disappeared from the political agenda together with the government that brought it forward, it remains a landmark in the discussion on environmental taxes for Italy, as Alberto Majocchi argues in his contribution to this book. The White Paper is the first official document to present the concept of a Pigouvian tax. It envisions the reduction in labor income taxation compensated by a reshaping of energy taxation on the basis of carbon content. Such a substitution would serve two out of three central goals set by the White Paper for tax reform: to lower the tax burden on labor and to simplify tax enforcement and compliance. The third goal, to decentralize revenues (and spending), could be attained with road pricing.

*Acknowledgements*
*I am grateful to Nicolas Wallart for his comments and suggestions.*

## References

**Poterba, J.** (1991) Tax policy to combat global warming : On designing a carbon tax. *In:* Global Warming. Economic Policy Responses. R. Dornbusch and J. M. Poterba (eds.), MIT Press, Cambridge, MA, 71-98.

**Sandmo, A.** (1975) Optimal taxation in the presence of externalities. *Swedish Journal of Economics, 77* : 86-98.

**Smith, S.** (1993) Distributional effects of a European carbon tax. *In:* The European Carbon Tax : An Economic Assessment. C. Carraro and D. Siniscalo (eds.), Kluwer Academic Publishers, Dordrecht, 49-65.

**von Weizsäcker, E. U.** (1989) Umweltsteuern sind unverzichtbar. *Wirtschaftsdienst, 69* (9) : 435-438.

Environmental Policy Between Regulation and Market
ed. by C. Jeanrenaud


# 2.2 Environmental Taxes in the Italian White Paper on Fiscal Reform

**Alberto Majocchi**
**University of Pavia, Pavia, Italy**

## 1. The role of environmental taxes

In recent years a broader utilization of environmental taxes has been included in the political agenda in many European States in an effort to overcome the limits of an environmental policy structurally based on command-and-control measures and also to provide the Treasury with the necessary means to cut other taxes with distortionary effects on the economy, such as income taxes with very high marginal rates or social security contributions, mainly those levied on unskilled or low-skilled workers. This kind of manoeuvre could provide a double benefit to the economy, that is known in the literature as the double dividend.[1]

Replacing -- or combining -- command-and-control measures with environmental taxes seems generally quite reasonable, especially in Italy where environmental policy shows the same limits as fiscal policy. Formally speaking, laws provide a set of very stringent conditions that represent a heavy burden for productive activity, but, for two main reasons, without any significant positive impact on environment:

---

[1] For a more skeptical view of this issue see Bovenberg and De Mooj (1993), Goulder (1995).

- It is very costly and largely unfeasible -- also due to the dreadful conditions in the public administration -- to check a very large number of polluting emissions. It follows that as far as polluters are concerned, the risk of punishment is very limited and certainly less than the advantages flowing from non-compliance with the constraining normative provisions.
- If the authorities try to force compliance with the law, firms are able to blackmail public opinion, supporting the idea that there is a trade-off between environmental protection and economic growth since environmental protection necessarily implies a cost in terms of employment and output.

If these ideas are correct, it seems that there are good reasons to implement environmental charges in lieu of pre-existing distortionary taxes. The double dividend argument appears sufficiently well-founded, on the basis of the existing literature and specifically on empirical grounds. In any case, it should be borne in mind that revenue from this source is substantially limited. Sweden is always -- and rightly so -- quoted as a country that has been able to implement a fiscal reform based on the lowering of marginal tax rates for income taxation, financed by a broadening of the VAT tax base and by new environmental taxes. But in this country as well the total revenue flowing from environmental charges represents only about 6% of the global revenue accruing to the Treasury.[2]

Even within these quantitative limits, environmental charges should be positively evaluated since -- unlike other taxes -- they do not cause new distortions, but:

- They eliminate some existing distortions in the economic system.
- They give firms an incentive to improve their productive choices, to look not only for maximum profit, but for the "optimal" level of exploitation of environmental resources.
- They can be phased in progressively in order to give firms the possibility to comply gradually with the more constraining level of environmental standards.
- They provide the Treasury with revenue that could be utilized to cut other distortionary taxes.

---

2 Ministry of Finance (1991).

## 2. Energy taxation and the double dividend

The debate in Italy has primarily focussed on the possibility of increasing the amount of energy taxation. From this perspective, it should be emphasized that even if a general energy tax represents a second-best solution in relation to a tax determined by the level of polluting emissions, it offers substantial advantages if the administrative costs of compliance with the tax are taken into account. A general tax on energy can be considered as a device to raise revenue and promote energy efficiency as well, providing further advantages in terms of reductions in other fields as far as air pollution is concerned. If the authorities wish to enhance the efficiency of an energy tax as a proxy to a charge on polluting emissions, fiscal differentiation could be implemented between the different kinds of fuels according to some characteristics, e.g. the amount of carbon content if the goal is to curb greenhouse gases emissions, as it has been the case in the past, with outstandingly positive results with regard to leaded and unleaded petrol.

The implementation of a general energy tax can also be considered as an opportunity to kick off a fiscal manoeuvre which, together with the positive environmental effects, will bring about an increase in the level of employment through a cut in social security contributions financed by the revenue of the energy tax. Following the European proposal for a Directive introducing a energy-carbon tax, simulation exercises on the economic impact of this kind of manoeuvre show that it is possible to increase employment significantly within the European Union. In Italy a manoeuvre implying a global revenue equal to 1% of GDP -- that means a tax rate for the energy tax equal to US$ 10 per barrel of oil -- spent to finance a general across-the-board cut in social security contributions, would increase GDP by 0.72%, while employment would rise by 0.79%, corresponding to some 165,000 new jobs.[3] These new jobs would mainly be created in the service sector (91,000), 24,500 jobs would be added in the manufacturing sector. Sectoral effects are linked to the energy and labour intensity in each sector. Hence, while added value would decline by 5.76% in the energy sector, it would grow by 1.8% in the sector producing intermediate goods, 0.41% in the investment goods sector, 0.61% in the consumption goods sector, 0.85% in the building sector, 1.31% for transport and communications and 1.62% for the other sectors.

---

3 Bureau du Plan-Erasme (1993).

## 3. The White Paper on fiscal reform

In Italy, the White Paper on fiscal reform[4] was presented by the Minister of Finance in December 1994. A few days later, the Italian government resigned. As a matter of fact, there has been no official discussion of this document within Parliament. In any case, it still seems useful to examine its content since the White Paper represents the culmination of a lengthy political and cultural debate and is considered an effective basis for a fruitful discussion about the future of the Italian fiscal system[5] -- even if the structure of the proposed reforms is sometimes controversial.

This remark is particularly true when environmental taxation is taken into account directly. The White Paper marks the first time in an Italian official document that a Pigouvian tax has been carefully considered as a mainstay of the fiscal system. In particular, it should be recalled that the White Paper defines the three main goals of the fiscal reform as follows:

1) Fiscal federalism. In Italy, there is currently substantial centralization of the tax system, and local and regional authorities derive the bulk of their financing from grants or revenue-sharing. The proposal is to shift a large amount of taxes from the centre to the periphery in order to enhance the efficiency of public expenditure and promote the responsibility of local authorities.
2) Limits on personal taxation. Due to the combination of a very progressive income tax and a high level of inflation in previous decades, the amount of revenue raised through direct taxation is now overwhelming. The proposal is to shift the bulk of taxation away from personal income and to tax goods to a greater extent. In this context, environmental taxation is clearly included.
3) Simplification. In Italy, four taxes (personal income tax, VAT, excise on gasoline and withholding tax on interests) provide with great efficiency and little equity some 72% of total revenue. Sixteen taxes -- included the four just mentioned -- supply some 97% of

---

4 Ministero delle Finanze (1994).

5 See the recent discussion within the "Italian Society of Public Economists", held in Pavia in October 1995, where the White Paper was at the centre of the debate.

total revenue. The proposal is to call a halt to the Italian "one hundred taxes"[6] and to limit the complexity of the tax system favouring the extremely widespread tax evasion which represents a special negative feature of the current tax system.

Before examining the proposed changes in the structure of environmental taxes, however, it is necessary to deal briefly with the current system of taxation of environmental resources.

## 4. The present structure of environmental taxes

Table 1 gives an overview of existing environmental taxes and charges in the Italian system. Apart from some small charges and the use of tax differentiation for promoting the use of unleaded gasoline, these levies represent a very limited share of total revenue. A fee was introduced in the past to limit the use of plastic bags, but it has been phased out and replaced by a charge on domestically produced or imported primary plastic material. Charges are also used to finance collection or recycling of specific wastes.

A landfill tax has been recently introduced in Italy with the provisions defined in Article 3 of the Law n. 549 (December 28, 1995). The goal of these provisions is to favour a lower production of wastes at the source and the recovery of energy and raw materials from wastes.

Starting from January 1, 1996 the landfilling site operators must pay the tax to the Regions and will be obliged to shift the tax by raising their charges to waste disposers. The taxable base is represented by the amount of wastes in weight and the rate of the tax is differentiated taking into account the environmental impact of different kind of wastes. The rate for inert wastes could be chosen by the Regions between 2 and 20 lire per kg; for special wastes between 10 and 20, for urban and assimilated wastes between 20 and 50.

One fifth of the expected revenue flowing from taxes should be devoted to a special regional fund that could provide incentives to favour a lower production of wastes, to recover energy and raw materials from wastes, to reclaim contaminated sites and to finance the creation and conservation of natural protected areas.

---

6 Tremonti (1986).

Table 1. Environmental taxes and charges in Italy.

| *Sector* | *Type of tax* | *Objective* | *Basis* | *Collector* | *Payer* |
|---|---|---|---|---|---|
| **Water** | Waste water tax | Partial financing collection and treatment | Volume and quality of water | Local water company | User |
| | Tax on release of polluted water in the environment | Partial compensation of damage | Volume and quality of water | Local water company | Polluting enterprise |
| **Waste**[a] | Municipal solid waste tax | Partial financing collection and treatment | Area of housing | Commune | User |
| | Plastic bag fee[c] (L 100 per bag) | Reducing consumption of plastic bags | Number of units on the national market | State (L 55 bn in 1990) | Producer Importer |
| | Plastic containers and packagings for liquid charge | Financing recycling | Primary material produced or imported | National consortium | Producer Importer |
| **Oil** | Lubricating oil charge (L 5/l) | Financing collection and recycling | Oil sale | Consortium for used oil | Enterprises |
| **Batteries** | Lead battery charge | Financing collection and recycling | Battery sale | Consortium for lead battery and lead waste | Producer Importer |
| **Unleaded gasoline**[b] | Tax reduction | Promoting use of unleaded gasoline | Gasoline sale | State | Benefit to the consumer |
| **Aircraft** | Noise tax | - | Noise | State | Airline |
| **CFC**[d] | Deposit/refund | Promoting CFCs collection and recycling or disposal | CFC volume | Consortium for CFCs | Buyer of equipment |

a) Waste exports to non-OECD countries can only take place after a deposit is made.
b) There are also taxes on motor fuels and on car registration.
c) Abolished in December 1993.
d) Created in December 1993.

Source: OECD, (1994).

One of the risks linked to the adoption of a landfill tax is the promotion of fly-tipping. In order to avoid the illegal disposal of wastes the law establishes a shared responsibility in the payment of the tax and the related fines of the waste-disposers together with the user and/or the owner of the land where the disposal takes place, while keeping the criminal sanctions against this type of activity.

The expected revenue for the Regions in 1996 amounts to about L 1'050 billion. In the environmental field, the only kind of taxation that is also important in terms of the revenue flowing to the Treasury is related to energy taxes. In Italy, excises are levied on all kinds of fuels except coal. A list of the different state or local taxes which apply to energy is given in Table 2. Quite clearly, energy taxation has been implemented on revenue -- not environmental -- grounds. But in any case, it is very important from a quantitative point of view.

Table 2. Energy taxes in Italy.

| *Mineral Oils* | *Methane* | *LPG* | *Electricity* |
|---|---|---|---|
| Excise on mineral oils | State consumption tax | Excise | State consumption tax |
| Regional surtax on gasoline[a] | Regional surtax | Value added tax | State surtax |
| Value added tax | Value added tax | | Provincial surtax[b] |
| | | | Municipal surtax[b] |
| | | | Value added tax |

[a] In 1993 introduced only in Piemonte and Puglia
[b] Starting in 1994, the revenue goes to the Treasury

The level is also relevant in an international comparison: 6.25% of total receipts of general government, 9.5% if VAT is included. The relative revenue is shown in Table 3. In terms of GDP, the revenue flowing from energy taxation -- excluding VAT -- is about 2.8 points, equivalent to some L 46,500 billion in 1993.

Finally, Figure 1 displays changes in revenue from energy taxation over time, showing clearly the steady increase in the amount of energy taxes up to 1991 and the relative stabilization of revenue in the following years.

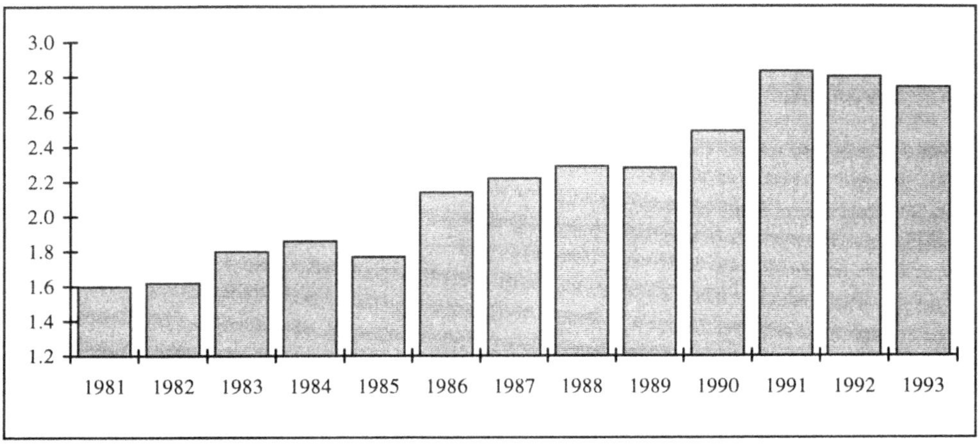


Source: Siniscalco, 1995

Figure 1. Energy taxation in Italy (as a percentage of GDP).

Table 3. Revenue from energy taxes[a].

| | | |
|---|---|---|
| *Mineral Oils* | Excises | 37,765 |
| | Border surtax | 318 |
| *Methane* | State consumption tax | 4,742 |
| | Regional surtax | 541 |
| *LPG* | Excise | 733 |
| | Border surtax | 328 |
| *Electricity* | State consumption tax | 487 |
| | Provincial surtax | 1,083 |
| | Municipal surtax | 566 |
| (1) | Total revenue | 46,563 |
| (2) | General Govt total revenue | 745,245 |
| (1) / (2) | | 6,25% |

[a] billion lire 1993

It should also be noted that per capita revenue from energy taxes varies widely according to region. As can be seen from Table 4, it is nearly twice as high in the North as in the South. The reason is linked to climatic factors (heating is only significant in the North) and to existing differentials in the industrial structure (firms are mainly located in the North). Since the distributive impact of energy taxation is generally considered one of the weakest features of energy taxation,[7] this point is quite important since income levels are lower in the South than in the North.

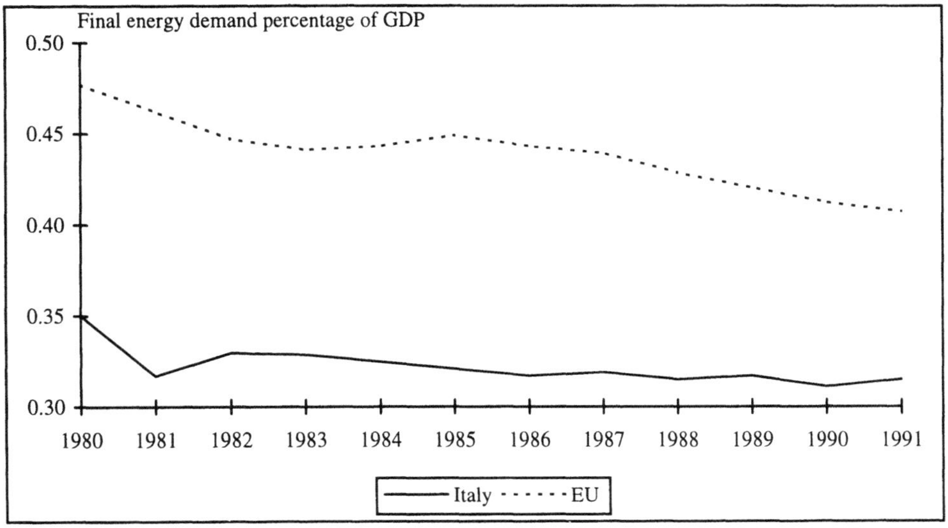


Source: Siniscalco, 1995

Figure 2. Energy intensity.

Following this high level of taxation, energy intensity in Italy is comparatively lower than in the other industrialized countries. Figure 2 shows the trend of energy intensity -- measured as final energy demand as a percentage of a unit of GDP -- in comparison with the other EU countries.

---

7 Smith (1992).

The high price of gasoline in Italy explains to a large extent the differences in gasoline consumption compared with other industrialized countries. Figure 3 shows more specifically the impact of the price of gasoline on its consumption. In Italy, the tax wedge amounts to 76.4%, one of the highest figures within the OECD countries (Table 5 shows the share of taxes out of total gasoline prices in some OECD countries in 1994).

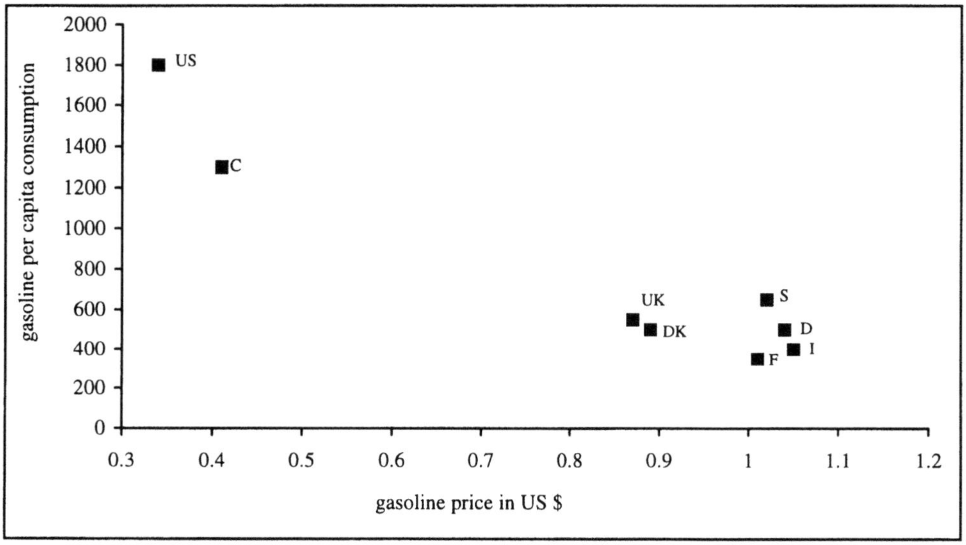


Source: Siniscalco, 1995

Figure 3. Prices and gasoline consumption, 1994.

In Italy, the level of energy taxes and consequently of final energy prices is higher in the civil and transport sector than in industry, since in the latter sector increases in the level of taxation are strictly constrained by the need to promote the international competitiveness of domestic production.

The fact that the high prevailing level of energy taxation in Italy is due to revenue generation rather than environmental grounds, can be easily proven since coal, that has a higher carbon content than mineral oil or natural gas, is taxed less even if it is the most polluting energy source.

Table 4. Energy taxes (without VAT). Revenue distribution among regions[a].

| | NORTH | CENTRE | SOUTH | ITALY |
|---|---|---|---|---|
| Electricity | 1,211 | 453 | 472 | 2,136 |
| Natural gas | 3,944 | 987 | 352 | 5,283 |
| Mineral oils | 19,390 | 8,368 | 10,007 | 37,765 |
| LPG | 401 | 108 | 224 | 733 |
| Total | 24,946 | 9,916 | 11,055 | 45,917 |
| Inhabitants (thousands) | 25,550 | 11,012 | 21,184 | 57,746 |
| per capita revenue (thousands) | 976 | 900 | 522 | 795 |

[a] billion lire 1993

Table 5. Tax share of gasoline prices. (Second Quarter 1994).

| | | | |
|---|---|---|---|
| France | 80.8 | Luxembourg | 67.9 |
| Germany | 76.6 | Denmark | 67.5 |
| Italy | 76.4 | Ireland | 67.5 |
| Sweden | 75.8 | Norway | 67.3 |
| Netherlands | 75.4 | Turkey | 66.6 |
| Greece | 74.6 | Austria | 61.5 |
| Portugal | 73.9 | Australia | 55.0 |
| Belgium | 73.8 | Canada | 50.9 |
| United Kingdom | 73.2 | Japan | 48.3 |
| Finland | 72.2 | New Zealand | 48.6 |
| Switzerland | 71.3 | United States | 34.6 |
| Spain | 68.5 | Mexico | 9.1 |

Source: OECD, 1994

## 5. The White Paper proposals

At present, given the very high level of existing taxes, increasing energy taxation in Italy could only be envisaged on environmental grounds. A change in energy taxes could be justified according to three different goals:

1) Improving air quality through lower energy consumption and the use of less polluting energy sources.
2) Strengthening the external competitiveness of the domestic industrial sector.
3) Providing new revenue sources that could be exploited for cutting income tax rates and in particularly offering tax relief for labour incomes.

This policy change should be included in the perspective of shaping a national plan for curbing $CO_2$ emissions following the obligations deriving from the Rio Convention and the EU decision to stabilize carbon dioxide emissions in the year 2000 at the 1990 level.

The White Paper accepts the suggestion put forward in the Proposal of a Directive presented by the European Commission[8] to introduce an energy-carbon tax with a revenue amounting to about L 10,000 billion. This additional revenue could be recycled back to the economy by cutting the level of other tax rates in order to exploit a double dividend in the perspective opened up by the Delors Report[9]: the first dividend is provided by the curbing of carbon dioxide emissions into the atmosphere, the second by the diminution of the deadweight loss associated with the existence of distortionary taxes.

This recycling of the revenue could be implemented through three different channels:

- A cut in the tax rates of the personal income tax. In this case, families' disposable income is fully compensated for the increased energy tax burden. At the macroeconomic level, limited expansion is achieved through an increase in the level of personal consumption.
- A cut in social security contributions. Firms are compensated for the increased tax burden. In this case firms with high labour intensity are favoured whereas energy-intensive firms are penalized. Expansionary effects on employment will probably

---

8 See Com (92) 226 in OJ No C 196 - 3/8/1992, p. 1

9 See European Commission (1993).

follow,[10] promoting a new model of sustainable growth, as described in the Delors Report.

- A cut in VAT. The price increase generated by the energy tax is compensated. At the macroeconomic level positive effects are derived due to the avoidance of a wage-price spiral and of the following inflationary impact. This means in particular the possibility to avoid an increase in the burden of interest payments on the existing stock of public debt, with negative consequences on public expenditure and the State budget-deficit.

The White Paper embraces the idea of using the revenue to cut personal income tax rates. This choice corresponds to the general goal to shift the burden of taxation away from persons and on to goods. A suggestion is put forward to cut the highest tax rate from the current level of 51% to 40% or 45% according to different options, and to reduce the average rate for the majority of taxpayers to below 25-30%.

A revision of energy taxes is envisaged, following the general ideas prevailing in the European debate, according to the carbon content of each energy source, while ensuring that an effort is made to keep the relative tax burden between the different sectors, i.e. households, industry, transport, generally constant.

In line with the overall inspiration of the fiscal reform project, existing energy taxes are included in a General Energy Tax that applies the same system to all the different energy sources and is strictly coordinated with VAT.

The tax increase will be applied to the industrial sector only if a decision is taken at the European level to implement a common energy-carbon tax. This provision is considered unavoidable since unilateral implementation of the energy-carbon tax would significantly weaken the overall competitiveness of the Italian industry.

Increasing existing energy tax rates would provide additional revenue amounting to L 10,000 billion that could be recycled back to the economy through a cut in income tax rates. This result could be achieved according to two different options that are considered in the

---

[10] For a survey of the likely effects of this kind of manoeuvre, see Majocchi (1996).

White Paper, each of which calls for an increase in the price of energy of about $ 7 for a barrel of oil equivalent, largely similar to the European proposal that results in a price increase of $ 10 per barrel of oil.

## 6. The economic and environmental effects of the manoeuvre

The economic impact of this manoeuvre is shown in Tables 6 and 7. In the first hypothesis (Table 6) all energy sources and all sectors -- households, industry and transport -- are subject to the new energy levy, that is shaped according to the carbon content of each energy source. In this case, it is assumed that an agreement has been reached at the European level on the harmonized implementation of the new energy tax. Furthermore, in the White Paper the new tax burden is modulated in a way that leaves relative prices in the different sectors more or less unchanged and is equivalent to a price increase of L 126 per litre of gasoline and L 142 per litre of diesel.

In the first year following the implementation of the manoeuvre -- and also taking into account the impact of the cut in income tax rates -- GDP increases by 0.56%, while in successive years it remains more or less stable. The consumption price level rises in the first year (+0.75%), but then the rate of increase diminishes rapidly, while the price impact on production costs in the industrial sector is more sensitive in the first year (+1%), but tends rapidly towards zero in the following years.

In the second hypothesis (Table 7), where it is assumed that no harmonization takes place at the European level, industry is not taxed and the new tax burden is concentrated on households and transport. It is shaped accordingly to the carbon content of each source, but the price changes following the manoeuvre are not proportional to the previous prices in the different sectors. In this case, the gasoline price increase is L 161 per litre, while diesel prices increase by L 182.

The impact on consumption prices is very similar, while it is naturally different as far as industrial prices are concerned since industry is exempted in this case. However, it is important to note that the growth rate of GDP is practically the same as in the first hypothesis. Exempting the industrial sector has no real impact on the growth of the economy.

Table 6. White Paper Proposal 1.

| | Civil | Transport | Industry |
|---|---|---|---|
| Gasoline | | X | |
| Diesel/fuel oil | X | X | |
| Heavy fuel oil | | | X |
| Natural gas | X | | X |
| Coal | X | | X |
| Electricity | X | | X |

| **CHANGES TOWARDS THE REFERENCE SCENARIO** | | | | | | |
|---|---|---|---|---|---|---|
| | 1995 | 1996 | 1997 | 1998 | 1999 | 2000 |
| | (billion lire) | | | | | |
| **ENERGY TAXES** | 11,167 | 10,144 | 9,893 | 10,079 | 10,241 | 10,319 |
| **GDP** | (changes in growth rates) | | | | | |
| | 0.56 | -0.10 | -0.16 | 0.03 | 0.01 | -0.07 |
| **PRICE INDEXES** | (changes in growth rates) | | | | | |
| Consumption prices | 0.75 | 0.29 | 0.07 | 0.04 | 0.03 | 0.03 |
| Energy price | 7.28 | 1.01 | -0.03 | -0.23 | -0.06 | -0.03 |
| Production prices | 1.00 | 0.01 | 0.03 | 0.04 | 0.04 | 0.05 |
| Industry energy prices | 7.01 | -0.21 | -0.09 | -0.07 | 0.24 | 0.29 |
| **ENERGY PRICES** | (price changes in percentage) | | | | | |
| **Civil sector** | | | | | | |
| Natural gas | 8.5 | 8.3 | 8.0 | 7.7 | 7.7 | 7.7 |
| Coal | 28.4 | 27.0 | 25.7 | 24.3 | 24.2 | 24.3 |
| Heating oil | 10.1 | 9.9 | 9.8 | 9.6 | 9.6 | 9.6 |
| Electricity | 4.5 | 4.3 | 4.1 | 3.9 | 3.9 | 3.9 |
| **Industry** | | | | | | |
| Natural gas | 13.6 | 13.0 | 12.4 | 11.8 | 11.8 | 11.8 |
| Coal | 27.6 | 26.2 | 25.0 | 23.6 | 23.6 | 23.6 |
| Heavy fuel oil | 22.2 | 21.5 | 20.9 | 20.2 | 20.2 | 20.2 |
| Electricity | 2.8 | 2.6 | 2.5 | 2.4 | 2.4 | 2.4 |
| **Transport** | | | | | | |
| Gasoline | 8 | 7 | 7 | 7 | 7 | 7 |
| Diesel | 11.8 | 11.6 | 11.4 | 11.1 | 11.1 | 11.1 |

Table 7. White Paper Proposal 2.

| | Civil | Transport | Industry |
|---|---|---|---|
| Gasoline | | X | |
| Diesel/fuel oil | X | X | |
| Heavy fuel oil | | | |
| Natural gas | X | | |
| Coal | X | | |
| Electricity | X | | |

| **CHANGES TOWARDS THE REFERENCE SCENARIO** | | | | | | |
|---|---|---|---|---|---|---|
| | 1995 (billion lire) | 1996 | 1997 | 1998 | 1999 | 2000 |
| **ENERGY TAXES** | 11,977 | 10,233 | 9,938 | 9,972 | 10,037 | 10,058 |
| **GDP** | (changes in growth rates) | | | | | |
| | 0.49 | -0.11 | -0.04 | 0.05 | 0.03 | -0.05 |
| **PRICE INDEXES** | (changes in growth rates) | | | | | |
| Consumption prices | 0.75 | 0.11 | 0.01 | -0.01 | -0.01 | -0.01 |
| Energy price | 10.43 | 1.38 | -0.05 | -0.33 | -0.09 | -0.05 |
| Production prices | 0.01 | -0.02 | 0.01 | 0.00 | 0.00 | -0.00 |
| Industry energy prices | 0.05 | -0.11 | 0.08 | 0.02 | 0.00 | -0.01 |
| **ENERGY PRICES** | (price changes in percentage) | | | | | |
| **Civil sector** | | | | | | |
| Natural gas | 13.6 | 13.1 | 12.7 | 12.2 | 12.2 | 12.2 |
| Coal | 45.2 | 43.0 | 41.0 | 38.7 | 38.7 | 38.7 |
| Heating oil | 16.0 | 15.8 | 15.6 | 15.3 | 15.3 | 15.3 |
| Electricity | 7.2 | 6.9 | 6.6 | 6.2 | 6.2 | 6.2 |
| **Industry** | | | | | | |
| Natural gas | 0.00 | 0.00 | 0.00 | 0.00 | 0.00 | 0.00 |
| Coal | 0.00 | 0.00 | 0.00 | 0.00 | 0.00 | 0.00 |
| Heavy fuel oil | 0.00 | 0.00 | 0.00 | 0.00 | 0.00 | 0.00 |
| Electricity | 0.00 | 0.00 | 0.00 | 0.00 | 0.00 | 0.00 |
| **Transport** | | | | | | |
| Gasoline | 10 | 10 | 9 | 9 | 9 | 9 |
| Diesel | 15.0 | 14.8 | 14.5 | 14.2 | 14.2 | 14.2 |

These are the main economic changes following the implementation of the fiscal reform regarding the increase in energy taxes, whose revenue is targeted to a cut in income tax rates. Notwithstanding this, the most important effect of the manoeuvre pertains to the environment, since it has a clear impact on air quality. The combined index of air quality -- including carbon dioxide emissions, $SO_2$, $No_x$ and particulates -- declines from 97.1 in 1994 to 95.4 in the year 2000 in the reference scenario and to 92.6 taking into account the impact of the manoeuvre, accompanied by a significant slowing in the rate of increase in emissions of carbon dioxide, that drops from the level of 110 in the reference scenario to 106.25 in the year 2000 -- 100 being the index level in 1990.

## 7. Road pricing and transport taxes in the White Paper

Energy taxation represents an important example of implementing the principle that requires internalization of the external costs linked to the use of natural resources, in so far as prices should not only mirror private costs but also social costs. Another extremely important sector where this efficiency principle should be applied is the use of public land within urban areas. This topic is one important aspect of the more general problem regarding internalization of the external costs linked to transports. In this sector, GDP growth has not kept up with the rate of increase of polluting emissions. Within the European Union, while the GDP growth rate was 25% during the eighties, $CO_2$ emissions linked to transport were increasing by 35%. The forecast up to the year 2000 is that energy demand in the transport sector will rise by 25.5%, with an equivalent increase in carbon dioxide emissions.

It is well known that there are important externalities connected to the transport sector. These externalities are linked to the problems of congestion -- due in urban areas to both traffic and parking, that imply significant costs for the community. In addition, the costs of accidents should also be considered, either in terms of deaths or medical treatment of injured persons. Lastly, there are the costs that arise from different types of pollution -- air, noise, landscape etc. -- included in the potential costs in the area of global pollution.

There are two main instruments that could be utilized to internalize the external effects flowing from the transport sector: (a) raising taxes on fuels used by transport vehicles; (b) exploiting the instrument of road pricing.

A recent study prepared by ECOTEC (1994) on behalf of the European Commission suggests the conclusion that it is possible to cover the external costs linked to the transport sector by charging for 80% of the congestion costs flowing from urban traffic through road pricing -- that will imply an average price of 0.12 Ecu/km -- and by paying the remainder through a fiscal levy raising the price of fuel, in a measure equal respectively to 53% for petrol and 77% for diesel.

This option must be carefully considered with regard to the Italian case,[11] particularly if a system of fiscal federalism is to be implemented. The problems of urban congestion are strictly linked to the shape of the city and to the territorial structure. Furthermore, it should be underlined that in this case as well -- as in the case of the energy tax -- the goal that must be achieved is not to dismiss the market, but rather to eliminate existing distortions, thereby enabling the market to reach an optimal solution. If the problem of urban traffic must be tackled, it is possible to use a tax on parking -- seen as a charge for using public property -- for residents, whereas for non-residents road pricing -- backed by the other necessary measures on the demand side -- could promote different alternative choices:

- using public transport;
- leaving vehicles in peripheral parking facilities and using the different public transport options to reach the city centre;
- going into the city and paying the tax for using the public streets and possibly the tax for using the parking facilities.

## 8. The general philosophy of the White Paper

All these options are founded on the principle that users should pay a price for exploiting natural and environmental resources. In general terms -- even if it could be unavoidable to strengthen some normative provisions in order to ban the consumption of goods that are particularly harmful to the environment -- it seems important to stress again that an effective environmental tax policy should be based on a limited number of taxes with a large tax base,

---

[11] On this topic see the papers presented to the recent Conference held in Bologna: "Road pricing: European experiences compared", October 9, 1995.

that would not pose any major compliance problems for taxpayers or entail high administrative and control costs.

The importance of these conditions has been underlined in the White Paper particularly with regard to Italian fiscal system, where some 5 million VAT taxpayers are now involved. Among these taxpayers, some 4.5 million have a turnover lower than L 500 million each year. This structure must be fully understood in order to analyse in depth the behaviour of the Italian lawmakers, who have tried for years to manage a mass fiscal system with the same criteria that could be utilized fruitfully in a system with a limited number of taxpayers. Consequently, compliance and administrative costs for firms have been raising rapidly, without any sizeable increase in the amount of revenue received from the Treasury and with no reduction in the amount of fiscal evasion.

A recent study shows that small firms have to bear very large costs -- totalling some L 35,000 billion -- for management of the accounting system and compliance with the fiscal provisions. These costs vary according to the level of turnover, ranging between 2 million each year for a craftsman with a turnover of 20 million and 8 million for a craftsman with a turnover of L 170 million.

The collection of taxes and social security contributions adds a supplementary burden to these indirect costs. Global revenue collection is currently very high indeed: for an income of 30 million it reaches some 45% -- with a total burden of about 13 million -- while for every additional million the marginal rate is above 57%. It is quite clear that these additional costs represent a heavy burden that diminishes the ability of small firms to build up new capacity and hence create jobs.

Since it is inconceivable -- given the size of the public debt -- to lower the level of taxation significantly in the short run, the only way out of this dilemma could be implementing on a priority basis the following policy choices that have been included as the main goals of the fiscal reform suggested by the White Paper:

a) Shifting the bulk of taxation away from labour. This is the proposal put forward by the European Commission, that suggests raising the level of energy taxation and more generally taxing the use of natural resources in order to use this additional revenue to finance a parallel cut in social security contributions, with special emphasis on unskilled or low-skilled workers. The new jobs that could be created through this manoeuvre has been estimated -- for the whole European Union -- at some 2-2,5%. The White Paper shares the idea of raising energy taxation, backed by an increased taxation on transport - especially in the more congested urban areas -- and by charges levied on tourism exploiting the Italian cultural and natural heritage while not contributing in a significant way to covering the huge costs of its conservation.

b) Shifting the bulk of taxation from the central to the local level. This policy choice is complementary to the first one since, while the income tax is typically a national tax, environmental taxes are generally -- by their very nature -- local taxes as they are linked to territory. Also from this perspective, housing taxation should be considered and -- reformed, simplified and coordinated with environmental taxation -- could represent the core of local taxation.

c) Simplifying the fiscal system. This implies a choice of taxes that should generate significant revenue with low collection and control costs. Consequently, compliance costs for taxpayers could be significantly diminished as well.

These policy choices, while offering promising prospects of an increase in employment and better conservation of environmental resources, are at the same time a necessary pre-requisite for improving Italian public finances since tax administration would be facilitated and each tier of government would be responsible for financing its own expenditures. A system of fiscal federalism implies that each level of government must provide its own revenue, leaving it up to the national level to redistribute resources through unconditional grants in order to equalize territorial income distribution. This system of fiscal federalism could provide different positive effects:

1) Efficiency would be ensured since each level of government would be made accountable for its behaviour and would be able to take on a new expenditure only if it could persuade its own citizens that the expenditure provided benefits higher than the cost of the taxes needed to finance it.

2) This mechanism represents the only effective curb on the expansion of public expenditure since at each level of government consensus would be maximized if -- given the amount

of services provided by the public authority -- the expenditure -- hence the levying of taxes on behalf of citizens -- was minimized.

3 ) Equity would be promoted since a regional redistribution of financial resources is needed so that the citizens all over the country are guaranteed -- given the same fiscal effort -- that they will receive comparable amounts of public goods.

## 9. Conclusions

Up to now, there has been no specific follow-up of the proposals included in the White Paper. In any event, however, they have yielded one important result. The debate about the use of environmental taxes has been revamped, as a result of which the Ministry of Environment has been able to include a proposal for a new landfilling tax in the draft budget for 1996 -- that has been recently approved by the Parliament.

The implementation of the proposed increase in the general energy tax -- including a share measured according to the different carbon content for each energy source -- is strictly linked to the final result of the European-wide debate on the energy-carbon tax. It is particularly difficult for reasons tied to its competitive position for Italy to take a unilateral decision to implement this new tax. But, at least, the White Paper has been able to put on the table the need to include environmental taxes in the broader context of a fiscal reform targeted to the achievement of a new model of sustainable growth, as intended by the Delors Report.

## References

**Bovenberg, A. L. and De Mooj, R. A.** (1993) Do environmental taxes yield a double dividend?. Erasmus University, Rotterdam.

**Bureau du Plan-Erasme** (1993) Un redéploiement fiscal au service de l'emploi. Réduction du coût salarial financée par la taxe $CO_2$/énergie. Bruxelles.

**ECOTEC** (1994) Cost-Benefit Evaluation of the Fifth Environmental-Action Programme: Transport Sector. Brussels.

**European Commission** (1993) Growth, Competitiveness and Unemployment. Brussels.

**Goulder, L. H.** (1995) Environmental taxation and the double dividend : a reader's guide. *International Tax and Public Finance, 2* : 157-183.

**Majocchi, A.** (1996) Green Fiscal Reform and Employment : A Survey. *Environmental and Resource Economics* (forthcoming).

**Ministero delle Finanze** (1994) Libro Bianco sulla riforma tributaria. Roma.

**Ministry of Finance** (1991) The Swedish tax reform. Stockholm.

**OECD** (1994) *Environmental Performance Review : Italy*. Paris.

**Siniscalco, D.** (1995) Note in tema di tassazione ambientale. SIEP, Pavia.
**Smith, S.** (1992) The distributional consequences of taxes on energy and the carbon content of fuels. *European Economy, Special Edition n° 1:* 241-268.
**Tremonti, G.** (1986) Le cento tasse degli italiani. Il Mulino, Bologna.

Environmental Policy Between Regulation and Market
ed. by C. Jeanrenaud


# 2.3 Cars and Environment in Switzerland: What Kind of Taxes?

**Nicolas Wallart,**
**University of Geneva, Geneva, Switzerland**

## 1. Introduction

Beyond their usefulness, motor vehicles are responsible, directly or indirectly, for pollution, greenhouse gases, waste, noise, and accidents. These are external costs, i.e. costs that are not borne by car drivers. Moreover, cars are also responsible for a significant part of the consumption of non-renewable natural resources. Polluting emissions vary according to place, time of day or year, but may also vary according to car, as well as fuel consumption. On the one hand, car-makers have insufficient incentives to reduce emissions because this implies an increase in production costs; on the other hand, car users fail to take these costs into account when making their choices.

Economists often propose taxes to internalize external costs of pollution or lower fuel consumption (see for example Maibach *et al*, 1992; Neuenschwander *et al*, 1992; Grosclaude and Soguel, 1992; Meier, 1993). The quoted authors advocate fuel taxes; nevertheless, other taxes could be used, such as "road pricing", taxes on new cars, annual taxes, etc. Accordingly, section 2 of this paper deals with the question of the most appropriate tax base to internalize external costs and lower fuel consumption. Section 3 deals with a particular tax base, new cars. It relies on simulations to show the impact of a tax on new vehicles, of a tax/subsidy scheme, of a tax on large polluters, and finally compares the results with stricter standards. A conclusion can be found in section 4.

## 2. Pollution, sustainable development and the tax base

*Cars and the environment*

Motor vehicles have varying impacts on the environment. They are responsible for a significant share of air pollution, noise, climate change and congestion. The main air pollutants from cars in Switzerland are:

- $NO_X$. The introduction of obligatory catalytic converters for new gasoline vehicles sold has decreased $NO_X$ emissions; however, immissions still remain higher than the maximum value defined in the Air Protection Statute [Ordonnance pour la protection de l'air], in cities and along main roads (OFEFP, 1994).
- *HC*, during the summer, combined with $NO_X$, produces ozone. There is no maximum value for HC immissions in Switzerland. Such values do exist for ozone, but are regularly exceeded (id.).
- *Particles* are emitted by diesel engines, which are not common in Switzerland. Although maximum values are sometimes exceeded, the trend is downwards. However, it is thought that particles may be carcinogenic.

In Switzerland, about 30% of the population suffers critical levels of *noise* (>60dB(A)) from road traffic (id.). *Congestion* also has an impact on urban pollution; congestion costs are high, especially in terms of time loss. For motor vehicles, *climate change* mainly originates in $CO_2$ emissions by fuel combustion.

In addition, there is increasing agreement that the economy has to take the direction of sustainable development. This is a major challenge for our societies. Motor vehicles, because of their responsibility in the consumption of non-renewable resources and the greenhouse gas emissions, are a major part of this challenge, perhaps the most crucial element if we consider the potential growth in world road traffic.

Many definitions and interpretations of the concept of sustainable development exist (see for example Pearce *et al*, 1989). All these definitions are normative and are open to criticism. However, most of them imply a strong decrease in world use of fossil energy. Motor vehicles are one of the largest consumers of fossil energy; moreover, the steady increase in number of cars worldwide is pushing future potential energy consumption higher and higher. From the standpoint of sustainable development, instruments to reduce fuel consumption of motor vehicles are becoming more and more necessary. Switzerland recently proposed a convention along these lines.[1]

To give an idea, we can use the definition of the Dutch report "Sustainable Netherlands", which stipulates a 50% reduction in fossil fuel use by 2010 (Friends of the Earth, 1994). Even if smaller reductions in the transport sector could be offset by larger reductions in other sectors, this definition implies a significant reduction in fuel use.

*What determines pollution and fuel consumption?*

Total pollution (or total fuel consumption) of motor vehicles depends on two factors: existing vehicles and their use. For each pollutant, this implies:

Total pollution = average emissions per kilometer • number of kilometres travelled

For some pollutants, localization is important, but for others (like $CO_2$ emissions), it is not. It is useful to carry out a detailed analysis of these two factors.

*Average emissions per kilometer*

Average emissions per kilometer depend on:

- driver behaviour;
- vehicle maintenance;
- the technological characteristics of vehicles and fuels used, i.e. what can be called the technology.

---

1 "Ordonnance sur la réduction de la consommation spécifique de carburant des automobiles" [Statute on Reduction of Automobile Gasoline Consumption]. This bill, which is dated 18 April 1995, states that the average gasoline consumption of new cars sold has to be reduced at a rate of 3% per year, to reduce the level in 2001 to 85% of the value for 1996. Car importing firms have to comply volontarily with this objective, otherwise new instruments will be introduced.

Driver behaviour can be influenced by transport policy. Slowing traffic, introducing harsher punishment for people who exceed the speed limit, better roads or driver education could influence behaviour. However, the effect on pollution is not straightforward: slowing the traffic can sometimes even increase pollution if it increases congestion or if engines have to turn at a more polluting speed. Moreover, even in the best of cases it is not possible to obtain a strong and lasting reduction of traffic emissions with measures influencing behaviour. This can only be a complementary action.

Likewise, improving vehicle maintenance cannot lead to a strong decrease in pollution, considering the good general condition of Swiss vehicles. Globally, these actions cannot be forgotten, but are not a long-term solution.

The technological characteristics of vehicles and fuels can be directly influenced by policy. For example, the introduction of compulsory catalytic converters has led to a significant reduction in new car pollution. Recent environment statistics reflect the success of such a measure: in Switzerland, despite increasing mobility, total emissions of $NO_X$ and HC from road traffic are decreasing. Unlike the two preceding measures, and in view of the success of previous measures, it seems that influencing the technological characteristics of vehicles is a promising approach in the long-term.

*The number of kilometres travelled by cars*

The number of kilometers travelled by motor vehicle depends on supply and demand for the service they offer. Although it is not easy to distinguish them, we can say that *demand* depends on:

- the price per kilometer;
- the price and quality of the substitute good (public transportation);
- income;
- non-monetary costs (travel time for a given distance);[2]
- behaviour of populations, determined for example by the housing structure and the location of economic activities.

---

[2] Travel time is a non-monetary cost for individuals but a monetary cost for firms.

Supply is determined by:

- the capacity and quality of the road network;
- the parking infrastructure;
- characteristics of vehicle used (comfort, safety, etc.).

Usually, transport policy aimed at reducing pollution is directed towards a reduction of supply and demand.

Measures to reduce demand have been tested in many places. In Switzerland, the supply of public transportation has been improved in big cities and agglomerations (especially in Zurich) without decreasing private traffic. Measures such as an increase in the travel cost through a supplementary fuel tax are generally not very effective: the price elasticity of fuel demand is much smaller than 1 (see also next point), and above all an increase in the fuel price is easily offset by an increase in the income of individuals. On the contrary, measures increasing travel time are more effective in reducing demand and encouraging substitution toward public transportation. Yet they raise acceptability problems. For example, a study about acceptability of an environmentally friendly transportation policy show that "less restrictive measures for motorists are best accepted" (Bütschi and Kriesi, 1994). Technological improvements to vehicles, car-pooling and electrical cars (that change supply) are well accepted, while supplementary taxes (that modify demand) or parking restrictions (that modify supply) are less well accepted.

This reflects the fact that individuals are not ready to change their use of cars. In England, a survey showed that 80% of motorists would find it difficult to do without their car, even if public transportation had to improve significantly (Lex report, 1995). As individuals are not ready to reduce their use of cars, policy proposals to reduce demand raise acceptability problems.

Supply measures have proved more effective. In cities, control and pricing of parkings, partial closing of city centres and physical and psychological barriers are generally effective. However, these measures seem to be restricted to cities and agglomerations, and applied only for local transportation policy. Outside of agglomerations, the opposite may be seen. In Switzerland for example, proposals for building a third lane for the Bern-Zurich highway or finishing the highway network mean an increase in supply. All in all, the total network length has remained approximately the same since 1984 (Bürgenmeier *et al*, 1994). Congestion costs imply that a diminishing supply is not the best way to reduce externalities. Moreover, it is hard to imagine that this could be a long-term solution.

*Which solution?*

Most likely, the number of kilometres travelled will continue to increase in the next years, especially with population growth, economic growth and the continuing decrease in the size of households. Measures to reduce traffic are effective, but mostly apply to cities and agglomerations. They have an impact on local pollution. For global pollution, an increase in car travel costs or a better supply of public transportation does not seem able to curtail growth in car traffic, hence pollution.

Moreover, all measures aiming at reducing car use give rise to significant political opposition (Witherspoon, 1995). Measures aiming at lowering the number of kilometres travelled are not popular, hence unlikely to be adopted. It thus seems that the only measures able to reduce durably car pollution and consumption pertain to the technological characteristics of vehicles. If the number of kilometres travelled increases, pollution and consumption per kilometer have to be reduced more in order to reduce total pollution and consumption. This raises the question if such a decrease is possible, and economically feasible.

Different studies converge and show that a reduction in fuel consumption is feasible. A report of IEA (1992) states "that most important technological possibilities to increase fuel efficiency are not used by the market. Therefore, market conditions have to change notably in order that these possibilities may be applied, even partly". In the same way, a US study (DeCicco and Ross, 1994) concludes that increasing the energy efficiency from 28 to 46 miles per gallon (i.e. decreasing consumption from 8.4 to 5.1 litres per 100 kilometres) would cost about $ 800 per car. The fuel savings over the lifetime of the car would add up to many times this amount.[3] But a car maker who reduces the consumption of his cars would perhaps lose

[3] $ 800 are worth CHF 1,000 if $ 1=CHF 1.25. If 1 liter of gasoline costs CHF 1.20, and the average car reaches 100,000 km, the gasoline saving is worth CHF 3,960.

clients to the benefit of his competitors, who would increase power or comfort for unchanged consumption. Ledbetter (1993) finds approximately the same value: reducing the average consumption of cars sold in the US from 8.3 to 5.3 litres/100km would cost $ 750 per car, and would thus be "cost effective". Likewise, Greene and Duleep (1993) say that "substantial fuel economy improvements can be made by using proven technologies already incorporated in at least one mass production vehicle somewhere in the world today". For the individual, the additional cost of more efficient cars is offset by the fuel savings (at US price; at European price, fuel savings are much higher than the additional investment). In the case of pollution, the situation is likely to be similar; we will show later that, even with a substitution between current vehicles, pollution can be reduced significantly.

The studies show that *today,* the possibilities of building more efficicent cars exist and are profitable, but the market does not use them. Fuel consumption is one buying criterion among many, and by far not the most significant one (McCarthy and Tay, 1989). Nowadays, car makers use the following principle: "maximum engine power for reasonable consumption" (TCS, 1992). Government has to step in if this principle has to become "minimum consumption and emissions for reasonable power".

*Analysis of demand elasticities*

Another way to handle the problem is to analyse the results of studies about price and income elasticities in the motor vehicle sector. Mattei (1994) calculated price and income elasticities for different goods in Switzerland (see Table 1 for car and motorbike fuels).

Table 1. Fuel price and income elasticities in Switzerland (1975-1989).

| | Short-term | Long-term |
|---|---|---|
| Price elasticity (a) | -0.011 | -0.16 |
| Income elasticity (b) | 0.089 | 1.33 |
| Multiplying factor (b/a) | -8 | -8 |

Source: Mattei (1994).

This study shows that the price elasticity of fuel demand in Switzerland is very low. Other studies estimate different values, but it seems that globally, price elasticity is lower in Switzerland than in other countries.[4] From the table, following remarks can be made:

- Income elasticities show that fuel is a luxury good in the long run.
- The relation of income elasticity to price elasticity (b/a) shows that, if Swiss economic growth amounts to a long-term average of 2% yearly, and without any technological progress, fuel price would have to increase 16% annually in order to keep consumption constant.
- Internalizing external costs with an increase of the fuel tax of 1.- per liter[5] (+83%) leads to a decrease of consumption of only 13%; five years of growth are enough to reach the initial levels of consumption and pollution.
- If we aim at a 50% reduction in long-term consumption of fuel, this implies a huge initial increase in prices, then a continuous increase to offset the effect of economic growth on demand.[6]

We can already conclude that internalization of external costs of cars through a fuel tax is neither enough to lower long-term pollution, nor to lead to sustainable development. Table 2

4 For example, Spierer (1988) finds a long-term price elasticity of -0.25 to -0.35. In other countries, Dahl & Sterner, summing up 175 studies, find an average long-term price elasticity of -0.5.

5 This number corresponds to average numbers proposed by the Swiss studies quoted in the introduction.

6 Price increases as high as to reduce demand by 50% have never been observed. Therefore, it is dangerous to evaluate price increases from elasticities that are best used for marginal changes.

sums up price elasticities of fuel demand and compares them with new motor vehicle elasticities (the study aggregated cars and motorbikes).

In this study, vehicle demand is much more sensitive to price than fuel demand (19 times more in the long-term). In the same way, Blomqvist and Haessel (1978) show for Canada that the short-term price elasticity of new car demand is fairly high.

This is part of a more global problem, the question of equipment (machinery, fittings, cars, etc.), that is given more and more priority when dealing with environmental problems. In fact, once somebody has purchased equipment, influencing his behaviour is not an easy task. It is easier for government to influence the purchase of the equipment than its use once bought. For instance with cars, it is comparatively easy for an individual to buy a low-polluting car. On the other hand, once he has bought his car, it will be very hard not to use it, hence not to pollute. It is easier for government to apply environmental policy to the equipment purchase, and not to the equipment *use* (which is often not effective and raises acceptability problems).

Table 2. Price elasticity of fuel and vehicle demand in Switzerland (1975-1989).

| | Short-term | Long-term |
|---|---|---|
| Fuel (a) | -0.011 | -0.16 |
| Cars + motorbikes (b) | -4.56 | -2.99 |
| Multiplying factor (b/a) | 415 | 19 |

Source: Mattei (1994).

These remarks suggest that a new car tax, aiming at influencing new car purchases, is an interesting alternative to a fuel tax, rather aiming at influencing use of existing vehicles. However, as could be observed during the oil crisis, fuel price increases also influence new car purchases. Therefore, a fuel tax should encourage purchase of fuel efficient vehicles (Table 3).

Taffe (1993) shows that fuel price has less effect than income on average consumption of cars sold. For the elasticity of average consumption relative to fuel price, we find in the swiss literature values as different as -0.02 (see table) or -0.7 (Bühler, 1993). In the international literature, we find values of MPG relative to fuel price of 0.23 (Witt, 1994), 0.21 (Greene,[7] 1990), of even greater than 1.0 (Atkinson and Halvorsen, 1984).

Table 3. Elasticities of average specific consumption of cars sold in 1990.

| | |
|---|---|
| Elasticity relative to fuel price (a) | -0.021 to -0.027 |
| Income elasticity (b) | 0.27 to 0.29 |
| Multiplying factor (b/a) | -11 to -13 |

Source: Taffe (1993).

In summary, we can conclude from the elasticities:

- An increase in fuel price has little effect on fuel demand. Large price increases are needed to offset the impact of economic growth on fuel demand.
- Fuel price has less influence on new vehicle choice than income.
- An increase in the new vehicle price has a strong influence on new vehicle demand.

What are the implications for taxation? A fuel tax has to reach *a high level* to influence significantly demand for fuel and vehicles. Moreover, economic growth implies a continuously rising price to stabilize consumption. If such a level or increase is not politically acceptable, then a tax on new car sales, aimed at influencing the composition of vehicles sold, becomes an attractive instrument.

*Analysis of correlations between pollutants and consumption*

In a classic Pigovian framework, internalisation of external costs through a fuel tax increases the marginal cost of car use, which implies a decrease in the use of cars, hence a reduction in the polluting emissions. This conclusion is valid in the short-term.

---

[7] Witt concludes that "the adjustment process within the new car market permits little substitution for more fuel efficient cars in response to petrol price changes". For Greene, "new energy and conservation standards are far more effective in influencing car manufacturers to increasing new car fuel economy than consumer responses to higher petrol prices".

In the long-term however, we have to take into account the replacement of old vehicles through new ones. If the fuel price increases because of a tax, people will buy new cars that consume less fuel. But: do more fuel-efficient cars pollute less?

To analyse the relation between fuel consumption and pollution, we calculated, for the 314 most sold models in Switzerland in 1993, the correlations between consumption and polluting emissions. We also looked at correlations between consumption and vehicle price. Polluting emissions are measured by the Swiss Federal Police Bureau (Office fédéral de la police) following the FTP-75 test procedure, and published by Association Transports et Environnement (1994). Fuel consumption is calculated by car makers following the same test, and published by Touring Club Suisse (1994) as well as prices (Table 4).

Table 4. Correlations between consumption, pollution and price.

| | Gasoline vehicles | Diesel vehicles | All vehicles |
|---|---|---|---|
| Correlations between fuel consumption and... | | | |
| ...Noise | 0.01 | 0.21 | -0.01 |
| ...CO | 0.26 | 0.54 | 0.36 |
| ...HC | 0.32 | 0.44 | 0.37 |
| ...$NO_X$ | 0.07 | 0.72 | -0.11 |
| ...Particles | --- | 0.70 | 0.70 |
| ...Price | 0.80 | 0.81 | 0.78 |

For *gasoline* vehicles (97.1% of motor vehicles in Switzerland in 1993[8]), $NO_X$ emissions and noise are not correlated with consumption. CO and HC emissions are very weakly correlated with consumption. The lack of correlation between consumption and polluting emissions is mainly due to nature and size of catalytic converters installed. Finally, price is

8 Swiss Federal Statistical Office (1995).

correlated with consumption. For taxation, this implies:

- With the existing technology, fuel is an inappropriate tax base to induce a long-term substitution towards vehicles emitting less noise and $NO_X$. For CO and HC, the question is open, but correlation is weak.
- As there is a good correlation between price and consumption, and as consumption and $CO_2$ emissions are perfectly correlated,[9] external costs of $CO_2$ emissions can be internalized by a tax dependent on the purchase price of the vehicle.

For *diesel* vehicles, all pollutants (except noise) are well correlated with consumption. For taxation, this implies:

- External costs of pollution can be internalized by a fuel tax, except for noise.
- Again, good correlation between consumption and price implies that external costs of $CO_2$ emissions can be internalized by a tax dependent on the purchase price of the vehicle.[10]

For all vehicles, the same conclusions can be drawn as with the gasoline vehicles. However, a consumption tax identical for gasoline and diesel vehicles[11] would increase the proportion of diesel vehicles sold: because diesel engines consume less fuel (for the same power), buying a diesel car instead of a gasoline one is a mean of avoiding the payment of too much tax.

Compared to comparable gasoline cars with catalytic converters, diesel cars emit less $CO_2$, CO and HC, but they emit more $NO_X$, particles and noise (TCS, 1994). Therefore, a fuel tax could lead to an increase in $NO_X$, particles and noise emissions, due to a substitution between

---

9 Perfect combustion converts all C in $CO_2$. In case of incomplete combustion, a small quantity of CO is emitted, but this quantity finally ends up in $CO_2$.

10 Further calculations show that most pollutants are also correlated with the purchase price of diesel vehicles. This implies that external costs of pollution can be internalized by a tax dependent on the price of the vehicle.

11 Or any other measure designed to reduce gasoline consumption, like the "Ordonnance sur la réduction de la consommation" quoted above.

gasoline and diesel vehicles! (this explains negative correlations in the table for noise and $NO_X$).

*Average consumption of new vehicles*

Owing to technological progress, new vehicle consumption should decrease. What happens in fact? In the USA, "the average mileage of new light vehicles has been stagnant for a dozen years" (DeCicco and Ross, 1994). In Germany, "average car energy consumption per kilometre has been quasi-constant for the two last decades" (OECD, 1993, our translation). In Switzerland, calculations by Bühler (1993) even show an increase in average consumption since the mid-80s; this is confirmed by McInnes (1993) for Germany and Japan (between 1985 and 1988).

There are many reasons for this. As their income grows, people tend to buy more powerful cars. Moreover, cars sold nowadays have more and more fittings and kits, that increase the weight and the energy consumption:

- car radios, catalytic converters for gasoline vehicles, power assisted steering or head-rests can be found on nearly all vehicles sold;
- electrical windows, airbags, ABS, electrical seats and rearview mirrors, central locking, sun roof but also air conditioning[12] are becoming more and more popular.

The fashion of recreational vehicles, land vehicles or big volume family cars is also responsible for this evolution, as well as the tendency of car-makers to increase the size of vehicles when building a new model. All in all, the increasing size, weight and power of vehicles sold offsets the increasing efficiency of engines. Technological progress is not directed towards lowering consumption, but rather towards increasing performance, comfort and security.

Although security (and to a lesser extent comfort and performance) is desirable, such a tendency is worrying from a sustainable development perspective. Global warming threats

[12] In Europe, 24% of new cars sold in 1994 had air conditioning, against 10% in 1987 (Journal de Genève, Jeudi Economie, 13.7.95).

imply that the increasing number of cars in the world is only compatible with much more efficient cars than now. As new car sales today determine the vehicle fleet tomorrow, smaller average consumption of cars in the future can only be attained by taking measures today. If no such measures are taken, average consumption may still increase.

*Which tax base to internalize external costs?*

We know that environmental taxes use three main channels (Bürgenmeier *et al*, 1994):

1) *Lowering demand.* This is the main short-term channel, a reduction in the use of cars.
2) *Substituting* less polluting alternatives for polluting goods. Present polluting cars can be replaced by either public transportation or cleaner private vehicles.
3) *Promoting research and development.* The tax gives car makers an incentive to do research and development, with a view to manufacturing cleaner vehicles.

Taxes are only fully effective using these three channels. In this respect, the choice of a tax base is very important to lower long-term pollution: the best tax base will use all three channels. In the Swiss case, the third channel is less important, since the country is too small to catch enough attention from world car-makers to modify their research and development programmes. Consequently, we will primarily cover the first two. For cars, different taxes are possible:

- fuel taxes;
- road pricing;
- changing fixed costs into variable ones;
- taxes on new cars;
- an increase in annual taxes on motor vehicles.

*Fuel taxes*

In Switzerland, most projects aimed at internalizing external costs propose an increase in fuel taxes.[13] The mechanism is as follows. In the short run, a higher fuel price lowers demand; in the medium -- and long-run, it also influences the choice of cars by purchasers, which has an

---

[13] The often quoted ecobonus is not about taxation itself, but about redistribution of the revenue of the tax.

impact on research and on development of more efficient vehicles. In the very long run, infrastructures and lifestyle are changed (Von Weiszäcker *et al*, 1992).

With the fuel tax, different problems arise. First, we have seen that price elasticity is low. For many people, fuel price is no longer a constraint, where travel time is a stronger constraint. If price is to become a significant constraint once again, large price increases are needed. These price increases meet *acceptability problems* that are particularly important if the tax level is high. Since the acceptance of a 20 centime increase in the fuel tax in 1993, plans for further increases have encountered strong opposition.[14] Consequently, even if a higher fuel tax is effective and necessary, it is not accepted by the people.

Second, we have seen that fuel consumption is a bad approximation of pollution. A fuel tax can lower the use of cars and modify the choice towards more fuel-efficient vehicles, but it does not lead to a shift towards vehicles which are quieter or emit less $NO_X$ or particles. The tax base is only appropriate for $CO_2$ emissions, perhaps also HC and CO emissions.

These remarks show that a higher fuel tax cannot bring about a significant decrease in long-term pollution. However, from a sustainable development perspective, it can lead to a reduction in fuel consumption if fuel prices rise high enough.

*Road pricing*

Different cities have adopted road pricing throughout the world (for example Singapour and Hong Kong). In Europe, cities are still testing the system (for example Stuttgart and Cambridge), but highway or tunnel tolls have long existed. With road pricing, car drivers must pay for their use of a road, for example every time they enter a city. Payment is intended to internalize both pollution costs and congestion costs. The primary advantage of road pricing is that it adapts the tax to the road or even to the time of the day, a feature which other taxes are unable to offer. Unlike fuel taxes, it is effective in fighting local pollution.

Two problems arise. First, road pricing requires substantial infrastructures that can be used even by cars from other countries. As the development cost of such a system is high, real

---

[14] The last proposal for an increase of 15 centimes per liter was rejected by the Parliament, and opinion polls show that the nation would not even accept a temporary increase of 15 centimes to finance the new tunnels crossing the Alps (Poll of Télévision Suisse Romande, 21.3.95). The country might well reject even a 10 centime increase.

agreement among the population is necessary. However, technological progress could lower the cost in the future. The most promising system, electronic road pricing, is able to adapt the price to the existing congestion at the moment the car enters the city, a factor which means great efficiency.

Second, it is not possible for the time being to use road pricing to differentiate vehicles according to the pollution they emit. Again, no incentive exists for people to buy cleaner vehicles, nor for car-makers to invest in R&D aimed at lowering pollution.

To conclude, road pricing seems quite effective in fighting congestion costs, hence also local pollution (cities, tunnels). However, it is not able *for the time being* to induce a shift towards cleaner or more fuel-efficient cars.

*Changing fixed costs into variable costs*

Some authors suggest changing fixed costs such as insurance, yearly taxes and technical controls, which are paid on a lump-sum basis, into variable costs, which are paid according to the number of kilometres travelled (see for example Schwartz, 1990, or Derenne, 1992). For these authors, this measure is quite effective in increasing the use cost of cars, hence lowering demand. The main objective is to increase the efficiency of the insurance and tax system, but this would also have an impact on pollution.

Letting fixed costs become variable may in the short-term reduce the number of kilometres driven by cars. However, the system needs reliable mileometers in order to avoid fraud. This can be expensive. If cheap and reliable mileometers were available, this system could be considered. It would be quite effective against pollution if at the same time the tax rules are changed so that each car pays according to the pollution it emits.

However, with existing technology, changing fixed costs into variable ones does not seem to be a good solution. Moreover, if tax rules are not changed and if only current fixed costs become variable, there will be no incentive for people to shift to cleaner vehicles. Accordingly, mileometer technology should be followed carefully; technology could change a bad tax base today into a good one tomorrow.

*Taxes on new cars*

We have seen that, although it is relatively easy to influence the purchase of a car, it is more difficult to reduce pollution after the car has been bought because this implies a restricted use. Since the price elasticity of new car demand is high, a tax on new vehicles must be considered carefully.[15]

In a way, purchasing a car means purchasing a *pollution permit* which is valid throughout the lifetime of the vehicle. A tax on new cars means nothing else than putting a price on this permit: the tax can be calculated to internalize the external costs of the car during its lifetime. This reasoning uses *the "presumption principle"* (International Energy Agency, 1993), which states that "ownership of equipment is tantamount to intent to use it". For example, ownership of a car means driving some kilometres and emitting some pollution. With such a system, cars that pollute little will be subject to a small tax, while cars that pollute a great deal will be subject to a large tax. This system should influence the long-term composition of the cars on the road, by encouraging a shift towards cleaner cars. Other systems can be considered, such a tax/subsidy system.[16] Many reasons militate in favour of a tax on new cars sold:

- (Fixed) depreciation costs are more than twice as high as fuel costs (TCS, 1992).
- We know that new car sales are sensitive to price: we have seen that the price elasticity of new vehicle demand in Switzerland is high. A US study shows that the price elasticities as well as the cross-price elasticities for individual models are much higher (Irvine, 1983).[17] The author concludes that "the size of these price elasticities also suggests that the relative mix of car sales by size-class could be altered substantially by a tax or subsidy which favored fuel-efficient cars".

---

[15] Here, we are interested in the long-term impact. In the short-term, a gasoline tax affects all car drivers, while a new car tax affects only new car buyers. In the long-term however, all car drivers are affected by both taxes.

[16] Denmark has introduced a rebate on the purchase tax for the cleanest cars (IEA, 1993).

[17] This study is to our knowledge the only one on individual new car models. Estimated price elasticities are very high (average -10.4), due to the fact that it is about individual models: there are more substitutes for an individual model than for all cars, or even for a class of cars.

- Pollution varies greatly according to vehicle, but is not dependent on size;[18] substitutions are potentially important.
- Even between different classes, models can be substitutes (Irvine, 1983).
- We know that new cars are driven more than old ones.[19] Consequently, they can rapidly lower the average emissions and consumption of the fleet.
- Taxes on new cars can have a psychological advantage over fuel taxes. When they choose a car, people tend to ignore future costs of fuel, but take more notice of a tax that reflects the future social costs of their choice (Levenson and Gordon, 1990).

However, a tax on new cars can have the adverse effect of giving people an incentive to keep their old vehicles instead of buying new ones. As a result, the average age of vehicles on the road tends to rise, which is a problem because old cars generally pollute more and are noisier than new ones (IEA, 1991). A potential solution against this adverse effect is to tax only cars which pollute more than a given level. We will study this later in the paper.

*Annual taxes on motor vehicles*

Different systems are applied at present. The annual tax may depend on the weight of the car (Denmark), cubic capacity (Germany, Switzerland[1]), power (France) or a mix of these criteria (Japan) (IEA, 1993). One potential solution for pollution is to change the annual car tax to take account of environmental criteria.

Theoretically, changing the yearly tax may encourage a move towards cleaner cars in the same way as levying a tax on new vehicles, if information is perfect and consumers rational. Compared to a tax on new cars, the advantages are:

- This system has no influence on the price of new cars, hence there is no incentive to keep old, polluting vehicles.

---

[18] For example, the two smallest cars tested by the TCS (Suzuki Swift 1.0 and Daihatsu Charade 1.0) emit more NOx than the largest one (Mercedes 560 SEL) (Touring Club Suisse, 1992).

[19] 16,800 km per year for a car aged 1 to 2 years, against 11,500 km if the car is aged 9 to 10 years (Swiss Federal Department of Transportation, Communication, and Energy, 1991)

[20] Different systems exist. For example, the Canton of Lucerne is considering a 50% tax reduction for vehicles which consume less than 7 litres per 100 km (Journal de Genève, 4.1.95).

- On the contrary, old polluting vehicles pay a higher yearly tax, which pushes their owners to replace them by new cars.

Other considerations argue against such a system:

- Car buyers often lack information about future financial consequences (fuel consumption, yearly tax, maintenance costs, etc.); moreover, car purchase often follows irrational motivations.
- Payment does not occur at the moment the decision is taken (the car purchase) but rather over many years; therefore, it is less painful and may be less effective.
- If a tax dependent on pollution is not very different (in size) from the current tax dependent on cubic capacity, the system will not be very effective.
- Finally, changing the yearly tax in Switzerland implies changing legislation in the 26 cantons, whereas a tax on new vehicles sold would necessitate only one federal law (on the contrary, a majority could be more difficult to obtain at the federal level).

Unfortunately, there are no estimates of elasticities of cars sold relative to the yearly tax. Consequently, the impact on polluting emissions is impossible to evaluate. However, we know that an increase in the yearly tax for vehicles without catalytic converters in the canton of St. Gallen had virtually no impact (Meier, 1993). In our opinion, changing the yearly tax according to polluting emissions may convince individuals to replace old polluting vehicles, for example those without catalytic converters. However, some administrative problems may occur in view of the diversity and the age of existing models, implying high costs. And above all, emission standards already exist at the federal level; new laws in some isolated Swiss cantons may not give a sufficiently strong signal to induce a decisive change towards cleaner cars.

*Conclusion*

Each tax has advantages and disadvantages. Economists have long viewed the fuel tax as a sine qua non for fighting pollution, but it raises efficiency problems, related to acceptability problems. Introducing road pricing, on which intensive research is still being conducted, or letting fixed costs become variable, may be future solutions. For the reasons explained above, introducing a new vehicle tax may be more effective than changing the yearly tax. Consequently, we will now make a more detailed analysis of an instrument that could be applied immediately, and may modify the long-term composition of vehicles on the road: the tax on new cars.

## 3. Taxes on new cars in Switzerland

In this part, we will estimate the impact of different taxes and stricter standards on new cars sold in Switzerland. Accordingly, we will first present the model used to estimate demand changes. Next we will present the data, and finally we will analyse the impact of three different taxes and stricter standards on new car demand.

*Model and data*

A demand function of the Cobb-Douglas type can be defined as:

$$\mathbf{q}_i = \mathbf{A}\mathbf{p}_i^{\alpha}\mathbf{p}_j^{\beta}\mathbf{p}_k^{\beta}\ldots\mathbf{Y}^{\delta}$$

where $q_i$ is the quantity sold of model i, $p_i$ its price; $p_j$, $p_k$, ... are the prices of substitute models; Y is income; $\alpha$ (<0) is the own price elasticity of demand, $\beta$ (>0) is the cross price elasticity of demand and $\delta$ the income elasticity. Dividing $q_{i1}$ by $q_{i0}$ and isolating $q_{i1}$, we easily obtain the following demand function ($q_{i0}$ being the quantity of good i at time 0, i.e. without tax, and $q_{i1}$ the quantity at time 1, i.e. with the tax):

$$\begin{aligned} q_{i1} &= q_{i0}\left(\frac{p_{i1}}{p_{i0}}\right)^{\alpha}\cdot\left(\frac{p_{j1}}{p_{j0}}\right)^{\beta}\cdot\left(\frac{p_{k1}}{p_{k0}}\right)^{\beta}\cdot\ldots \\ &= q_{i0}\left(\frac{p_{i1}}{p_{i0}}\right)^{\alpha}\cdot\left[\left(\frac{p_{j1}}{p_{j0}}\right)\cdot\left(\frac{p_{k1}}{p_{k0}}\right)\ldots\right]^{\beta} \end{aligned}$$

which we will use for our simulations.[21] This function only necessitates informations about quantities sold, prices and own and cross price elasticities. We suppose that the substitute

---

[21] Greene & Duleep (1993) use such a function for cars, but without including prices of other goods (j, k, ...).

models are the six car models whose prices are closest to the price of the estimated car model.[22]

Data on new cars sold in Switzerland in 1993 comes from the Swiss Federal Statistical Office [Office fédéral de la statistique, 1994]. Data on pollution of different models is taken from an unpublished listing of the Swiss Federal Bureau of Police, Vehicle Certification Department [Office fédéral de la police, Section des homologations]. Data on fuel consumption is provided by vehicle importers and is published by the Touring Club Suisse (1994). The 244,559 vehicles under consideration[23] feature the technical characteristics described in Table 5.

No standard for fuel consumption exists in Switzerland. There are two kinds of standards for pollution, one for ordinary cars (group I), the other for land vehicles or those with a carrying capacity of more than 760 kg or 9 people (group II). For land vehicles, *this distinction is no longer valid today*, because these vehicles are increasingly used like ordinary cars, i.e. mainly in cities and agglomerations. The table also shows that, for all pollutants as well as for consumption, there are major differences between the lowest and the highest values, thus suggesting good possibilities for substitution.

In order to estimate the impact of changes in the price of cars, we need price elasticities for each model of car. Yet such data does not exist. Consequently, for the own *price elasticity* of each model, we took a value of -10.4, from a US study of 7 models (Irvine, 1983). The value for cross price elasticity has been calculated so that the global price elasticity of the whole model is equal to -3, which has been estimated for Switzerland (see before).[24] We obtain a cross price elasticity equal to 1.22.

---

22 Classification by vehicle class could be preferable, but raises the problem of categorizing these classes. Moreover, models that are very different are in fact substitutes, like big cars with land vehicles, or with vans. As a result, we considered price to be a better variable and an objective one.

23 of total 256.917, i.e. 95%. We didn't include small vans, nor models of which less than 100 vehicles were sold during the year.

24 This approach may be open to criticism, but we are obliged to use estimates, because no figures exist for Switzerland. However, for what follows, direction of variation is more important than size of this variation. Therefore, these numbers are not so important as they look here.

Table 5. Main vehicles sold in 1993 in Switzerland.

| | Average | Standard deviation | Min | Max | Standard (group I) | Standard (group II) |
|---|---|---|---|---|---|---|
| Fuel consumption (litres/100km) | 9.3[a] | 2.6 | 4.5 | 19.6 | --- | --- |
| Noise (db(A)) | 74.3 | 1.5 | 69 | 77 | 75 / 77 | 75 / 77 |
| CO (g/km) | 0.93 | 0.42 | 0.04 | 3.7 | 2.1 | 6.2 |
| HC (g/km) | 0.14 | 0.05 | 0.03 | 0.37 | 0.25 | 0.50 |
| $NO_X$ (g/km) | 0.22 | 0.16 | 0.01 | 1.1 | 0.62 | 1.1 |
| Particles (g/km) | 0.10 | 0.04 | 0 | 0.24 | 0.124 | 0.162 |

[a] This value is bigger than the one published by the TCS, which only includes cars with a cubic capacity of less than 2000 cm3 and is not weighted according to sales (DFTCE, 1995). Yet the percentage of vehicles with high cubic capacity is increasing. The microcensus of 1989 shows that vehicles >1951 cm3 accounted for 34.2% of the vehicles on the road (DFTCE, 1991). Therefore, our value, weighted according to sales, is higher than that of the TCS. Moreover, as the TCS (1994) suggests, we took the urban fuel consumption (FTP75), that "more or less corresponds to actual fuel consumption".

We will now use our model to evaluate the impact of three kinds of taxes (a tax equal to total external costs occuring during the life of the car, a tax/subsidy system, a tax on large polluters) as well as stricter standards.

*A tax internalizing external costs (Pigovian solution)*

The tax is equal, for each car, to the evaluation of total external costs occuring during the probable lifetime of the car.[25] We took the hypothesis that a vehicle bought in 1993 will be driven an average of 120,000 km during its lifetime.[26] The external costs result from a compilation of existing estimations, in particular Greene and Duleep (1993), Swiss Federal

[25] We suppose here that the tax is entirely shifted forward onto consumers. If this is not the case, the impact on demand is reduced.

[26] No statistics exist in Switzerland. The TCS uses an average value of 100,000 km. This value seems too low if we consider only new cars, the quality of which is increasing every year.

Office for Business Cycles [Office fédéral des questions conjoncturelles] (1994), Kageson (1993), and Pearce (1991). We took the following figures.[27]

- $NO_X$: CHF 30/kg
- HC: CHF 10/kg
- CO: CHF 1/kg
- Particles: CHF 30/kg
- $CO_2$: CHF 125 per ton of carbon (=$ 100/t C),[28] for the costs of climate change. Other external costs such as energy dependency have been neglected.
- For noise, we use the following hypothesis: on average, vehicles sold cause external costs worth CHF 1,500 during their lifetime.[29]

Each vehicle is taxed an amount equal to the external costs over 120,000 km. The following table shows the impact such a tax would have had on car sales in 1993 in Switzerland. We evaluated the impact of a tax which internalizes total external costs, then of a tax which internalizes only external costs of $CO_2$, then only external costs of noise and finally only external costs of $NO_X$.

As can be seen from Table 6, in every case, the reduction in external costs mainly comes from a reduction in the number of vehicles sold. Yet we saw that a significant reduction in the number of vehicles sold could have adverse effects, through an ageing of the vehicles on the road. As old vehicles often pollute more, this has a negative impact on pollution. We do not know the total effect on pollution (our model handles only new cars and provides no data on total pollution).

---

27 We do not take account of congestion costs, because they are virtually independent of the type of vehicle bought.

28 i.e. CHF 34 ($ 27) per ton of $CO_2$.

29 The noise values, expressed in db(A), have been linearized for the calculus.

Table 6. A tax internalizing external costs.

| | Type of tax | | | |
|---|---|---|---|---|
| | (1) Total external costs | (2) Only costs of $CO_2$ | (3) Only costs of noise | (4) Only costs of $NO_X$ |
| Impact of the tax (variation in %): | | | | |
| Number of vehicles sold | -25.8 | -9.2 | -12.3 | -5.3 |
| Average consumption | +1.0 | 0.0 | +0.3 | +0.5 |
| Average noise | -4.8 | 0.0 | -5.6 | +0.4 |
| Average $No_X$ | -7.9 | +0.1 | +0.2 | -8.6 |
| Total external costs | -28.7 | -9.2 | -14.3 | -7.1 |

To avoid this adverse effect, it is better to change the composition of vehicles sold than to lower the quantity. Accordingly, a tax should lower the average emissions of vehicles. Yet the tax in column (1) *increases* the average consumption of cars.

To understand this, let us examine column (2), which shows the impact of a tax equal to external costs of $CO_2$ emissions during lifetime of cars. Such a tax has no effect on average consumption. In fact, because they use more fuel, large expensive cars will be subject to a higher tax than small vehicles, but one which is proportionally smaller compared to their respective price. If the price elasticity of demand is the same for these two kinds of vehicles, demand for small cars will fall more than for large cars, thus leading to an increase in the average size of the vehicles on the road! For instance, a CHF 15,000 vehicle is subject to a tax of CHF 1,500, i.e. 10% of the purchase price, while a CHF 50,000 vehicle is subject to a tax of CHF 3,000, i.e. 6%. The price increase due to the tax is proportionally larger for the small vehicle (10% as against 6%). If the price elasticity of demand is the same for these two vehicles, demand will fall more for the small vehicle than for the large one, despite a smaller tax!

This implies that even if people choose more fuel-efficient vehicles for a given size, this is offset by an increase in the size of the vehicles on the road. This effect is even stronger if we consider, using the findings of Blomqvist and Haessel (1978), that price elasticity of demand is higher for small cars than for large ones: even if price increases were the same for all cars, demand for small cars would fall more than demand for large cars. Overall, the average size of vehicles on the road increases.

Thus, a tax equal to the external costs of $CO_2$ has no effect on the average consumption of vehicles sold. Taxes on noise (3) or $NO_X$ (4) are more effective in reducing noise and $NO_X$ respectively, because these emissions are not directly related to the price of vehicles. However, adverse effects cannot be excluded, for example an increase in the average emission of other pollutants.

*A budgetary neutral tax/subsidy system*

Let us consider now the following system: we take the same taxes (Pigovian taxes), but tax revenues are used for a subsidy which is the same for every car. For example, government imposes a tax of CHF 2,000 on one car, and CHF 5,000 for another one. The subsidy amounts to (2,000 + 5,000)/2 = 3,500. For the less polluting vehicle, the price will be:

$$\text{price} + \text{tax} - \text{subsidy} = \text{price} + 2{,}000 - 3{,}500 = \text{price} - 1{,}500$$

i.e. a bonus equal to 1,500. For the big polluter:

$$\text{price} + \text{tax} - \text{subsidy} = \text{price} + 5{,}000 - 3{,}500 = \text{price} + 1{,}500$$

If the subsidy is calculated so as to equal the tax revenues, we have a budgetary neutral tax/subsidy system (Table 7).

With a tax/subsidy system, the model shows a large increase in the sales of new cars. Again, the explanation comes from elasticities: most small (and cleaner) cars are subsidized, so their sales increase. Large polluting vehicles are taxed, so their sales decrease, but the reduction is lower than the increase for small cars. All in all, more cars are sold.[30] If these do not replace old vehicles but add to existing cars, pollution may *increase*. Moreover, an

---

30 Here again, this effect is stronger if we consider the fact that price elasticities are higher for small cars than for big ones (see preceding section).

increase in the number of vehicles means more congestion, hence once again an increase in external costs.

Table 7. A tax/subsidy system.

| | Type of tax | | | |
|---|---|---|---|---|
| | (1) Total external costs | (2) Only costs of $CO_2$ | (3) Only costs of noise | (4) Only costs of $NO_X$ |
| Subsidy (CHF/vehicle) | 3,260 | 876 | 1,392 | 728 |
| Impact of the tax (variation in %): | | | | |
| Number of vehicles sold | +8.8 | +0.7 | +3.6 | +3.3 |
| Average consumption | -1.9 | -0.7 | -1.0 | -0.2 |
| Average noise | -6.6 | -0.3 | -6.4 | +0.2 |
| Average $NO_X$ | -9.5 | -0.1 | -0.1 | -9.0 |
| Total external costs | +2.4 | +0.4 | +0.4 | +1.0 |

*A tax on large polluters*

We have seen the adverse effects of a tax on all new vehicles sold, as well as those of a tax/subsidy system. To avoid these problems, another solution is to tax only cars that pollute more than a given level. The tax and the exemption level are calculated so as to attain a given reduction in pollution (second best solution). In the US, such a system exists for cars that use a great deal of fuel, and is called the "gas-guzzler tax".[31] We extended the system to noise and $NO_X$ emissions (Table 8).

[31] It concerns all cars that consume more than 10.4 litres per 100 km (Ledbetter, 1993).

Table 8. A tax on large polluters.

| | Type of tax | | | |
|---|---|---|---|---|
| | (1) Gas-guzzler tax | (2) Noisy vehicles | (3) Large $No_X$ emitters | (4) Large polluters and consumers |
| | CHF1,000/litre >8 l/100km | CHF1,000/dB >73 dB(A) | CHF1,000/0.1g > 0.2 g/km | All 3 taxes together |
| Impact of the tax (variation in %): | | | | |
| Number of vehicles sold | -8.6 | +1.8 | +4.5 | -4.4 |
| Average consumption | -3.6 | -1.7 | -0.1 | -4.8 |
| Average noise | -0.5 | -12.2 | +0.7 | -11.8 |
| Average $NO_X$ | +0.1 | +0.9 | -13.3 | -10.6 |
| Total external costs | -9.6 | -4.0 | +1.2 | -13.2 |

Here, substitutions are larger. Average $NO_X$ emissions and average noise are significantly reduced. In the same way, average consumption falls with such an instrument, even if a significant reduction seems difficult to attain. In the case of an isolated measure, some adverse effects may occur, as the reduction in one pollutant may be attained to the detriment of another (for instance, replacing gasoline engines with diesel engines). For this reason, it is of interest to combine all three measures. We have done this in column (4), which combines a tax on noisy vehicles, a tax on large $NO_X$ emitters and a gas-guzzler tax. The results are the best of all taxes we have analysed, for all pollutants (there is also a significant reduction in pollutants not covered by the table). The effectiveness of the measure comes from high tax levels; however, for 12 models (out of 314) there is no tax at all.

In columns (2) and (3), the number of vehicles sold or the external costs increase is due to a particularity of the model.[32]

[32] This comes from the structure of the model. As no budget constraint exists, price increases for models of which few cars are sold may lead to a high increase in the sales of models of which many are sold. As a result, a tax can increase demand.

*Comparison with stricter standards*

To date, the approach used to fight against car pollution in Switzerland has been based on emission and immission standards. Instead of changing or introducing taxes, another possibility is to reinforce existing standards or introduce new ones.

To evaluate the impact of stricter standards, the methodology used is the following. Among vehicles sold in 1993, we eliminate those which do not comply with the stricter standards, and we distribute the eliminated quantity equally among the 6 substitute models, i.e. the models whose price are closest. With such a method, the total impact on sales is zero.

A consumption standard for new vehicles (for example 13 litres/100km) would imply a ban on most luxury cars, land and sport vehicles. The impact on average consumption of new vehicles sold would be small (1.5%). To avoid this ban, more accurate standards have to be defined, for example a standard for each vehicle category. However, this would imply a complex law, and above all would cause problems (now or in the future) for vehicles inbetween two categories.

Stricter $NO_X$ emission standards (for example 0.40 g/km) would imply a ban on almost all diesel cars, whereas gasoline vehicles would remain virtually unaffected. Moreover, total impact on pollution would be small.

On the contrary, with noise, CO and HC emissions, stricter standards seem possible. With the methodology described above, we tested the impact of stricter standards for noise, CO, HC, and $NO_X$[33] emissions (Table 9).

As with the tax on large polluters we analysed before, the impact on emissions is significant. Differences are as follows:

- As a consumption standard implies a ban on entire categories of cars, no consumption standard has been imposed here; as a result, the reduction in average consumption is small.
- Here, we used the hypothesis that the number of cars sold remains constant, unlike in the case of the tax.

---

[33] A $NO_x$ emission standard equal to 0.50 g/km would eliminate most diesel cars, except some low polluting models.

- As a standard implies a ban on all cars that fail to comply, it cannot be fixed at a very low level, unlike the exemption level for the tax; for example, in the case of the tax, the exemption level has been set at 0.2 g/km for $NO_X$; in the case of the standard, a level of 0.5 g/km has been chosen, as a lower level would imply a ban on diesel vehicles.
- As a consequence, the tax will have an impact on all vehicles emitting more than 0.2 g/km, i.e. most cars, while the standard prohibits only some models which emit more than 0.5 g/km but has no impact on those between 0.2 g/km and 0.5 g/km.
- Once compliance with the standard has been achieved, there is no incentive to market cleaner vehicles, whereas, in the case of the tax, the incentive exists until the exemption level, which can be set very low, has been reached.

Table 9. Stricter standards.

| | Existing standards (group I) | Stricter standards g/km | | Impact in % |
|---|---|---|---|---|
| | | | Average consumption | -1.1 |
| Noise (db(A)) | 75/77 | 76 | Average noise | -5.5 |
| $NO_X$ (g/km) | 0.62 | 0.50 | Average $NO_X$ | -9.8 |
| CO (g/km) | 2.1 | 1.5 | Average CO | -4.2 |
| HC (g/km) | 0.25 | 0.20 | Average HC | -4.5 |
| | | | Total external costs | -5.4 |

We see that a strengthening of noise, CO and HC standards is possible. For $NO_X$, the limit where diesel vehicles are prohibited is rapidly reached; for gasoline consumption, the imposition of a standard would eliminate most luxury, sport and land vehicles. Stricter $NO_X$ and consumption standards have a high cost, for a modest environmental result. A possible improvement in current legislation would be to transfer land vehicles (and possibly other vehicles) from group II to group I, to take into account the fact that these vehicles are more and more like ordinary cars, both from a technological point of view and with regard to use. Apart from this case, limits to standards are rapidly reached. On the contrary, the tax proposed here is more flexible, does not prohibit any vehicle, and above all gives the vast majority of car buyers an incentive to take pollution into account when the norm applies only to some car models.

## 4. Conclusions

We have seen in the first part that, as the number of kilometres travelled by cars in Switzerland is not likely to diminish, the only way to significantly lower pollution and fuel consumption is to introduce a technological change in vehicles used. This change is both technologically possible and economically feasible.

Price elasticities show that, for a fuel tax to influence significantly fuel consumption of vehicles on the road as well as average fuel consumption of new vehicles sold, it has to reach a high level. Such a level does not seem politically acceptable at present. If political opposition to higher fuel taxes continues, other instruments will be necessary. We have seen that there is a poor correlation between pollution and consumption for gasoline vehicles. Higher gasoline taxes are therefore not likely to induce people to buy less polluting vehicles.

New car sales can be easily influenced by economic instruments because of their high price elasticity. Accordingly, we have analysed taxes on new vehicles sold in greater detail. In the second part, we have thus estimated the impact of different taxes on new car demand, and compared the results with a tightening of standards.

It may be recalled that the price elasticities we used, hence our results, cannot be taken literally. Moreover, the model is partial insofar as it does not include used cars. Our findings however suggest that:

- A tax equal to total external costs occuring during the lifetime of the cars is possible but induces a significant reduction in demand, especially for small cars; the average size of vehicles sold tends to increase.
- A tax/subsidy system induces higher total vehicle sales, which is not desired.
- On the contrary, a tax imposed only on large polluters is effective and achieves a significant reduction in vehicle emissions if its level is high enough.
- A tightening of standards is only possible for some pollutants, otherwise entire categories of vehicles are prohibited; limits are rapidly reached.

The "gas-guzzler tax" played a significant role in the US, increasing fuel economy of cars (Ledbetter, 1993). In Switzerland, compulsory catalytic converters success in reducing pollution shows that already in the past, influencing the choice of new vehicles has paid off. In view of the problems with other taxes and especially standards, it would appear that a tax on new cars which pollute and consume more than a certain level can be recommended today:

- To stabilize or reduce fuel consumption (if supplementary increases in the fuel tax are not accepted).
- To reduce polluting emissions which are not correlated with fuel consumption.

A tax on new cars is not incompatible with higher fuel taxes. Indeed, the transport sector is complex because demand for mobility is high and inelastic and significant opportunities for substitution exist: between public and private transportation, between vehicle models, between new and used vehicles, between gasoline and diesel cars, between pollutants by tuning the engine, etc. Consequently, a combination of instruments is generally necessary. The same remark applies to command and control instuments. We have seen that a reduction in one pollutant often implies an increase in another. In our opinion, the number and size of possible substitutions and adverse effects imply that an approach based on economic instruments cannot replace the command and control approach. However, it may well complement it by achieving further pollution reductions at lowest cost.

The tax proposed here applies only to new vehicles, and only new vehicles that pollute and consume a great deal. The tax is fair: cars not subject to the tax are all among the cheapest on the market. As some cars will continue to be free of tax, political acceptability should be better than that of a fuel tax. In order to increase acceptability, the exemption level could be set high, and then gradually decreased; moreover, tax revenues could be used in a way that raises political acceptability. From a legal point of view, a change in the Constitution is not necessary; rather a federal law would suffice.

Finally, it should be noted that fuel consumption of vehicles is most important from a long-term perspective. Cars sold today will be driven for many years. Doing nothing today already has a long-term effect, for example on future fuel consumption and on the acceptability of future fuel taxes. Indeed, fuel taxes are more acceptable if cars consume little fuel. Doing nothing about vehicle fuel consumption today means that future fuel tax increases, if they are necessary, will be more painful.

*Acknowledgements*
*The author thanks the Swiss National Science Foundation (project 5001-34605) that made this work possible. He also thanks B. Bürgenmeier, Y. Harayama, and J. Krishnakumar, (University of Geneva), Ph. Thalmann (Swiss Federal Institute for Technology) as well as the participants at the conference "Environmental policy between regulation and market" (Neuchâtel, 19 and 20 June 1995) for numerous remarks and suggestions.*

## References

**Association Transport et Environnement** (1994) Tableau comparatif pour 250 voitures de tourisme: Bruit, gaz d'échappement, consommation. Delémont, ATE.

**Atkinson, S. E. and Halvorsen, R.** (1984) A new hedonic technique for estimating attribute demand: an application to the demand for automobile fuel efficiency. *The Review of Economics and Statistics, 66*: 417-426.

**Blomqvist, A. and Haessel, W.** (1978) Small cars, large cars, and the price of fuel. *The Canadian Journal of Economics, XI( 3)* : 470-489.

**Buchanan** (1962) Politics, policy and the pigovian margins. *Economica,* 29: 17-28. Reprinted in R.J. Staaf and F.X. Tannian (1970) Externalities: theoretical dimensions of political economy. Dunellen, New York.

**Bühler, G.** (1993) Augmentation de la taxe sur l'essence: assainissement de l'environnement et/ou des finances fédérales? [Increase in the fuel tax: to clean up the environment and/or the Federal budget?]. *In:* Analyses et prévisions 1992-1995, Institut CREA de macroéconomie appliquée, avril 1993.

**Bürgenmeier, B.; Butare, T.; Harayama, Y. and Wallart, N.** (1994) Taxes écologiques et réforme fiscale: premier rapport intermédiaire [Ecological taxes and tax reform: first intermediate report]. Project 5001-34605, Swiss National Science Foundation.

**Bütschi, D. and Kriesi, H.** (1994) Dispositions des Suisses à l'égard d'une politique des transports respectueuse de l'environnement [State of mind of the Swiss towards an environmentally friendly transportation policy]. Working Paper (University of Geneva), presented at the Conference of the Association Suisse de Science Politique [Swiss Political Science Association], Balstahl.

**Dahl, C. and Sterner, T.** Analysing fuel demand elasticities: a survey. *Energy Economics.*

**DeCicco, J. and Ross, M.** (1994) Improving automotive efficiency. *Scientific American, December*: 30-35.

**Département fédéral des transports, des communications et de l'énergie** (1991) [Federal Department of Transportation, Communication and Energy] Verkehrsverhalten in der Schweiz 1989: Mikrozensus 1989 [Traffic behaviour in Switzerland 1989: microcensus 1989]. DFTCE, Bern.

**Département fédéral des transports, des communications et de l'énergie** (1995) [Federal Department of Transportation, Communication and Energy], Consultation relative à l'ordonnance sur la réduction de la consommation spécifique de carburant des automobiles [Consultation on the statute on reduction of automobile fuel consumption], 18.4.1995, Bern.

**Derenne, B.** (1992) Fiscalité indirecte et environnement: la problématique des transports terrestres de personnes [Indirect taxation and environment: the question of transportation of persons by land]. IRES report presented to the King Baudoin Fondation, 5 November 1992, Bruxelles.

**Friends of the Earth** (1994) Sustainable Netherlands. Friends of the Earth, Amsterdam.

**Greene, D.** (1990) CAFE or price? An analysis of the effects of federal fuel economy regulations and gasoline price on new car MPG, 1978-89. *The energy journal, 11*: 37-57.

**Greene, D. L. and Duleep, K. G.** (1993) Costs and benefits of automotive fuel economy improvement: a partial analysis. *Transportation Research, 27A (3) :* 217-235.

**Grosclaude, P. and Soguel, N.** (1992) Coûts externes du trafic routier: évaluation en milieu urbain [External costs of road transport: evaluation in an urban milieu]. *Swiss Journal of Economics and Statistics, 128 (3)* : 453-469.

**International Energy Agency** (1991) Fuel efficiency of passenger cars. OECD, Paris.

**International Energy Agency** (1992) L'utilisation rationelle de l'énergie et de l'environnement [Rational use of energy and the environment]. OECD, Paris.

**International Energy Agency** (1993) Taxing energy: why and how ? OECD, Paris.

**Irvine, F.** (1983) Demand equations for individual new car models estimated using transaction prices with implications for regulatory issues. *Southern Economic Journal, 49 (3)* : 764-782.

**Kageson, P.** (1993) Getting the prices right: a European scheme for making transport pay its true costs. T&E 93/6, European Federation for Transport and Environment.

**Ledbetter, M.** (1993) Overcoming barriers to fuel-efficient automobiles. *In:* Toward clean and fuel efficient automobiles, OECD/IEA, Paris.

**Levenson, L. and Gordon, D.** (1990) DRIVE+: promoting cleaner and more fuel efficient motor vehicles through a self-financing system of state sales tax incentives. *Journal of Policy Analysis and Management, 9 (3)* : 409-415.

**Lex service plc** (1995) Lex report on motoring. London.

**Lipsey, R. and Lancaster, K.** (1956) The general theory of second best. *Review of Economic Studies, 24* : 11-32.

**Maibach, M.; Iten, R. and Mauch, S.** (1992) Internalisieren der externen Kosten des Verkehrs: Fallbeispiel Agglomeration Zurich [Internalization of external costs of transport: the example of the Zurich agglomeration]. *Swiss Journal of Economics and Statistics, 128 (3)* : 471-493.

**Manski, C. and Sherman, L.** (1980) An empirical analysis of household choice among motor vehicles. *Transportation Research, 14A* : 349-366.

**Mattei, A.** (1994) Estimation de fonctions de demande désagrégées en utilisant les données des budgets des ménages [Estimation of disagregated demand functions using data on household budgets], *Economics research paper* 94.17, Econometrics and Economics Department, University of Lausanne.

**McCarthy, P. and Tay, R.** (1989) Consumer valuation of new car attributes: An econometric analysis of the demand for domestic and Japanese/Western European imports. *Transportation Research, 23A (5)* : 367-375.

**McInnes, G.** (1993) The effectiveness of policy measures to promote energy efficiency in road transport. *In*: Toward clean and fuel efficient automobiles, OECD/IEA, Paris.

**Meier, R.** (1993) Umweltgerechte Verkehrsabgaben, Vorschläge für eine Neuorientierung [Environmentally friendly traffic taxes: proposals for a reorientation]. Rüegger, Chur/Zurich.

**Neuenschwander, R.; Sommer, H.; Suter, S. and Walter, F.** (1992) Externe Kosten des Agglomerationsverkehrs und Internalisierung [External costs of agglomeration traffic and internalization]. *Swiss Journal of Economics and Statistics, 128 (3)* : 437-451.

**OECD** (1989) Economic instruments for environmental protection. OECD, Paris.

**OECD** (1993) Germany: OECD environmental performance reviews. OECD, Paris.

**Office fédéral de l'environnement, des forêts et des paysages** (1994) [Swiss Federal Office for Environment, Forestry and Landscape] Rapport sur l'état de l'environnement 1993 [Report on the state of the environment 1993]. OFEFP, Bern.

**Office fédéral de la police** (1994) [Swiss Federal Police Office] Listing FAV1 Gr.1 on polluting emissions of cars.

**Office fédéral des questions conjoncturelles** (1994) [Swiss Federal Office for Business Cycles] Coûts externes et surcoûts inventoriés du prix de l'énergie dans les domaines de l'électricité et de la chaleur [External costs and identified additional costs of energy prices in the field of electricity and heat]. Publications PACER, OFQC, Bern.

**Office fédéral de la statistique** (1994) [Swiss Federal Statistical Office] Véhicules à moteur en Suisse: Véhicules à moteur neufs mis en circulation, 1993 [Motor vehicles in Switzerland: new motor vehicles put into circulation]. OFS, Bern.

**Office fédéral de la statistique** (1995) [Swiss Federal Statistical Office] Annuaire statistique de la Suisse, 1995 [Swiss Statistical Yearbook]. OFS, Bern.

**Pearce D.** (1991) The role of carbon taxes in adjusting to global warming. *The Economic Journal, 101* : 938-948.

**Pearce, D.; Markandya, A. and Barbier, E.** (1989) Blueprint for a green economy. Earthscan Publications, London.

**Rosen, H. S.** (1992) Public Finance (3rd ed.). Homewood IL, Irwin.

**Schwartz, J.-J.** (1990) Variabilisation de certains éléments de coût des véhicules à moteur [Variabilization of some elements of the cost of motor vehicles]. *Revue fiscale, 2.*

**Spierer, Ch.** (1988) Modélisation économétrique et perspectives à long-terme de la demande d'énergie en Suisse [Econometrical model and long-term perspectives for energy demand in Switzerland]. Département fédéral des transports, des communications et de l'énergie [Federal Department of Transportation, Communication and Energy], Bern.

**Stählin-Witt, E.** (1991) Konflikte und Widerstände [Conflicts and opposition]. *In:* Mit Oekonomie zur Oekologie [With economics to ecology]. Frey, R.L. et al, Helbing und Lichtenhahn, Basel.

**Taffe, P.** (1993) Modèle de choix multinomial et application au choix d'une cylindrée [Multinomial model and application to the choice of the cubic capacity of an engine]. Master's degree thesis, Econometrics Department, University of Geneva.

**Touring Club Suisse** (1992) [Swiss Automobile Association] Besteuerung von Motorfahrzeugen [Taxation of motor vehicles]. TCS, Emmen.

**Touring Club Suisse** (1994) [Swiss Automobile Association] Consommation de carburant [Fuel consumption]. TCS, Emmen.

**Von Weizsäcker, E.; Jesinghaus, J.; Mauch, S. and Iten, R.** (1992) Oekologische Steuerreform: Europäische Ebene und Fallbeispiel Schweiz [Ecological tax reform: At the european level and with Switzerland as a case study]. Rüegger, Chur.

**Wallart, N. and Bürgenmeier, B.** (1996) L'acceptabilité des taxes incitatives en Suisse [The acceptability of environmental taxes in Switzerland]. *Swiss Journal of Economics and Statistics, 1.*

**Witherspoon, S.** (1995) Deep fears and free riders: public knowledge, attitudes and behaviour relevant to carbon use and the greenhouse effect in advanced industrial economies. Paper presented at conference: $CO_2$ - A challenge for mankind, Interlaken, 20-21.4.95.

**Witt, R.** (1994) The demand for car fuel efficiency: an hedonic price approach. Surrey Energy Economics Discussion papers Series, No. 77 University of Survey.

# *3. Greening of the Tax System*

Environmental Policy Between Regulation and Market
ed. by C. Jeanrenaud


# 3.1 Principles of an Ecological Fiscal Reform

**Klaus A. Vallender and Reto Jacobs**
**Hochschule St-Gallen, St. Gallen, Switzerland**

## 1. Introduction and definitions

Ecological fiscal reform is concerned with the utilization of taxes in order to meet environmental objectives. Entirely or in part, taxes are therefore functionally applied as market-based instruments to influence behaviour. From a legal point of view, public contributions relevant in this context can be classified according to the following definitions (see Figure 1). In this classification, taxes and causal contributions constitute the public contributions.[1] Whereas tax liability is unconditional, causal contributions are linked to a service provided to the contributor. Depending on whether the contributions are aimed at raising revenues and/or influencing taxpayer behaviour, a distinction is made between revenue-raising contributions and incentive contributions. Contributions with an effect on public revenues and without any intention to modify the behaviour of taxpayers are genuine revenue-raising contributions. Contributions are of a mixed nature if they raise additional revenues but also aim at an incentive-based target and have an incentive effect (e.g. tobacco

---

1 Vallender (1976) 29 ff.; Höhn (1993) 57 ff.; Häfelin/Müller (1993) 488 ff.; in contradiction Auer (1980) 91 ff.; more differentiated Blumenstein/Locher (1995) 5 ff.

tax [art. 41^bis para. 1 lit. c of the Federal Constitution (FC)], alcohol tax [art. 32^bis para. 2 BV]). Contributions with absolutely no effect on revenue but which aim at an incentive-based target and have an incentive effect are genuine incentive contributions. If these are due unconditionally, they are genuine incentive taxes. If the revenue-raising objective were to be considered an indispensable element of the definition of taxes,[2] the relevant term (instead of genuine incentive taxes) would be the genuine incentive contributions unconditionally due.

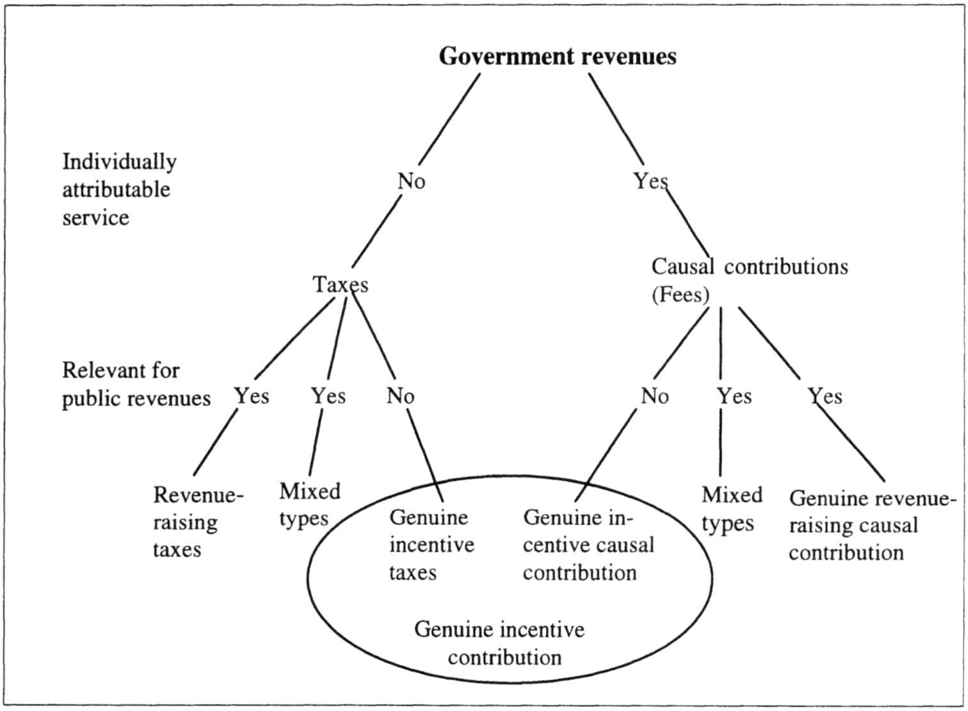


Figure 1 : Classification of public contributions

2 Blumenstein/Locher (1995) 4 f.; Tipke/Lang (1994) 43 ff. In German law the tax definition is laid down in § 3 para. 1 Abgabeordnung: "Taxes are monetary contributions, that are not a remuneration for a specific service, that are imposed on all by a public body in order to raise revenues and of which can be said that they are linked with a legal obligation to contribute. The raising of revenues can be a secondary purpose."

An ecological fiscal reform is based on the objective of utilizing fiscal policy to pursue extra-fiscal goals, i.e. for the protection of the environment. Thus, there is no « ideal » ecological fiscal reform, but rather a number of approaches, the most significant and promising of which we shall discuss from a legal point of view.

Although a multitude of possible concepts for reform exists, three main strategies can be distinguished typologically:

- extending the range of the tax system to include ecological taxes;
- modifying certain existing taxes;
- introducing a comprehensive reform of the tax system.

*Extending the range of the tax system* means introducing new taxes, not in order to raise revenues, but to influence behaviour by changing related prices (internalization). *Modifications* of certain existing taxes aim at giving mainly fiscally motivated taxes an ecological orientation. Comprehensive *reform* of the existing, fiscally motivated tax system could imply replacing certain elements of the existing tax system (e.g. introducing a compensatory energy tax) or altering their relative importance. Such a reform could also consist of a transition to consumption-oriented income taxation.

From a legal point of view, the three reform types raise a number of issues, in which the emphasis is on questions concerning the power of the Confederation to levy taxes, the compliance of the new regulations with basic principles of tax law (generality, capacity, economy of taxation), their formal legal basis (sufficiently precise basis on the level of legislation) and their conformity with constitutional rights.

## 2. Principles

An ecological fiscal reform must respect the basic principles of the legal system, mainly for reasons of rationality, consistency and unity of the legal system. Accordingly, an ecological fiscal reform must consider the principles of environmental law and the principles of taxation. Like all principles in general, those of tax law and those of environmental law, respectively, may be in a harmonious, neutral or antinomic relation to each other. Based on common sense, clashes between these principles should be resolved by achieving practical concordance, i.e. by optimizing rather than maximizing particular principles to the detriment of other principles.

*Principles of environmental law*

The most important principles of environmental law, which have a restricting impact in the context of an ecological fiscal reform, are the precautionary principle, the principle that the party responsible is liable for damages (the « polluter pays » principle) and the principle of equivalent burden.

a) The *precautionary principle* requires that pollution which may become damaging or bothersome be limited at an early stage (art. 1 para. 2 USG). Ecological damage must therefore be prevented in advance. In order to realize this principle, the legislator primarily applies a technological approach by defining minimum standards (emission limits) for preventive action (best available technology approach, art. 11 para. 2 USG).
b) The *« polluter pays » principle* in current legislation (art. 2 USG) primarily consists of a cost allocation rule. Based on this, the costs incurred in order to protect the environment (prevention costs, reduction costs, removal costs) should generally be borne by "whoever the environmental pollution to be prevented or removed is attributable to".[3] However, the « polluter pays » principle as laid down in art. 2 USG does not -- for each case -- define who the costs are attributable to, i.e. who is responsible.[4] In particular, the principle cannot be understood as the undiluted codification of the « polluter pays » principle as established in welfare theory.
c) For environmental law, the Federal Supreme Court has developed the *principle of equality of burden*. As a specification of the principle of equality, this principle requires equal (non-discriminatory) treatment of those responsible for excessive pollution. Each of those responsible must contribute proportionally to improving the situation.[5]

---

3 Rausch (1985) art. 2, N 13.

4 BGE 118 Ib 410.

5 BGE 119 Ib 484 f.; 118 Ib 34 f.; cf. the criticism of the principle of equality of burden Jacobs (1994) 341 ff.

*Principles of tax law*

As far as tax law is concerned,[6] the principle of generality of taxation, the principle of proportional taxation based on economic capacity ("ability to pay principle") and the principle of competitive neutrality must be emphasized.

a) The principle of *generality of taxation* requires that the public revenues needed by the state to perform its tasks should whenever possible be provided by all its members and not just by certain groups; originating from the principle of equality this is a judicial principle of non-preferentiality and non-discrimination.[7] As a result, the principle specifically requires adequate selection of the taxpayers.[8]

b) The principle of taxation according to economic *capacity* (ability to pay principle) is a measure for allocating the tax burden. It determines that the burden should be allocated amongst the taxpayers based on their economic capacity. Ability to pay indicators are mainly income, wealth and consumption.[9] At present, the applicable measure for economic capacity in the area of direct taxation on the federal and cantonal level is generally the net income obtained ("theory of net income growth").[10] In the context of the value-added tax (VAT), the capacity is partially taken into account by means of exemptions and zero-rating.[11]

c) In addition to the ability to pay principle, the *principle of equivalence* must be noted. According to this, the public services provided to an individual and the contributions payable by this individual should in principle be equivalent, the main emphasis being on equivalent costs. This principle reaches its limits whenever the costs or benefits connected with the public services are no longer attributable to specific taxpayers or

---

6 With regard to the principles of levying taxes cf. Reich (1990) 97 ff. with additional references; further Höhn (1989) 125 ff.; for case law BGE 118 Ia 3, 116 Ia 323, 110 Ia 14 f.

7 Hangartner (1995) 91 ff.

8 Blumenstein/Locher (1995) 91 ff.

9 Kirchhof (1981) 243 ff.; Tipke/Lang (1994) 77 ff., 81 ff.; Lang (1995) 21 ff.

10 Cf. Höhn (1993) 176 ff.; Blumenstein/Locher (1995) 153 ff.

11 Art. 14 MWSTV, Art. 15 MWSTV; Art. 8 UeB.

groups of taxpayers. The equivalence principle is particularly important with respect to causal contributions (fees) and for the taxation of taxpayers who benefit from specific public services as a group. If the amount of the contributions is based on the costs incurred by the state and/or the benefit obtained by the recipient of the services, this corresponds with the equivalence principle.

d) According to the principle of *competitive neutrality,* taxation should not put economic actors who are competing in the same market at an advantage or disadvantage with their competitors in this market. Consequently, the principle of competitive neutrality may also be interpreted as a principle of decision neutrality,[12] according to which taxation should not prevent or impede rational business decisions and should not encourage irrational business decisions. From an economic point of view, decision neutrality demands that market-based allocations of resources should whenever possible not be altered as a result of taxation. Any deviations from this principle must be justified by significant public interests. Obviously, the principle of competitive neutrality is particularly emphasized in the field of corporate taxation.

*Relationship between the principles*

In answering the question of what the *relationship between the principles* of tax law and the principles of environmental law is, the principles should not be regarded separately but rather in connection to each other. Strong links exist between the principle of equality of burden and the « polluter pays » principle on the one hand and on the other the principle of generality as well as the principle of proportional taxation based on economic capacity (ability to pay principle).

According to the principle of equality of burden all polluters, specifically new ones and existing ones, should contribute to improving environmental quality. As a rule, the « polluter pays » principle laid down in art. 2 USG provides the fundamental rule for distributing the burden. Accordingly, any legal entities who significantly pollute the environment should to the largest possible extent be obliged to share the burden in proportion to the extent of their

[12] Cf. Tipke/Lang (1994) 622.

responsibility for the costs incurred. A similar relationship exists between the principle of generality of taxation on the one hand and the ability to pay principle on the other hand. The principle of generality of taxation requires as many citizens as possible to be made to contribute to financing the state. The ability to pay principle and the equivalence principle have measurement functions. Tax burdens should be allocated to taxpayers according to their respective economic capacity and/or based on the costs or benefits of public services for the taxpayers.

Both categories of principles have a similar legal basis : they are derived from the equality principle. However, their relationship is antinomic, since not all taxpayers significantly pollute the environment and since economic capacity has nothing to do with environmental pollution. For example, if environmental charges are determined based on the economic capacity of the polluters instead of on their contribution to pollution, the results may be arbitrary.[13]

To a certain extent, the technologically influenced precautionary principle laid down in art. 11 para. 2 USG contradicts the utilization of tax legislation as a means of influencing behaviour. The technological approach to precaution (best available technology) can only be implemented with rules and bans (e.g. with emission limits) on the federal level. On the other hand, being market-based instruments, the indirectly effective incentive taxes should leave the taxpayers with different options (payment of the contribution, total or partial substitution of the taxable facts). The purpose of this is to achieve reductions of emissions mainly where abatement costs are low.

In the context of an ecological fiscal reform, a solution must be found which is compatible with the legal principles and which takes the principles into account in an optimal way, i.e. which leaves as much room as possible for principle-driven reasoning in both areas of law.

---

[13] Cf. as an example -- from the area of toll law -- the decision of the Zurich cantonal government in: URP 1994, 90.

## 3. Main strategies

Below, the three most important reform types or reform strategies -- extending the range of the tax system to include ecological taxes, modifications of certain existing taxes, comprehensive reform of the tax system -- will be briefly characterized and assessed from a legal point of view.

*Extending the range of the tax system to include ecological taxes*

Extension of the range of the tax system leaves the various existing taxes basically unchanged, which means that the structure of the tax system as a whole will hardly be changed. According to the extension concept, new taxes, the primary purpose of which is to influence behaviour, are added to existing taxes. We shall first introduce examples which illustrate the extension concept, followed by a legal assessment. The examples refer to the VOC-charge and the charge on « extra-light » heating oil, as planned in the context of the revision of the USG, as well as to the $CO_2$-charge. The two (genuine) incentive contributions to be outlined are considered to be a "legislative première", an "actual test for market instruments";[14] they are elements of "pilot tests".[15]

*Examples*

- VOC-CHARGE: VOCs are volatile organic compounds, in particular compounds which are used as solvents, propellants or cooling substances. Together with nitrogen oxide they are preliminary substances for the formation of ozone. The levying of the charge aims to reduce VOC-emissions by approximately 75,000 tons in order to reduce ozone formation. In its version of December 21, 1995,[16] the USG defines the VOC-charge as follows: payers of the charge are the importers and the producers of the volatile organic compounds as well as the importers of paints and lacquers which contain VOCs (art. 35a para. 1 and 2 USG). Subject to tax (chargeable subject) are the import and production of the volatile organic

---

14 Reporter in the National Council, Sten. Bull. NR 1995, 1251.

15 Reporter in the State Council, Sten. Bull. StR 1994, 461.

16 BBl 1996 I 233 ff., 244 f. (Referendum proposal); cf. also Botschaft Revision USG, BBl 1993 II 1519 ff.

compounds including paints and lacquers with VOC-components. Art. 35a para. 2 alinea 2 USG empowers the Federal Council to subject the import of VOCs and other compounds and objects to the charge if the amount of the materials is relevant for environmental pollution or if the materials' proportion of the costs is significant. Art. 35a para. 3 stipulates certain exemptions: utilization of the VOCs as fuel (due to the relatively low VOC-emissions), utilization or treatment of the VOCs without their being able to enter the environment (concerns the utilization of VOCs as synthesis components as well as VOCs which are retrieved after use and burned), transit and export of VOCs. Art. 35a para. 4 USG allows the possibility of offsetting the charge if investments are made which are able to help to reduce the VOC-emissions significantly below the required level. VOCs which do not harm the environment may be exempted from the charge by the Federal Council (art. 35a para. 5 USG). The criteria applied in determining the amount of the charge are the level of environmental pollution through using VOCs, the environmental risk connected with the taxed materials, and the price level of the VOCs and of ecologically suitable substitution materials (art. 35a para. 7 USG). During their parliamentary debates the Swiss Chambers of Parliament fixed CHF5 per kg as the highest admissible rate (plus inflation since the introduction of the regulations). This upper limit considerably restricts the administration's discretion to act. In order to allow the industries affected by the charge to adapt to the burden, art. 35a para. 8 USG prescribes that the charge be introduced in stages and that a time frame be established which allows the applicable percentages for each of the introductory stages to be known in advance. This regulation improves the predictability of state action.

The revenues from this charge shall be fully « reimbursed » to the population (art. 35a para. 9 USG). The Federal Council is responsible for determining the method and allocation of « reimbursement ».

- CHARGE ON "EXTRA-LIGHT" HEATING OIL: The purpose of this charge is to reduce $SO_2$-emissions by 6,000 tons. The charge on "extra-light" heating oil (ELH) covers heating

oil with a sulphur content exceeding 0.1 percent (percentages).[17] According to Art. 35b USG the charge is characterized as follows: Payers of the charge are the importers or the producers of ELH with a sulphur content exceeding 0.1 percent. Importation and production are subject to the charge, even if these are for the purpose of self consumption. The amount of the charge shall be fixed by means of an Ordinance. As with the charge on VOCs, a maximum rate was decided in the parliamentary debate, according to which the charge may amount to no more than CHF 20 per ton of heating oil (art. 35b para. 3 USG). As with the VOC-charge, the legislator defines criteria for determining the amount of the charge (environmental pollution with $SO_2$, additional costs of producing ELH with a sulphur content of no more than 0.1 percent, requirements of public supply).
After deduction of the costs of imposition, the revenue from this charge is to be « reimbursed » (art. 35b para. 5 USG), as is the case with the VOC-charge. The Federal Council will determine how this will be done.

- (POSTPONED) $CO_2$-CHARGE: In March 1994 the Swiss Federal Department of Internal Affairs (EDI) started a public inquiry with regard to a new federal law for a $CO_2$-charge on fossil energy sources. The purpose of this charge would have been to stabilize $CO_2$-emissions and total consumption of fossil energy sources at the level of 1990 up to the year 2000 and subsequently to reduce them further. The preliminary draft stipulates that all fossil energy sources (fuels) which are imported, extracted or produced in Switzerland be subject to the charge (art. 1 para. 2 VE $CO_2$). The law would only allow a small number of usages to be exempt (art. 6 para. 1 VE $CO_2$). The Swiss Federal Council may grant additional exemptions, in particular for aircraft fuels (art. 6 para. 2 VE $CO_2$). Payers of the charge are the inland producers and those liable for customs duties on importation. Exports of fossil sources of energy and those fossil energy sources not utilized for energy production are exempt. The charge shall be increased in stages: CHF 12 until the end of 1997, CHF 24 until the end of 1999 and subsequently CHF 36 per ton of $CO_2$-emissions (art. 3 para. 1 VE $CO_2$). Based on these charges, the Federal Council is responsible for

[17] Botschaft Revision USG, BBl 1993 II 1525.

determining the charge for the various categories of fossil energy sources. There is a reduced charge for energy-intensive industries (art. 11 para. 1 VE $CO_2$).

According to the preliminary draft, one-third of the revenue is to be used to support international ecological measures (art. 14 para. 2 VE $CO_2$). The remaining revenue is to be « reimbursed » to the population and to the economy (art. 15 para. 1 and art. 16 para. 2 VE $CO_2$). The modalities of payment will be decided by the Federal Council (art. 15 para. 2 VE $CO_2$).

In the course of the public inquiry there was criticism that fiscal neutrality was absent as a result of the partial dedication to a specific purpose (earmarking of revenue). Furthermore, the possibility that Switzerland might be the only state to impose such a charge meant that the proposal was looked upon with scepticism. By mid 1996, the Federal Council is expected to present a report on a law for the reduction of $CO_2$-emissions which should include the $CO_2$-charge as well as additional instruments for the reduction of $CO_2$.[18]

*Legal assessment*

- EMPOWERMENT OF THE FEDERATION: The Federation may only levy revenue-raising taxes to the extent it is empowered to do so by the Federal Constitution. This point of view is mostly unchallenged.[19] There is, however, dispute over the issue of whether the « Environmental Protection Article » in the Federal Constitution (FC) empowers the federal legislator to introduce incentive taxes.[20] Agreement is likely to exist on the fact that art. 24$^{septies}$ FC empowers the Federation to extend the existing tax system by adding environmental taxes (power to extend), as long as the charges do not affect the fiscal quota (genuine incentive taxes) and are adequate instruments for protecting human health and the environment[21]. The VOC-charge and the charge on « extra-light » heating oil generally comply with these requirements. Both are genuine incentive taxes, the net revenues being fully reimbursed. It must be noted however that "reimbursed" in the meaning of neutrality

18 Cf. NZZ no. 125 of June 1, 1995, 13.

19 Saladin (1987) N, 128; Höhn (1993) 71.

20 Vallender (1995) 208 with additional references.

21 Fleiner (1987) N 76-80.

of the tax quota implies that no new state goals (financing of public duties, distribution policies) are introduced, which could otherwise only be financed with additional revenues. If this should occur, the tax would be transformed from a genuine incentive tax to a mixed type in the definition of a « control and utilization driven tax »,[22] comparable with the tax on tobacco. There is no empowerment with regard to the $CO_2$-charge, since the Federation procures additional revenues (earmarking, dedication to a specific purpose). The $CO_2$-charge should therefore also be constructed in such a way as to ensure the neutrality of the fiscal quota. Another option would be to introduce a specific basis for the $CO_2$-charge in the Federal Constitution.

- LEGALITY: The principle of legality demands a sufficiently precise definition on the level of the formal law. The law[23] must contain a definition of the taxpayers and the amount of the charge to be paid. The essential elements of the fiscal relationship (taxpayer, subject of taxation, tax base and tax rate) are sufficiently defined as far as the VOC-charge and the charge on « extra-light » heating oil are concerned. Delegations are within the scope of the Constitution. The increasingly precise definitions which were laid down by the Chambers of Parliament in the course of the debates (maximum rate and introduction in stages of the VOC-charge, maximum rate for the ELH-charge) are welcome refinements.
  The principles with respect to the legality of public contributions developed by scholarship and practice are met by the $CO_2$-charge. The essential elements of the fiscal relationship are sufficiently precise on the level of the law, even though the Federal Council is empowered to fix the exact amounts of the charge for the various categories of fossil sources of energy.

- CONFORMITY WITH BASIC PRINCIPLES AND WITH BASIC RIGHTS: The charges are within the context of the extension strategy. They are based on the « polluter pays » principle and do not interfere with existing fiscal taxes. The technological approach of the

---

22 Vallender (1976) 36.

23 Höhn (1993) 78 f.; Blumenstein/Locher (1995) 12; Häfelin/Müller (1993) 38; BGE 120 Ia 266; 118 Ia 323; 114 Ia 11; 112 Ia 43.

precautionary principle may hinder the optimal application of the extension strategy. The principles of levying taxes remain untouched. These statements apply, provided the Federal Council introduces modalities for « reimbursing » the revenues from the charges which ensure the neutrality of the fiscal quota and which do not finance state tasks for which other revenues would have to be utilized.

The extension strategy conforms with the constitutional rights. The selection of the subjects of taxation and the stipulated exemptions are factually adequate. There is no interference with the freedom of trade and commerce clause; the measures are ecological in character and qualify as means of risk deterrence (police measures in a wider meaning). Nor are there any conflicts with public interest or with the principle of proportionality. Since there are possibilities of substitution for both VOCs and ELH, the charges seem adequate for meeting the objectives. Contractual liberty, a prerequisite for freedom of trade and commerce, is limited by charges to a lesser extent than it would be by alternative measures, such as rules or bans. The impacts resulting from the charges are adequate and their intensity is not exaggerated in relation to the objective of reducing VOC- and $SO_2$-emissions.

*Modification of existing taxes*

In the context of modifications of existing taxes the question is whether and to what extent existing taxes should be modified in such a way that they may be used to meet ecological objectives. The following possibilities are most frequently discussed:

- changes in the income taxation of employees;
- changes in corporate income taxation;
- amendments to the value-added tax;
- increases in existing consumption taxes.

Art. 24^septies^ FC empowers the Federation in the area of environmental protection. Furthermore, the Federation is explicitly empowered to levy individual and corporate income taxes and turnover taxes (art. 41^ter^ FC). Consequently, the federal legislator may in principle also consider ecological objectives in connection with establishing taxes.[24] As far as empowerment to legislate is concerned, the cantons may pursue ecological objectives when stipulating direct taxes, insofar as this is not in conflict with federal legislation, especially federal environmental and tax law.[25]

However, the Federation and the cantons are required to comply with constitutional law. Further to the requirements laid down in the empowerment norms, constitutional rights and legal principles must be considered.[26] This applies to the equality principle in particular and to the principles derived therefrom. It therefore leads to misunderstandings if the Federal Council demands a material and fiscal empowerment in the Constitution, "because otherwise there is no justification for the unequal treatment regularly resulting from fiscal encouragement".[27] The material empowerment and the power of taxation are necessary but insufficient prerequisites for modifying income taxation on ecological grounds. In view of the fundamental principles of fiscal legislation and considering the rationality and consistency of the tax system, modifications of existing taxes are only conceivable within relatively restricted limits. In the past, ecological proposals in the area of direct taxation referred especially to amendments (privileges, discriminatory measures) designed to consider ecological objectives when determining tax base, tax amount as well as exemptions.[28] With respect to indirect taxes, especially the value-added tax, varying tax rates depending on the ecological impacts of consumption are theoretically conceivable. Some examples will be elaborated below.

Examples in the area of individual income taxation are limitations on certain deductions for professional expenses of *employees* or, for the self-employed, preferential depreciation rates

---

24 Cf. Yersin (1990) 52 f.

25 This should always apply, unless something else can be deduced from the Constitution. According to the view of Danielle Yersin, the legislator is very restricted in this area. She demands a clear constitutional basis, which explicitly mentions the pursuit of objectives with the means of fiscal measures and which states expressly that the unequal treatment is intended. Yersin (1990) 59.

26 Höhn/Vallender (1995) N 93 ff.; Oberson (1995a) 22.

27 Botschaft über die Steuerharmonisierung 45.

28 Cf. Rodi (1994) 204 ff., 208 f.

for investments that facilitate ecologically adequate production. Professional expense deductions for employees could, for example, be linked to ecological considerations in connection with transportation costs. Existing tax legislation has already moved a bit in this direction by accepting the costs of transportation to the place of work as deductible costs,[29] but limiting the amount of this deduction to the costs of the available public transportation, if the taxpayer can reasonably be expected to use this service. This rule regarding the deduction of transportation costs, that is already applied by the Federation and most cantons, is reflected in proposals made in neighbouring countries.[30]

Analogous reflections to those already made with respect to employees could also be made for the *self-employed*. On the one hand, certain (ecologically undesirable) costs, although justified by commercial practice, could be excluded from deductibility; on the other hand, certain investments could be promoted by way of particularly advantageous depreciation possibilities. However, the adequacy and especially the economic efficiency of such differentiated measures are questionable in view of their objectives.[31]

The taxation of income from privately held real estate also contains certain ecological elements.[32] For example, art. 32 para. 2 DBG empowers the Federal Finance Department to determine the extent to which investments made for the purpose of saving energy and protecting the environment can be treated as deductible maintenance costs. In the Federal Council Ordinance, investments made for the purpose of saving energy and protecting the environment are defined as measures which contribute to the rational utilization of energy or to the utilization of renewable sources of energy.[33] The Federal Finance Department has defined which measures are considered to contribute to the rational utilization of energy and

---

29 Cf. e.g. art. 26 para. 1 lit. a DBG.

30 Sprenger/Körner/Paskuy/Wackerbauer (1994) 420 ff. (summary 433).

31 Cf. the pro and contra arguments in Sprenger/Körner/Paskuy/Wackerbauer (1994) 510 ff.

32 Agner/Jung/Steinmann (1995) 119 ff.

33 Ordinance of the Federal Council of August 24, 1992, on the deduction of costs of privately held real estate for direct federal taxes, SR 642.116 (in force since January 1, 1995).

to the utilization of renewable sources of energy.[34] Costs such as maintenance costs relating to heating and to the production of hot water, specific energy costs, are excluded from the deduction.[35]

The incorporation of ecologically motivated incentive-based norms in income tax law raises questions with regard to widely differing aspects. It is especially important to ensure that the rationality of the tax system is not unduly disturbed. However, there will inevitably be some disturbance if the taxpayer is taxed on more than his net income for ecological reasons. Likewise, if he is taxed on less than his net income, through benefiting from deductions justified only by ecological concerns, disturbance of the system's rationality will also occur. If this were to result in a significant alteration of the tax base, the ability to pay principle would not be fully adhered to, which would affect horizontal and vertical tax justice. Regulations concerning deductible professional expenses (transportation costs), rules concerning depreciation and the acceptance of investment costs as maintenance costs in the category of income derived from real estate, all of which are motivated by ecological objectives, result in an amendment of the tax base for direct taxes for ecological reasons. The extent of the resulting tax relief (indirect subsidies) or additional tax burden varies greatly for each of the taxpayers concerned, depending on their income bracket. Because of the progressive tax rates for direct taxes for individuals stipulated in practically all tax laws (Federation/cantons) and progressive rates for corporations which are still widespread (often based on the corporations' profit/capital ratio),[36] regulations which change the tax base are unpredictable as far as their effect is concerned and create new inequalities, which cannot be justified by the incentive-based target. Consequently, these regulations clash with the general equality principle, specifically with the ability to pay principle derived from it, moreover these inequalities can hardly be justified by the incentive-based target, because the discriminatory treatment of the different taxpayers results from the progression and has nothing to do with the incentive-based

---

[34] Ordinance of the Federal Finance Department of August 24, 1992, on the measures for the rational utilization of energy and the utilization of renewable sources of energy, SR 642.116.1 (in force since January 1, 1995).

[35] Ordinance of the Federal Tax Administration of August 24, 1992 on the various deductible costs of privately held real estate for direct federal taxes, SR 642.116.2.

[36] Höhn (1993) 325 ff., 374 ff.

target itself. When viewed from this angle, the tax subsidies or additional tax burdens also clash with the « polluter pays » principle. From a tax point of view the reduction or increase of taxable income with incentive-based regulations, in connection with the progressive tax rates, leads to an inadequate discrimination between the polluters themselves as well as in the comparison of the polluters to third parties. We do not by any means wish to state that the amount of the motivational subsidies should never be linked to the capacity of the recipients of these subsidies, but it should be noted that the reduction (or increase)[37] of the tax base for individual or corporate income taxes may result in a considerable reduction (or increase) of the progressive tax rate. We stress therefore that there is no adequate justification for subsidizing taxpayers with higher capacity to a larger extent due to the mechanism of progressive tax rates.[38] Recent studies on the legality of indirect financial aid conclude that this type of subsidy is unacceptable -- if judged according to the principles of taxation.[39]

The definition of those professional or business expenses which should be encouraged and those which should be liable for taxation also raises the information issue, in particular if the dynamic and rapid changes in the highly differentiated process of producing goods and services are taken into account. Furthermore, application of the tax laws would be rendered more complicated. It should also be remembered that present income tax law already includes a number of incentive-based norms and allocation norms. A recent examination of the reform of income tax law in Germany follows similar arguments and concludes that incentive-based norms should generally be extracted from income tax legislation and abolished wherever possible. If incentive-based measures seem indispensable, they should -- according to the

---

[37] The tax base for individual and corporate income taxes is e.g. - as laid down above - increased as a result of a total or partial inacceptance of certain professional expenses (e.g. transportation costs) for employees or for certain types of business justified costs.

[38] There would probably be no consensual approval if the progression effect were to be justified by stating that taxpayers with higher economic capacity require greater economic incentives to change their behaviour. Nonetheless, Rasenack (1981) 21 f., seems to believe so; sceptical opinions: Rodi (1994) 213; Rodi (1995) 88 ff., 94 f.

[39] Cf. Yersin (1990) 56 and passim; also Richli (1995) 291 ff. (summary 322). Richli concludes that direct financial aids are in conflict with the subsidizing principles and with the general principle of the Budget Law.

author -- be implemented as subsidies in order to make public budgets more transparent and to significantly relieve them.[40]

Scepticism also seems appropriate when considering the question of whether or not the *value-added tax* should include ecological components. Depending on the desirability of consumption from an ecological point of view, theoretical possibilities are exemptions and/or zero-rating and/or varying rates. In this context the following must be considered: the value-added tax is a general consumption tax, created as an all-phase-tax with input tax recovery, which is levied in order to finance state costs and which is based on the principle of competitive neutrality, the state of destination principle, on the economy of taxation and on Euro-compatibility.[41] Furthermore, taking current law into account, this area of legislation is already relatively complicated due to a number of exceptions.[42] Mainly because of the complications connected with special rates, there is a tendency in the OECD countries to reduce the number of varying tax rates to a minimum.[43] Within the EU, Member States may not introduce more than two (reduced) rates in addition to the regular rate.[44] The ecologically motivated differentiation of the tax base and the tax rate for the value-added tax raises similar problems to those in direct taxation. Here too, the information problem is acute; a specific example is the dynamic changes with respect to the production and distribution of goods and services.

A separate question is whether certain special consumption taxes should be maintained in addition to the general value-added tax or whether such taxes should be integrated into a general turnover tax with special tax rates.[45]

---

[40] Einkommensteuer-Kommission (1995) 32 ff.

[41] Cf. Höhn/Vallender (1995) N 33 ff.

[42] Cf. e.g. art. 14 MWSTV, art. 13 of the 6. VAT-directive, § 4 UStG (Germany), art. 261 ff. CGI (France).

[43] OECD, Consumption Tax Trends (1995) 15: "... there is general agreement that in the case of VAT, having several tax rates adds significantly to administrative and compliance costs, and inevitably creates absurdities with similar goods being taxed at different rates".

[44] Art. 12 of the 6the VAT directive.

[45] For the reasons to levy special consumption taxes cf. Böckli (1975) 248-252.

Finally, there is a need to examine the desirability and acceptability of an *increase* in existing *special consumption taxes*, which include motivational elements in the areas of health and ecology. Some examples are the tobacco tax, the tax on distilled spirits and the tax on mineral oils, which admittedly have to be considered in view of their constitutional context.[46] The effect of such an increase specifically depends on the elasticity of demand.[47]

*Reform concepts*

The emphasis is on two concepts: the introduction of a revenue-raising energy tax and the transformation of the individual and corporate income tax legislation into a consumption-oriented income tax system.

The most relevant reform concept is currently the concept of *energy taxation*. According to this, two objectives should be met with one instrument: improvement of the environment is one, while the other is the raising of significant tax revenues which are to be fully reimbursed (neutrality of the fiscal quota) to the private sector (private households and enterprises). Revenue from energy taxation at the level of enterprises should flow directly back into the economy to reduce supplementary payroll costs such as pension scheme contributions (win-win-strategy). Revenue from the taxation of private households would be reimbursed to the population on a per capita basis, e.g. as a corrective to the regressive effect of the energy tax as an object-linked tax. Other proposals envisage the reduction of existing taxes.

There are varying ideas on how to strucutre a tax on energy. There are, for example, proposals which suggest that the consumption of any kind of energy fossil and nuclear energy, etc. should be subject to tax. Energy from renewable sources could be exempted.[48] As a result of such a comprehensive circumscription of the taxable subject, considerable revenues could be expected. Mauch, Iten, Banfi and von Stokar estimate that energy prices would double within about 16 years if the energy tax were increased by 5 per cent each year (real

---

46 Cf. Vallender (1991) 278 ff.

47 Cf. from an economic viewpoint Kirchgässner (1993) 153 ff.

48 As a representative of the multitude of publications cf. Binswanger/Geissberger/Ginsburg (ed.), Der NAWU-Report (1979) 186 ff.; Mauch/Iten/von Weizsäcker/Jesinghaus (1992) 124 ff.

increase).[49] For the same timespan, they estimate a decrease of total energy consumption of approximately 20 percent. According to this calculation, revenue from the energy tax would amount to CHF 20 billion after 16 years.

Win-win strategies such as the one described are not new in legal practice. Smoking and consumption of hard liquor are already subject to special taxes (tobacco tax, charges on distilled spirits) due to their negative effects on health.[50] Because demand has proven to be relatively inelastic in the past, these taxes have had a (rather modest) incentive effect but have led to considerable tax revenues which are used for financing important state budgeting costs. Whereas the envisaged energy tax should reduce supplementary payroll costs in the manner described above, the tobacco tax and the alcohol tax (partially) finance the state old age pension scheme (AHV).

Currently, constitutional law does not allow revenues to be raised by means of an energy tax, mainly for two reasons: first, neutrality of the fiscal quota is to be obtained by utilizing the revenues to achieve other public goals -- e.g. reduction of the supplementary payroll costs -- in order to obtain results which have nothing to do with environmental protection. Second, the financing structure would be altered. Contrary to the extension of the tax system, which is also neutral with regard to the fiscal quota, questions of constitutionality arise in this case, even if the reform implies neutrality of the fiscal quota and the revenues are used (for instance) to reduce other taxes (e.g. direct federal taxes) or to reduce supplementary payroll costs (mandatory old age pension contributions). The Federal Constitution determines with which taxes and charges the Federation has to finance its activities. If the Federation were empowered to levy any kind of taxes simply on the basis of material empowerment (environmental protection), provided the revenue is used to reduce other contributions, it would be able to determine the subject of taxation and the amount of the contributions more or less freely and in a different manner than the one envisaged by the Federal Constitution. However, the financial regulations as laid down in the Constitution determine which contributions shall cover which financial needs. If this is to be fundamentally changed, the

---

49 Mauch/Iten/Banfi/von Stokar (1995) 28 ff.

50 Cf. Diriwächter (1981) 1 f.

discussion must take place on the constitutional level, i.e. the ecological fiscal reform in the meaning of the reform outlined above requires a corresponding amendment to the Constitution.

Placing the discussion on the constitutional level corresponds to existing practice. For the "double-dividend taxes" mentioned above -- alcohol tax and tobacco tax -- the issue of the Federation's legislating power is clearly settled (art. $32^{bis}$ para. 1 FC, art. $41^{bis}$ para. 1 lit. c FC).

With respect to alcohol taxation ("fiscal charge on distilled spirits") the legislator explicitly mentions the incentive-based target: according to art. $32^{bis}$ para. 2 al. 1 FC, legislation is to be fashioned such "that it reduces the consumption of distilled spirits and consequently decreases its importation and production". In connection with the tobacco tax as well as with the fiscal charge on distilled spirits, the Federation uses its proportion of the revenue to finance the old age pension scheme (art. $32^{bis}$ para. 9 FC, art. $34^{quater}$ para. 2 lit. b FC).

For an energy tax, which is expected to raise considerably more revenue, the constitutional level would be the adequate level of regulation. The need for widespread consensus is also an argument which favours choosing this level of regulation.

The same conclusion is adequate in view of the fact that the Federation's power to levy ecological taxes on the federal level is an implied power ("Annexkompetenz") based on art. $24^{septies}$ FC. However, the objective of changing the relative burden of federal taxes or the goal of reducing supplementary payroll costs for employers cannot be justified with reference to the "environment article" in the Federal Constitution.

A proposal currently under discussion which refers to the constitutional level is the federal initiative "for a secure old age pension scheme -- tax energy instead of labour!".[51] This proposal envisages a new art. $41^{quater}$ in the Constitution and an amendment to art. 24 of the transitional regulations of the Constitution (art. 24 UeB FC). According to the proposed art. $41^{quater}$ FC the Federation should levy a tax on non-renewable sources of energy and on electricity from hydroelectric power stations which generate more than one megawatt of energy in order to partially or wholly finance social insurances. Art. 24 UeB FC determines the utilization of the revenues. First, the revenue from the energy tax according to art. $41^{quater}$ is

---

[51] BBl 1994 V 407 ff.

used to cover the additional costs (para. 1). In addition, the revenue is used to reduce the employees' and employers' contributions to the social insurance schemes (AHV, IV, EO and ALV) as well as the contributions of the self-employed to AHV, IV and EO. The reduction should be carried out in a socially acceptable manner.

A *consumption-oriented individual and corporate income tax* is expected to have a positive impact on the environment as a result of the incentives for saving and investing related to the tax. If the concept is to be implemented, the indicators for measuring economic capacity must be reconsidered. The measure for this capacity would no longer be the net income obtained, but rather the portion of this income which is spent. There are various ways in which a consumption-oriented income tax could be implemented. A savings-corrected income tax would be relatively simple to implement in practice.[52] In Switzerland, various similar approaches already exist; e.g. the various types of mandatory savings and at least part of voluntary savings (provisions for old age) are exempt from taxation.[53] Thus, premiums are deducted from the tax base, whereas pensions received are taxed as income. This basic idea could be generalized: "As long as income is saved, it remains untaxed. Only withdrawals from the accumulated savings are taxed".[54] Expectations connected with these rules are not only an improvement regarding regularity of taxation, support of inflation neutrality and the principle of taxation of life-long income, but also -- as mentioned -- considerable positive effects on the environment.[55]

In the sphere of corporate taxation, a reform seeking the implementation of a "cash-flow-tax" seems to be particularly promising: steps in this direction would be the acceptance of immediate depreciation[56] and/or the introduction of an equity-linked interest exemption, at

---

52 Cf. Rose (1990) 3 ff.; McLure/Zodrow (1990) 335 ff., 363 ff.; Tipke/Lang (1994) 89.

53 Höhn (1993) 296 ff.; Vallender (1994) 658.

54 Tipke/Lang (1994) 89.

55 Lang (1993) 145 ff.; Tipke/Lang (1994) 91; Lang (1993a) 285 f. with a detailed legislation proposal.

56 McLure/Zodrow (1990) 363 ff.

least up to the inflation rate.[57] In Switzerland there are also -- though not as a result of ecological concerns -- approaches to tax exemption of net profits which are allocated to certain reserves: The Federal Law on the Formation of Tax-exempt Reserves for Creating Employment Opportunities[58] allows companies to accumulate tax-free reserves up to a limited amount.

The restrictions concerning income taxation laid down in art. $41^{ter}$ para. 1 lit. c and para. 5 FC do not present any obstacles to an adequate implementation of the concept.[59] Nonetheless, we must ask if the issue should not be discussed on the constitutional level due to the fundamental impact of the amendments involved. The Constitution does not stipulate in detail how the legislator should determine economic capacity in the context of direct taxation. Certain questions arise in connection with the consumption orientation of taxation in view of ecological efficacy. Specifically, the question arises of whether a higher charge on consumption as compared to capital accumulation results in ecological progress.[60] It is up to economists to seek answers to this question. In our view, an approach which leads to a consumption-oriented direct taxation law should not be limited to environmental issues but should rather be extended. Many arguments indicate that, compared to the present structure of direct taxation which is mainly based on the theory of incoming net wealth, a consumption-oriented income tax legislation is advantageous in many respects.[61]

The advantages and disadvantages of a consumption-oriented income tax require further examination. It is worth reflecting on whether these postulates should not be treated separately from environmental issues in the context of a fundamental discussion on an adequate reform of individual and corporate income taxation.

---

57 Lang (1993) 152 f.

58 ABRG, SR 823.33.

59 With respect to the obligation for the legislator defined in art. $41^{ter}$ para. 1 lit. c and para. 5 cf. Höhn/Vallender (1995) N 92 ff.

60 Cf. in agreement Lang (1993), Rodi (1993) 131 ff. is sceptical.

61 Cf. the pioneering contributions to the Heidelberg Congress in 1991, where the advantages and disadvantages of a transformation of the tax system to a consumption-oriented tax system were discussed under a multitude of aspects (Rose Manfred [ed.], Heidelberg Congress on Taxing Consumption, Berlin/Heidelberg/New York etc. 1991).

*Prospects*

The characterization and legal assessment of the three major strategies for ecological fiscal reform demonstrate that, from a legal point of view, the extension of the existing tax system with genuine incentive taxes, the revenues of which are fully reimbursed to the private sector without any allocative bias, faces fewest obstacles.

The material and legal limitations facing a modification of the existing general taxes in order to pursue ecological objectives are relatively restrictive. This is primarily because an ecological reform of the current tax system clashes with fundamental principles underlying the tax system. A modification strategy which strives to introduce a revenue-raising energy tax must be embedded on the level of the Federal Constitution, because this strategy implies significant modifications of the structure of federal revenues as currently defined in the Constitution.

The reform strategy in the sense of a consumption orientation of direct taxes should not be pursued or solely pursued in the context of ecological issues, but rather in the context of establishing a rational tax system.

*Acknowledgements*
*This contribution presents partial results of a more comprehensive investigation on the legal basis of an ecological fiscal reform, which is supported by the Swiss National Research Fund and which is due to be published in 1996. The authors thank Dr. phil. et lic. oec. Philipp Robinson for translating the manuscript.*

## References

**Agner P. / Jung B. /Steinmann G.** (1995) Kommentar zum Gesetz über die direkte Bundessteuer. Zürich.

**Auer R.** (1980) Sonderabgaben. Ein Beitrag zu besonderen Erscheinungsformen öffentlich-rechtlicher Geldleistungspflichten. Bern / Stuttgart.

**Binswanger H.-Chr. / Geissberger W. / Ginsburg Th.** (Hrsg.) (1979) Wege aus der Wohlstandsfalle. Der NAWU-Report: Strategien gegen Arbeitslosigkeit und Umweltzerstörung, Frankfurt am Main.

**Blumenstein E. / Locher P.** (1995) System des Steuerrechts. 5. Auflage, Zürich.

**Böckli P.** (1975) Indirekte Steuern und Lenkungssteuern. Basel / Stuttgart.

**Bundesrat** (1983) Botschaft zu Bundesgesetzen über die Harmonisierung der direkten Steuern der Kantone und Gemeinden sowie über die direkte Bundessteuer (Botschaft Steuerharmonisierung). *Bundesblatt III*, 1-381.

**Diriwächter H.B.** (1981) Die Zuständigkeit zur Erhebung und Regelung von Lenkungsabgaben, St.Galler Diss., Bern / Stuttgart.

**Einkommensteuer-Kommission** (1995) Thesen der Einkommensteuer-Kommission zur Steuerfreistellung des Existenzminimums ab 1996 und zur Reform der Einkommensteuer. Schriftenreihe Bundesministerium der Finanzen, Heft 55, Bonn.

**Fleiner Th.** (ab 1987) Kommentar BV. Art. 24$^{septies}$; *in :* Kommentar zur Bundesverfassung, J.-F. Aubert et al (Hrsg.), Basel / Zürich / Bern, Loseblatt.

**Häfelin U. / Müller W.** (1993) Schweizerisches Bundesstaatsrecht. 3. Auflage, Zürich

**Hangartner Y.** (1995) Der Grundsatz der Allgemeinheit der Besteuerung, *In:* Ausgewählte Probleme am Ende des 20. Jahrhunderts. Festschrift zum 65. Geburtstag von Ernst Höhn, F. Cagianut / K.A. Vallender (Hrsg.), Bern / Stuttgart / Wien, 91-107.

**Höhn E.** (1989) Verfassungsgrundsätze über die Besteuerung. *In:* Aktuelle Probleme des Staats-und Verwaltungs-rechts, F. Cagianut / W. Geiger / Y. Hangartner / E. Höhn (Hrsg.), Bern/Stuttgart, 125-138.

**Höhn E.** (1993) Steuerrecht. 7. Auflage, Bern/Stuttgart/Wien.

**Höhn E. / Vallender K.A.** (1995) Kommentar BV. Art. 41$^{ter}$; *in :* Kommentar zur Bundesverfassung, J.-F. Aubert (Hrsg.), Basel / Zürich / Bern, ab 1987, Ergänzungslieferung.

**Jacobs R.** (1994) Lastengleichheit - ein sinnvolles Prinzip bei der Anordnung verschärfter Emissionsbegrenzungen?, *Umweltrecht in der Praxis (URP) 8*, 341 ff.

**Kirchgässner G.** (1993) Akzeptieren die Bürger Steuererhöhungen? Einige Bemerkungen im Zusammenhang mit der Erhöhung des Treibstoffzolls, *Aussenwirtschaft 48*: 153-174.

**Kirchhof P.** (1981) Besteuerung und Eigentum. *Veröffentlichungen der Vereinigung der Deutschen Staatsrechtslehrer 39:* 213-285.

**Lang J.** (1993) Verwirklichung von Umweltschutzzwecken im Steuerrecht. *In:* Umweltschutz im Abgaben-und Steuerrecht, P. Kirchhof (Hrsg.), Köln, 115-160.

**Lang J.** (1993a) Entwurf eines Steuergesetzbuches. Schriftenreihe Bundesministerium der Finanzen, Heft 49, Bonn.

**Lang J.** (1995) Über das Ethische der Steuertheorie von Klaus Tipke. *In:* Festschrift für Klaus Tipke, J. Lang (Hrsg.), Köln, 3-23.

**Mauch S. / Iten R. / Banfi S. / von Stokar Th.** (1995) Ökologisch-ökonomische Fiskalreform: mehr Beschäftigung und weniger Umweltzerstörung?, *Volkswirtschaft 68/Heft 6*: 25-32.

**Mauch S. / Iten R. / von Weizsäcker E.U. / Jesinghaus J.** (1992) Ökologische Steuerreform, Europäische Ebene und Fallbeispiel Schweiz. Chur / Zürich.

**McLure Ch.E. / Zodrow G.R.** (1990) Administrative Advantages of the Individual Tax Prepayment Approach to the Direct Taxation of Consumption. *In:* Heidelberg Congress on Taxing Consumption, M. Rose (ed.), Berlin / Heidelberg / New York etc., 335-359.

**Oberson X.** (1995) Environmental Taxes on Products, *in:* Environmental Taxes and Charges, International Fiscal Association (ed.), The Hague / London / Boston, 73-78.

**Oberson X.** (1995a) Some Legal Issues raised by Eco-Taxes, *in:* Environmental Taxes and Charges, International Fiscal Association (ed.), The Hague / London / Boston, 21-23.

**OECD** (1995) Consumption Tax Trends, Paris.

**Rasenack Ch.** (1981) Steuersubventionen oder direkte Finanzhilfen?, *Der Staat.* 20, 1-30.

**Reich M.** (1990) Von der normativen Leistungsfähigkeit der verfassungsrechtlichen Steuererhebungsprinzipien, *In:*Steuerrecht im Rechtsstaat. Festschrift für Francis Cagianut zum 65. Geburtstag, E. Höhn / K.A. Vallender (Hrsg.), Bern / Stuttgart, 97-124.

**Richli P.** (1995) Über die rechtliche Fragwürdigkeit von indirekten Finanzhilfen durch Abzüge vom steuerbaren Einkommen - Das Beispiel der Wohneigentumsförderung. *In:* Ausgewählte Probleme am Ende des 20. Jahrhunderts, Festschrift zum 65. Geburtstag von Ernst Höhn, F. Cagianut / K.A. Vallender (Hrsg.), Bern / Stuttgart / Wien, 291-326.

**Rodi M.** (1993) Umweltsteuern. Das Steuerrecht als Instrument der Umweltpolitik, Baden-Baden.

**Rodi M.** (1994) Steuervergünstigungen als Instrument der Umweltpolitik. *Steuer und Wirtschaft 71:* 204-213.

**Rodi M.** (1995) Tax Expenditures for Environmental Purposes, *in:* Environmental Taxes and Charges, International Fiscal Assocation (ed.), The Hague / London / Boston, 87-95.

**Rose M.** (1990) The superiority of a consumption-based Tax System. *In:* Heidelberg Congress on Taxing Consumption, M. Rose (ed.), Berlin / Heidelberg / New York etc., 3-26.

**Saladin P.** (ab 1987) Kommentar BV. Art. 3; *in :* Kommentar zur Bundesverfassung, J.-F. Aubert et al (Hrsg.), Basel / Zürich / Bern.

**Sprenger R.-U. / Körner J. / Paskuy E. / Wackerbauer J.** (1994) Das deutsche Steuer- und Abgabesystem aus umweltpolitischer Sicht - eine Analyse seiner ökologischen Wirkungen sowie der Möglichkeiten und Grenzen seiner stärkeren ökologischen Ausrichtung. München.

**Tipke K. / Lang J.** (1994) Steuerrecht. 14. Auflage, Köln.

**Vallender K.A.** (1976) Grundzüge des Kausalabgabenrechts. Bern/Stuttgart.

**Vallender K.A.** (1987) Die Auslegung des Steuerrechts. 2.Auflage, Bern/Stuttgart.

**Vallender K.A.** (1991) Sondersteuern des Verbrauchs, *In:* Schweizerisches Steuerlexikon Bd. II, K.A. Vallender / E. Hess / C. Stockar (Hrsg.), Zürich, 277-280.

**Vallender K.A.** (1994) Gedanken zur Ökologisierung des Steuerrechts, *Archiv für Schweizerisches Abgaberecht 62*: 641-668.

**Vallender K.A.** (1995) Wirtschaftsfreiheit und begrenzte Staatsverantwortung, Grundzüge des Wirtschaftsverfassungs- und Wirtschaftsverwaltungsrechts. 3. Auflage, Bern.

**Yersin D.** (1990) Les buts extra-fiscaux assignés aux impôts directs, au regard de quelques principes constitutionnels, *In*: Steuerrecht im Rechtsstaat. Festschrift für Francis Cagianut, E. Höhn / K.A. Vallender (Hrsg.), Bern / Stuttgart, 47-65.

Environmental Policy Between Regulation and Market
ed. by C. Jeanrenaud


# 3.2 Testing the Double Dividend Hypothesis for a Carbon Tax in a Small Open Economy

**Stefan Proost and Denise Van Regemorter**
**Centre for Economic Studies, Catholic University of Leuven, Leuven, Belgium**

## 1. Introduction

In this paper, we analyse the introduction of the proposed EC energy-carbon tax in the tax and transfer system of a small open economy like that of Belgium. The European energy-carbon tax can be considered as an international agreement of the minimum national tax type.[1] In order to reach the stabilization of emissions promised by the EC in Rio de Janeiro, all EC-Member States have agreed to introduce a national carbon tax of a given magnitude in their tax system but can decide themselves on the use of the extra tax proceeds.

The way these tax proceeds are used is important and is at the heart of what has become known as the double dividend debate (Goulder, 1994). The double dividend hypothesis states that the introduction of an environmental tax entails gross costs (abatement expenditures and price distortions between energy sources) that can be compensated by two types of dividends. The first dividend is the environmental benefit. The second benefit is the efficiency gain that

---

[1] Alternative types of international agreements are studied in Hoel (1992) and empirically in Eyckmans, Proost, Schokkaert (1994).

can be made by using the tax revenue to substitute existing distortionary taxes. The strong double dividend hypothesis claims that this second dividend is even larger than the gross costs so that precise knowledge of the environmental benefits is not necessary to justify the environmental tax. Bovenberg (1994) surveys the different theoretical models that can be used to discuss the double dividend hypothesis and shows that the strong double dividend hypothesis can only occur under very strict and rather unlikely assumptions about the functioning of the economy.

In this paper, we add the equity dimension in discussing the costs and benefits of alternative recycling options. The welfare function for the evaluation of policies includes explicitly environmental benefits and equity concerns. Furthermore, the importance of the macro-economic regime for the welfare ranking of policies is tested.

We study this question using a dynamic (two periods) Applied General Equilibrium (AGE) model. A first analysis with a one-consumer AGE-model for Belgium that examined a related question was reported in Proost and Van Regemorter (1992a). The one-consumer model was used to study the effects of a unilateral Belgian carbon tax. This exercise yielded insights into the sectoral impacts of a carbon tax. The model was clearly not suited for studying the problem of recycling the tax revenue as there could hardly be any conflict of interests with only one type of individuals in the model. Consequently, the model is extended to several income groups. It is a dynamic, multi-sectoral model that makes it possible to study fixed price regimes for the labour market. In this case study, labour supply is fixed.

Section 2 surveys the different elements that enter our tax reform discussion. In section 3, we present very briefly the model structure. Section 4 presents results and section 5 concludes.

## 2. Insights into optimal directions of tax reform

We have to analyse an existing tax system to which the EC energy-carbon tax is added. The existing tax system is the result of a series of historical compromises and is probably not optimal. Even if it was optimal, the introduction of an additional tax might require reoptimizing the whole system in order to arrive at some bliss point designated by optimal tax theory. Our ambitions are more limited: we compare only three alternative tax recycling strategies that all include the same energy-carbon tax.

The three tax reform strategies are indexed r= LT, SS, DI and differ in their uses of the net tax revenue of the same EC energy-carbon tax ($ 10 per equivalent barrel of oil):

r= LT : revenue is used to increase welfare payments - this is a lump-sum type of tax reform that benefits more particularly the poor;

r= SS : revenue is used to finance a reduction of social security contributions by employers of approx. 5 percentage points (from 29% to 24 %);

r= DI : revenue is used for a reduction of all direct income taxes (including profit taxes) of approx. 15 %.

In all simulations, payment of unemployment benefits receives absolute priority in relation to payment of other welfare payments. This is important when the labour market does not clear because the total value of unemployment benefits is endogenous.

Much more refined tax reform strategies could be elaborated. The three we selected are more or less archetypes used in policy discussions. A strategy of indirect tax reform is missing because the EU framework does not allow large unilateral VAT or excise changes. Of course, one could also discuss the appropriate rate of the carbon tax itself -- here, the tax rate is taken as given and it is assumed that other European countries will also introduce the energy-carbon tax.

The three tax reform strategies will be compared using a welfare function that takes account of three types of effects : environmental benefits, efficiency effects and equity effects. We further test the sensitivity of the welfare ranking to the degree of income inequality aversion and to the macro-economic regime of the economy.

Throughout this paper, we will use the following notations for the welfare function of policy-makers (here expressed for the reference state 0):

$$W(0/s,e) = WP(0/s,e) + WE(0/s,e) \tag{1}$$

This function consists of two parts: the welfare associated with the quantity and distribution of private goods ("private welfare") and the welfare level associated with the quality of environmental goods ("environmental welfare"). The quality of the environment enters the welfare function in an additive way so that there is no feedback from the quality of the environment to the allocation of private goods.[2]

We will examine the effect on welfare rankings of two exogenous parameters: first, the macro-economic regime of the economy s (s= FL, FI) : flexible (FL) or fixed wages (FI), and second, the degree of inequality aversion e. Two values will be used for e.[3] When only efficiency is used as a criterion, e=0 : the marginal utility of income is always equal. However, when e=1, utility of income is measured on a logarithmic scale (marginal utility of income equals the inverse of real income). We assume that both terms of the welfare function are normalized by the marginal utility of income so that for small welfare changes, the normalized changes in welfare represent equivalent income changes.

The reference state will be compared to a new equilibrium (index 1 instead of 0) obtained with the EC energy carbon tax of $ 10 per barrel of oil and a recycling strategy r= LT, SS or DI. The difference in welfare with the reference state can be written as the sum of four effects:

$$
\begin{aligned}
W(1, r=SS/s,e) - W(0/s,e) = & \\
& + WP(1, r=LT/s,e) - WP(0/s,e) \\
& + WE(1, r=LT/s,e) - WE(0/s,e) \\
& + WP(1, r=SS/s,e) - WP(1, r=LT/s,e) \\
& + WE(1, r=SS/s,e) - WE(1, r=LT/s,e)
\end{aligned} \tag{2}
$$

Using the terminology of Goulder (1994), the first effect represents the "gross costs" of the carbon tax: namely, the loss in private welfare when an environmental tax is used and all revenue is recycled via a lump-sum transfer. The second effect is the first dividend of the carbon tax : the improvement in the environmental welfare given that the tax is recycled on a lump-sum basis. The third effect is the second dividend of the carbon tax: the improvement in

---

2 This assumption can be easily generalized but makes the exposition of double dividend notions more complex, see Mayeres and Proost (1994).

3 The private goods welfare function is of the utilitarian type. The degree of inequality aversion is introduced by choosing a particular cardinalization of the utility function, by analogy to the iso-elastic utility function used by Atkinson (1970).

welfare when we replace the revenue recycling via transfers by a reduction of distortionary taxes. Here, social security contributions are reduced; the same reasoning could just as well have been applied to a reduction of income taxes. The fourth effect is the change in the environmental welfare that can accompany a change in recycling strategy.

The *strong form of the double dividend hypothesis* states that the sum of the gross costs plus the second dividend is positive. This means that as long as you recycle the environmental tax revenue by reducing existing distortionary taxes, you will have a welfare gain and you do not need to determine the environmental benefits. This hypothesis is now rejected by most theoretical models (Bovenberg and Van der Ploeg, 1994a).

The *weak form of the double dividend* is now widely accepted: reducing a lump-sum subsidy in order to decrease distortionary taxes always generates a positive efficiency effect (Goulder, 1994).

We shall now discuss the impact of three important factors on the double dividend hypothesis: the valuation of environmental benefits, the macro-economic regime and the degree of income inequality aversion.

*The shift to greener preferences or the importance of the first benefit*

The main reason why an energy carbon tax has been proposed by the EC is the concern for global warming that can be interpreted as a shift to a greener social welfare function. We introduce environmental benefit considerations in our selection of alternative tax reform strategies because, even for a fixed energy-carbon tax rate, the recycling strategy can have a significant impact on total emission of pollutants. We distinguish two main ingredients of these green preferences: the direct $CO_2$-related effects and the side effects on other pollutants.

*The valuation of $CO_2$ emission savings*

There are three types of difficulties. The first difficulty is that $CO_2$ emissions are at the origin of an environmental problem on a world scale. Do we have to take into account the reduced damages at world level, at EC level or only at national level? We value reductions in the emissions of $CO_2$ at the total marginal damage of at least all the EC countries.

The second difficulty also has to do with the transfrontier character of greenhouse gas emissions. Some recycling strategies might induce a larger carbon leakage,[4] which should not be counted as a net gain. In this case study, no correction has been introduced for differences in carbon leakage.

The third difficulty has to do with the valuation of the damages of climate change. This is a highly speculative exercise (Cline, 1992). We reverse the argument and assume that the EC has set the energy-carbon tax ($ 10/bbl of oil) equal to the total marginal damage of its member countries and that this total marginal damage is constant for the variations in carbon emissions considered.

*The presence of other externalities*

Climate change is not the only source of environmental externalities. The introduction of an energy-carbon tax affects the level of other externalities and these variations have to be valued at their marginal damage. Here as well, valuation of these externalities is difficult due to the lack of a market mechanism to value public goods and because of their transboundary character. There is an additional problem because an energy-carbon tax is only a very indirect policy instrument for tackling the other externalities. The main externalities affected are probably air pollution, road safety and road congestion. Reduced congestion and safety externalities were not valued in this exercise because they were considered as being too indirectly linked to energy use and carbon emissions. Correcting these externalities requires the use of better instruments (De Borger, Mayeres, Proost and Wouters, 1994). For a small country like Belgium, air pollution damages (including those linked to car use) are to a large extent transboundary so that we face the same valuation problem as for carbon emissions. Belgium has subscribed to international agreements for $NO_x$ and $SO_2$ that limit the absolute total of emissions of both pollutants per country. When certain polluting activities are reduced due to the EC carbon tax, this implies that smaller emission reduction efforts will be needed

4 By carbon leakage, we mean the transfer of carbon-emitting activities to the rest of the world. This is a mere relocation of emissions, not a net reduction.

for $SO_2$ and $NO_x$ in other sectors. We use these "avoided costs" to value the differences in $SO_2$ and $NO_x$ emissions.

The data used in the valuation of externalities are presented in Table 1. The valuation was not made dependent on the degree of income inequality aversion e.

Table 1. Marginal damage of pollutants.

| | $CO_2$ | $NO_x$ | $SO_2$ |
|---|---|---|---|
| ECU1988/ton of emission | 18 | 2225 | 325 |
| Emissions 1990/1000 tons | 100,000 | 364 | 301 |

Source for the $NO_x$ values and $SO_2$ values : Mayeres (1993)

*The importance of the macro-regime of the economy for the second dividend*

Drèze (1985) has shown that the tax reform policy must in principle take into account the regime of the economy : is there equilibrium on all markets and can all markets be cleared by price adjustments? We concentrate on only one kind of fixed price equilibria: classical unemployment where the real wage is fixed at too high a level. In the European context and certainly in the Belgian context, where wages (including the social security contributions of the workers) are indexed proportionally to the consumer price index, real wage cuts may be hard to negotiate when they are necessary. We discuss environmental tax reform for two regimes of the economy: flexible wages (s=fl) and fixed real wages (s= fi).

In the case of *full flexibility of the labour market*, Bovenberg and De Mooij (1994) and Bovenberg and Van Der Ploeg (1994a) have shown that, in a small open economy, there is no strong second dividend. The intuition is that labour is the only production factor that is not mobile and always ends up bearing the extra pollution abatement costs of the production sector. This will happen in the form of a lower net wage and consequently a lower supply of labour. There are two ways to increase labour supply and employment : levying high environmental taxes on non-active population groups (living on transfers) or decreasing other public expenditures and using the additional public revenue to decrease taxes on labour supply. Both strategies have a welfare cost as either income distribution or supply of public goods is negatively affected.

In the case of *fixed real wages*, the results are different. Bovenberg and Van Der Ploeg (1994b) use a model for a small open economy with a fixed factor, an imported polluting input and a fixed real wage. An environmental tax reform could at first sight bring the user cost of polluting resources closer to the social marginal cost and simultaneously make it possible to reduce the gap between the demand and supply prices of labour. The authors show that this can only be the case if several conditions are met. A first crucial condition for a reduction of the producer tax on labour is that the public revenue effect of an increased taxation of the polluting inputs must be positive : this is the tax base effect. This approach requires a combination of low initial taxes on pollution, a high input share for polluting resources and for profits and a low tax on profits. Profits (the return on the fixed factor) will indeed pay the increased resource tax to a large extent. A second condition is that labour demand can indeed expand: this requires sufficient substitutability with the polluting input and absence of strict complementarity with the fixed input. When these conditions are met a second dividend can exist. Goulder (1994) has discussed several possible extensions of this model that can improve the chances for a double dividend.

In our case study, the labour supply is fixed and we compare tax reform strategies in the case of flexible and fixed wages. A fixed labour supply combined with flexible wages implies that there are no efficiency gains associated with a reduction of the labour taxes, so that, as we will see, other elements dominate the choice between tax recycling strategies. In the case of fixed real wages, lowering taxes on the use of labour can result in significant efficiency gains.

Assuming fixed real wages is a simplistic explanation of unemployment. Rigid real wages have to be explained by the functioning of the labour market, and this richer structure should be included in the model. Bovenberg and Van der Ploeg (1994b) use a model with job search and hiring costs. Through a bargaining process between unions and firms, the rents are distributed between employed and firms. The "equilibrium" unemployment rate can then be indirectly affected by lowering unemployment benefits through pollution taxes. The same mechanism can operate if there is an unofficial labour market that can only be taxed via higher consumer taxes.

*The income distribution dimension*

In any tax reform, the effects on the distribution of income are probably a dominant concern. A reform will only be accepted if the benefits and costs are to some extent distributed in an equitable way so as to obtain sufficient political support. The income distribution concern is also at the heart of the existing tax system: in the absence of income distribution concerns, the optimal tax system would be limited to a combination of a head tax (in as far as this is possible) and externality taxes so that all the second dividend concerns about the reduction of distortionary taxes would be irrelevant.

Van De Gaer, Schokkaert and De Bruyne (1992) have studied optimal marginal tax reform as a function of the degree of inequality aversion. In their examination of both the flexible and the fixed real wage case, they showed that the marginal welfare impact of an indirect tax change on the welfare of a particular income group is a sum of four effects: effect on relative good prices weighted by the consumption bundle of that group, effect on wage income via the change in the real wage, effect on profit income, and finally, the welfare impact of changes in involuntary unemployment (measured by the change in employment times the gap between consumer wage and the sum of reservation wage and unemployment benefit). In our model, we could add a fifth element: the change in environmental quality that could be weighed by the marginal willingness to pay of the income group concerned.

Returning to the double benefit terminology of Goulder (1994), the introduction of income distribution concerns requires that the gross costs, the environmental benefit and the recycling benefit all be weighted by the welfare weights of the different economic agents affected. Now even the weak form of the double dividend proposition no longer necessarily holds. If lump-sum subsidies accrue relatively more to the poor, this could offset the efficiency gains of a lower distortionary tax and invalidate the weak double dividend proposition, as we will see in the empirical section.

## 3. The model used

For this study, the model developed by C. Dewatripont, S. Erlich, V. Ginsburgh and D. Van Regemorter (1989) has been extended to include several income groups, a more elaborate description of the tax and transfer system, a distinction between all energy forms and emissions of different air pollutants. A model description with only one type of consumers can be found in Proost and Van Regemorter (1992b).

The model used is a dynamic perfect foresight AGE model for an open economy which covers two periods : a first period representing medium-term adaptations (say 5 to 7 years) and a second period, capturing the full adjustment taking place in the long term (say another 7 years). The institutional framework is perfect competition with however the possibility of labour market disequilibrium in the medium run.

There are two types of goods: tradeable and non-tradeable. All commodities are in principle considered as tradeable, while factors of production (labour and production capacities) are non-tradeable. Hence there will, at equilibrium and in each period, be only one price for each commodity and one price for each country for each factor of production. The model is solved as a Negishi welfare optimum. The model contains four types of agents: households, firms, the government and the rest of the world.

*Households*

Four homogeneous groups of consumers are used in the model. It is assumed that they have the same utility functions but differ in labour productivity, in the probability of becoming unemployed, in their share of the initial capital stock and in their share of the welfare payments.

The four groups represent different socio-economic groups. The first group contains mainly retired persons living from transfers and those in other low-income categories. The second group is a mixed one. The third group contains mainly households where both husband and wife are working, while the last group contains persons in the higher wages bracket and those living from large capital incomes. The lowest income group has the highest probability of becoming unemployed, the lowest share of capital income but the highest share of welfare payments. Preferences are intertemporally separable, and are explicitly defined for only two periods. The rest of the planning horizon is summarized in the value of future utility streams consistent with the terminal capital stock.

Table 2. Data by income class.

| | CLASS 1 | CLASS 2 | CLASS 3 | CLASS 4 |
|---|---|---|---|---|
| | Distribution of consumers in % | | | |
| Consumers | 11.2 | 25.8 | 35.7 | 27.3 |
| Active population | 5.0 | 23.3 | 42.6 | 29.1 |
| Unemployed | 14.6 | 33.1 | 41.6 | 10.7 |
| | Distribution of income | | | |
| Relative labour efficiency | 0.49 | 0.77 | 0.91 | 1.33 |
| Share in capital income | 3.1 | 18.2 | 16.1 | 62.6 |
| Share in transfer income | 18.7 | 26.3 | 29.5 | 25.5 |
| | average taxation rates | | | |
| Labour income | 0 | 6.4 | 22.6 | 34.8 |
| Capital income | 24.1 | 24.1 | 24.1 | 24.1 |

Consumer preferences are represented via a Linear Expenditure System with 15 aggregates of goods. A dedicated model is used for the allocation of the energy aggregate of private consumers. We assume that the quality of the environment enters the utility functions in an additive way. Furthermore, we assume a fixed labour supply. In the classical unemployment case, the individuals, as suppliers of labour, can be confronted in the first period with rationing on the labour market due to sticky real wages. In this case, effective labour supply is smaller than desired labour supply for some individuals. Total unemployment U is distributed among individuals via exogenous rationing parameters.

*Producers*

Producers maximize profits. Production possibilities are described by a dynamic Input-Output framework with constant returns to scale but flexible coefficients. The flexibility of the input-output coefficients is modeled differently for the 25 sectors and differs also between periods as shown in the following table.

Table 3. Structure of production parameters.

| | 11 industrial sectors (excl. Energy) | other sectors |
|---|---|---|
| FIRST PERIOD | exogenous klemo substitution effects | exogenous substitution effects |
| SECOND PERIOD | exogenous klemo substitution effects | exogenous substitution effects |
| | + ENDOGENOUS CAPITAL/LABOUR SUBSTITUTION | |

In the first period, coefficients are adapted exogenously, using detailed information from demand models. The input-output coefficients for the 12 industries (excluding the energy sectors) were adapted using sectoral Capital Labour Electricity Materials Other fuels (KLEMO) demand functions. These were estimated on the period 1966-1986 using a CRESH production function (Italianer, Strumelle and Van Regemorter, 1990). For the other sectors, ad hoc models were used to adapt the energy coefficients in response to the change in relative prices. As regards the energy sectors, only the electricity sector has any real degrees of freedom in the fuel mix. For this sector, an exogenous investment and fuel allocation model was used that represents substitution possibilities more accurately than a traditional econometric approach.

In the second period, capital/labour substitution is endogenized using a constant elasticity of substitution equal to 0.25, the other substitution effects remaining exogenous. This substitution process was endogenized rather than the others because the prices of these non-tradeables are more likely to change than the prices of the other commodities that are largely determined by the world market.

*Savings and investment*

Savings determine total investment in each period ("classical closure rule"). The allocation of savings between sectors ensures that, in each period and in every sector, replacement investments or expansion investments only take place if the cost of investment is entirely covered by the present and future earnings generated by that investment.

*Government*

The government collects indirect taxes (VAT and excises), direct taxes on labour income and on capital income and employers, social security contributions on labour. An energy-carbon tax can be levied on households and on the production sector. The tax proceeds are used to pay for exogenous expenditure on government goods and services, unemployment benefits and welfare payments to the different income classes. In each period, the budget of the government must balance.

*Rest of the world*

The behaviour of the rest of the world (ROW) is given by a trade welfare function representing willingness to trade imports and exports with the country concerned. Export and import behaviour of the ROW is obtained by maximizing the trade welfare function with the intertemporal trade balance as a constraint.

Given the current European institutional structure, electricity is not considered as a tradeable good. Crude oil, natural gas and coal cannot be produced in Belgium at a reasonable price and can therefore only be imported. These imports occur at a fixed price. Accordingly no account is taken of the gains in terms of trade that can be obtained with an energy tax. Price elasticities for Belgian exports vary between -1 and -2.5.

## 4. Results

*Scenario description*

The EC energy-carbon tax takes the form of an increase in excises on all energy products consumed. It is an excise and is applied to all sectors of activity and to households. The excise equals $ 10/bbl of oil equivalent. Half of the excise is proportional to the carbon content of the fuel, while the other half is proportional to the energy content. This excise is kept constant in real terms and is added to the energy prices of a given reference scenario. This implies an average increase of 50% for industrial energy prices and of 20% for the energy prices of private consumers.

In all scenarios considered, it is assumed that an identical energy-carbon tax of $ 10/bbl is imposed on trade partners, that these trade partners have a flexible wage market and that they recycle revenues via increased welfare payments. Because we only dispose of an AGE model for Belgium, an artefact had to be used to mimic the behaviour of the foreign economies. The only mechanism by which the actions of the trade partners affect the open economy under study is via the demand functions for Belgian exports and via the supply functions of imports to Belgium. We assume that the import and export behaviour of our trade partners reacts in the same way to carbon taxes and tax reform strategies as the Belgian economy would do.

This assumption can be justified when the major trade partners have a similar type of economy as the home country, adopt similar tax reform policies and when the home country has a very open economy. This is indeed the case for Belgium : 80% of its total export and import (except energy products imported at a fixed price) is with EU member countries, and its exports and imports each account for some 80% of its GDP.

*Results*

In Table 4, alternative domestic tax reform policies are compared for two macro-economic regimes: flexible wages and fixed wages at home.

*The case of flexible wages*

We shall first concentrate in greater detail on the first column of Table 4 that reports results for the *transfer scenario* where all variations in net public revenue are redistributed through a proportional variation of the welfare transfer payments (exclusive of unemployment benefits). Only in this first column do we report results for the two periods. In the rest of the tables, we only compare results for the first period. Welfare computations are however always reported jointly for both periods.

The results shown in the left half of Table 4 (first column) are in line with the results of the literature: low aggregate economic effects of a carbon tax and a restructuring to a less energy-intensive industry. This restructuring requires extra investment in the first period (+2.4%). The extra investment efforts are moderate because the energy-intensive sectors are the most capital-intensive sectors.

As regards the welfare effects, there is a clear efficiency loss when no environmental benefits are taken into account (line private welfare e=0 shows -2.95% loss). This is expected because in the production sector, an extra tax on an imported input (energy) is introduced and this is a new distortion (see Dixit (1985) and Diamond and Mirrlees (1971)). Because the import price of energy is fixed, there is also no role for a disguised import tax. The same holds for the tax on private energy use: this is an extra distortion that, together with the carbon tax on industry, is used to increase a lump-sum transfer.

The welfare results can only be reversed partially by attaching more importance either to the income distribution dimension or to environmental benefits. When the government is not only interested in total equivalent income variations but also attaches a higher weight to poor households (e=1), the welfare results improve. This is because higher energy taxes are paid more or less proportionally to income, while the welfare payments are more concentrated in the lowest-income group (cf. Table 2). The welfare ranking can be improved further by taking into account the environmental benefits (the line private welfare + environmental welfare in Table 4). There is no welfare loss when income distribution effects and environmental benefits are taken into consideration. Of course, these results depend to a large extent on the valuation of environmental benefits and on the degree of inequality aversion.

Table 4 also indicates the effects of the tax reform on the composition of public revenues (all variations measured as shifts in shares of GDP). The major effect is the increase in indirect tax revenue: +2.7% (the energy tax base decreases by some 10%). This is accompanied by small decreases in other revenues, such as income tax (-0.7%) and social security contributions (-0.5%) because the decline in real wages decreases the tax base, resulting in a 1.5% net increase in the welfare transfers share.

Due to the assumption of fixed labour supply and flexible wages, reducing social security payments (r=SS) or reducing direct income taxes (r=DI) (the next two columns in Table 4) does not entail a reduction in distortions. Consequently most economic effects are more or less comparable to the transfer scenario. The major difference between the three tax reform scenarios lies in the income distribution effects: the rich benefit much more from a reduction in social security contributions or a decrease in income taxes.

Finally, we would like to point out the important decrease in the real wage cost of firms in order to restore equilibrium on the labour market. This decrease is necessary because of the pollution taxes levied on the production sector.

The results are summarized in terms of the *double dividend discussion* in the upper half of Table 5. This table has been constructed by using the type of welfare evaluations reported in Table 4. As defined by equation (2), the total welfare effect of a tax reform can be broken down into three effects: gross costs, first benefit and second benefit (representing the sum of the effects on private and on environmental welfare), so that the sum of the three effects in Table 5 (selecting only one of the two second dividends) gives the total welfare effect.

When only efficiency counts (e=0), we see that there is no strong double dividend: the gross costs are always higher than the second dividend (the gain due to substituting distortionary taxes). The second dividend is almost zero here because the labour supply is fixed and this implies that there are no first-order efficiency gains associated to lower taxes on labour. It is negative, because of second-order effects on distorted markets.

When income distribution matters and the government attaches a higher weight to poor households, welfare results improve (right column of Table 5) because gross costs are born to a larger extent by the higher income groups whereas the increased lump-sum transfers are more directed to the poor. The environmental benefit remains positive. As expected, the second benefit is still more negative because the higher income classes are the main beneficiaries of the switch from increased welfare transfers to lower taxes on labour or income.

Table 4. Choice of tax reform strategy as a function of macro-economic regime.

| % difference reference scenario for period 1 | FLEX at home FLEX abroad and "TRANSFERS" abroad | | | | FIX at home FLEX abroad and "TRANSFERS" abroad | | |
|---|---|---|---|---|---|---|---|
| | TRANSFERS 1 per LT | 2 per | SOC-SEC SS | DIR-TAX DI | TRANS-FERS LT | SOC-SEC SS | DIR-TAX DI |
| GDP | +0.2 | -0.6 | +0.2 | +0.2 | -7.0 | -5.1 | -7.1 |
| CONSUMPTION | -0.9 | -1.0 | -0.9 | -0.9 | -4.7 | -3.7 | -4.7 |
| INVESTMENT | 2.37 | -5.2 | 2.3 | 2.3 | -20.6 | -14.2 | -21.1 |
| ENERGY INT.SECTORS | -2.5 | -1.7 | -2.5 | -2.3 | -11.9 | -8.8 | -11.9 |
| REAL WAGE COST | -5.2 | -6.2 | -4.5 | -5.2 | -0.2 | -3.3 | -0.2 |
| UNEMPLOYMENT (1000) | 0 | 0 | 0 | 0 | 238 | 176 | 243 |
| $CO_2$ emissions | -12.2 | -10.7 | -13.3 | -13.3 | -18.5 | -17.0 | -18.6 |
| PRIVATE WELFARE e=0 | -2.95 | | -3.12 | -3.1 | -15.7 | -12.4 | -16.5 |
| PRIVATE WELFARE e=1 | -0.6 | | -1.35 | -1.03 | -8.6 | -7.1 | -11.6 |
| PRIVATE WELF. +ENVIR WELF. e=1 | 0.31 | | -0.53 | -0.19 | -7.8 | -6.2 | -11.0 |
| Priv.cons Class 1 | 2.3 | 2.3 | -0.5 | 0.5 | -6.4 | -6.8 | -11.5 |
| Priv.cons Class 2 | -0.9 | -0.9 | -1.1 | -1.5 | -6.2 | -5.2 | -8.6 |
| Priv.cons Class 3 | -1.8 | -1.8 | -1.2 | -1.7 | -5.5 | -4.3 | -4.9 |
| Priv.cons Class 4 | -1.8 | -1.9 | -1.4 | -0.8 | -6.4 | -4.5 | -4.0 |
| Employer SS contrib.(Δ share in GDP) | -0.48 | -0.43 | -2.08 | -0.48 | +0.07 | -1.96 | +0.11 |
| Direct tax (Δ share in GDP) | -0.67 | -0.69 | -0.13 | -1.56 | -0.41 | +0.04 | -3.01 |
| Indirect tax (Δ share in GDP) | +2.68 | +2.49 | +2.63 | +2.68 | +2.80 | +2.72 | +2.8 |

Table 5. Carbon taxes and the double dividend hypothesis.

| Differences in welfare (income equivalents) % difference with reference scenario | FLEXIBLE WAGES AT HOME - flexible wages abroad and "transfers" as recycling strategy abroad | |
|---|---|---|
| | No inequality aversion, e=0 | inequality aversion, e=1 |
| GROSS COSTS (including recycling via lump-sum transfers and without environmental benefits) | -3.26 | -0.61 |
| 1st DIVIDEND (increase in environmental benefits) | +1.35 | +0.92 |
| 2nd DIVIDEND (via reduction of soc.sec.contributions employers) | -0.19 | -0.84 |
| TOTAL (via reduction soc.sec.contrib.) | -2.1 | -0.53 |
| 2nd DIVIDEND (via reduction of income taxes) | -0.16 | -0.50 |
| TOTAL (via reduction of income taxes) | -2.07 | -0.19 |
| | FIXED WAGES AT HOME - flexible wages abroad and "transfers" as recycling strategy abroad | |
| | No inequality aversion, e=0 | inequality aversion, e=1 |
| GROSS COSTS (including recycling via lump-sum transfers and without environmental benefits) | -17.3 | -9.3 |
| 1st DIVIDEND (increase in environmental benefits - HIGH assumption) | +2.0 | +1.4 |
| 2nd DIVIDEND (via reduction of soc.sec.contributions employers) | +3.6 | +1.7 |
| TOTAL (via reduction soc sec) | -11.7 | -6.2 |
| 2nd DIVIDEND (via reduction of income taxes) | -0.9 | -3.1 |
| TOTAL (via reduction income taxes) | -16.2 | -11.0 |

*The case of fixed real wages*

On the right side of Table 4, tax reform strategies are compared under the assumption of fixed real wages in the first period. In this case, the tax reform strategy matters much more because there is an initial important distortion on the labour market where there is an excess supply. The strategy of *reducing transfer payments or reducing direct tax payments produces catastrophic economic results*: GDP is reduced by 7% and there are many more unemployed. This should not come as a surprise : the increased energy tax on households is shifted to firms via constant real consumption wages. As a result of higher real wage costs (via the increased energy taxes on households) and higher energy costs of firms, the production is no longer sufficiently profitable. Thus, investments decline, production capacity is insufficiently adapted and unemployment rises sharply. The decrease in activity is particularly important in the energy-intensive sectors because they have the highest cost handicap compared to the rest of the world. It is assumed that the rest of the world has flexible wages and uses a transfer scenario, so the cost handicap may be significant.

The negative welfare results are somewhat mitigated by the better environmental performance: carbon emissions are reduced 50% more in the fixed real wage scenario compared to the case with flexible wages. Part of this effect has to do with a higher carbon leakage for which no correction was applied. As regards income distribution aspects, it appears that the net gain in public revenue (+2.4 % of GDP) is absorbed completely by the increase in unemployment benefits so that ultimately, due to the tax base effects, other welfare payments decrease rather than increase.

Reducing *social security contributions for employers* can at first sight be considered as a win-win scenario. It is a better tax reform strategy than the two others but overall welfare still decreases rather than increases. The reduction in social security rates is however insufficient to bring about the decrease in employer's labour costs that is needed to improve employment (real wage costs decrease by 3.3% instead of the necessary 4.5 to 5%). Possibilities of reducing wage costs are limited because the energy-carbon tax base is eroded (by some 18% rather than 10 to 12% in the flexible wage case). After the reduction of social security rates, nothing is left to pay for the increased unemployment benefits, hence welfare payments must be reduced. If we had protected unemployment benefits and welfare payments in real terms, the results would be even worse as there would be no room for a reduction of the social security contributions of employers.

This confirms the theoretical results of Bovenberg and Van der Ploeg (1993b), who state that a necessary condition for a positive employment effect is that pollution taxes must be born by the fixed factor when real wages are protected. In our model, there is no fixed factor as almost all sectors need at least replacement investments, accordingly profits in these sectors cannot be depressed without negative effects on production capacity and output.

We can reframe these results in terms of the *double dividend discussion* in the lower part of Table 4. Gross costs now dominate all other effects. The first benefit is larger (though partly through migration) and the second benefit is positive when the social security contributions are reduced because there is an efficiency gain associated with a reduction in the payroll taxes when there is an excess supply on the labour market. It remains positive when income distribution matters. The second benefit is negative when income taxes are reduced because with fixed real wages (before income tax) and a constant labour supply the labour market is not affected. It is more negative still when income inequality matters.

## 5. Conclusions

This paper analyses alternative ways of using the revenue (tax reform strategies) of an EC energy-carbon tax for a small open economy. This is a test of the double dividend hypothesis. It is shown that in order to test the double dividend hypothesis in a theoretic welfare framework, concepts have to be broadened to include the income distribution dimension. Once this is done, efficiency effects of recycling can be dominated by equity effects and the weak double dividend proposition does not necessarily hold: increasing lump-sum welfare payments can be more appropriate than reducing distortionary taxes mainly borne by the rich, even if the latter strategy entails clear efficiency gains.

The double dividend hypothesis is tested with the help of a dynamic, multisectoral AGE that can simulate two macro-economic regimes (flexible wages and fixed real wages) and contains several income groups but a fixed labour supply. To this model were added estimates of environmental benefits for the major air pollutants.

As long as there are no price rigidities on the labour market, the choice of tax reform strategy is not important for the aggregate economic results. The choice of strategy matters mainly for the income distribution: net tax proceeds can be used to increase welfare payments or to decrease direct tax rates.

When real wages are fixed, gross costs of an energy-carbon tax are very high. Reducing employer's social security taxes is a better strategy than increasing welfare payments or reducing direct taxes. The main problem is the real wage costs which cannot be decreased sufficiently.

In conclusion, the regime of the labour market together with the aversion to income inequality should be the main determinants of the choice of a tax reform strategy. The strong version of the double dividend assumption was rejected everywhere. The weak version of the double dividend was only verified in those cases where efficiency gains dominate distribution effects.

*Acknowledgements*
*This work has been supported by the Belgian Ministry of Economic Affairs, Energy Programme under contract No. 91-05. S. Proost has been supported by the National Fund for Scientific Research. The views defended here are the authors' views and are not necessarily those of the Ministry. We thank L. Bovenberg, L. Bui, I. Mayeres, R. McDougall, E. Schokkaert and T. Rutherford for their comments on a previous version.*

## References

**Atkinson, A. B.** (1970) On the measurement of inequality. *Journal of Economic Theory*, 2 : 244-263.

**Bovenberg, L. A. and de Mooij, R. A.** (1994) Environmental Levies and Labour-Market Distortions. *American Economic Review*, 94 (4).

**Bovenberg, L. A. and Van der Ploeg R.** (1994a) Environmental policy, public finance and the labour market in a second best world. *Journal of Public Economics*, forthcoming.

**Bovenberg, L. A. and Van der Ploeg, R.** (1994b) Consequences of environmental tax reform for involuntary unemployment and welfare. *CentER D.P.* No. 9408, Tilburg University ,The Netherlands.

**Cline, W. R.** (1992) Global Warming. Institute for International Economics, Washington D.C.

**De Borger, B.; Mayeres, I.; Proost , S. and Wouters, S.** (1994) Social cost pricing of urban passenger transport. D.P, CES - KULeuven.

**Dewatripont, C.; Erlich, Z.; Ginsburgh, V. and Van Regemorter, D.** (1991) The effects on unemployment of reducing social security contributions - A general equilibrium analysis for Belgium. *In*: Applied General equilibrium modelling, H. Don ; T. Van de Klundert and J.Van Sinderen (eds), Kluwer Academic Publishers, Dordrecht, Boston, London, 135-153 .

**Diamond, P. A. and Mirrlees, J. A.** (1971) Optimal taxation and public production I : production efficiency and II : tax rules. *American Economic Review*, 61 : 8-27 and 261-278.

**Dixit, A.** (1985) Tax policy in open economies. *In*: Handbook of public economics, Auerbach A.J. and Feldstein M. (eds), Elsevier Science Publishers B.V., 313-374.

**Drèze, J. H.** (1985) Second-best analysis with markets in disequilibrium. *European Economic Review*, 29 : 263-301.

**Eyckmans, J.; Proost, S. and Schokkaert, E.** (1994) A comparison of three international agreements on carbon emission abatement. *In*: International Environmental Economics, Van Ierland E. (ed.), Elsevier, Developments in Environmental Economics, 4, Amsterdam.

**Goulder, L.** (1994) Environmental taxation and the double dividend. Paper presented at the 50th Congress of the I I PF, Harvard University.

**Hoel, M.** (1992) Carbon Taxes- an international tax or harmonized taxes. *European Economic Review*, 36 : 400 - 406.

**Italianer, A.; Strumelle, P. and Van Regemorter, D.** (1990) Estimation of CRESH production functions with an application to Belgian industry. CES manuscript.

**Mayeres, I.** (1993) The Marginal External Costs of Car Use - with an Application to Belgium. *Tijdschrift voor Economie en Management*, XXXVIII : 225-258.

**Mayeres, I. and Proost, S.** (1994) Optimal Tax Rules for Congestion Type of Externalities. D.P., CES, KULeuven, Belgium.

**McDougall, R. and Meagher, G. A.** (1994) Distributional effects of a carbon tax in Australia. Paper presented at the 50th Congress of the I I P F, Harvard University.

**Proost, S. and Van Regemorter, D.** (1992a) $CO_2$ emissions and taxation: macro-economic, public finance and welfare impacts in the EC. *European Economy*, special issue n°1 : 91-124.

**Proost, S. and Van Regemorter, D.** (1992b) Economic effects of a carbon Tax - with a general equilibrium illustration for Belgium. *Energy Economics*, April 1992.

**Van De Gaer, D.; Schokkaert, E. and De Bruyne, G.** (1992) Marginal tax reform in different macro-economic regimes. *Public Finance*, 47 (Supplement) : 256-270.

**Vennemo, H.** (1994) Welfare and the environment : implications of a recent tax reform. Paper presented at the 50th Congress of the I I P F, Harvard University.

# *4. Tradeable Permits*

Environmental Policy Between Regulation and Market
ed. by C. Jeanrenaud


# 4.1 Designing Efficient Treaties to Protect the Global Environment[1]

**Peter Bohm**
**Department of Economics, University of Stockholm, Stockholm, Sweden**

## 1. Introduction

The global environment is threatened by anthropogenic factors in at least two respects: the ozone layer and the global climate. There are ways to reduce both these threats. If the options are selected and designed in a globally efficient manner, i.e., with a minimum loss of goods and services for a given reduction of ozone-depleting potential (ODP) and greenhouse gas (GHG) emissions, the chances increase that actions will in fact be undertaken. Efficiency also allows for actions to be made more stringent than if policy implied a waste of resources.

The purpose of this paper is to highlight a set of efficiency aspects of international policy to reduce threats to the global environment. Most of the discussion is relevant to global pollution in general, i.e., for environmental problems that affect all countries and are caused by emissions regardless of geographical origin. In the terminology of economics, this means that the problem is one of the environment as a global public good and pollution as a global public

---

1 This is a thoroughly revised version of a paper presented at the 1991 conference "Economy and Environment in the 1990s" in Neuchâtel. The revision was made in October 1995.

bad. As with other public goods or bads, costly actions taken by individual agents (here including most national governments) have a strictly limited effect on the public bad, e.g., the ODP or the GHG emissions. Thus, there are incentives for individual countries to avoid such actions in the hope of free-riding on actions taken by others. These incentives complicate the process of international cooperation to solve global environmental problems and to identify and implement globally efficient measures.

We will explicitly deal with $CO_2$ and CFC emissions, both of which are GHG emissions, the latter also the dominant cause of ozone depletion. The observation that ozone itself is a GHG means that reduced emissions of CFC will have a negative as well as a positive effect on GHG concentration in the atmosphere. Recently, there has been a tendency to leave out CFC emissions from the context of climate change assuming that these two effects approximately cancel each other out. This approach will be followed here as well. Focusing on $CO_2$ as a GHG means that we will not try to tackle one of the most important efficiency problems, that of a comprehensive approach to global warming; attempts to follow this approach imply observing all sources of a given type of environmental change and undertaking action to reduce emissions from one source (or increase the capacity of one carbon sink) as long as this is more cost-effective than actions concerning other sources or sinks. However, since the feasibility of the comprehensive approach is as yet uncertain, we will focus on $CO_2$ which is the dominating greenhouse gas.

The first sections of the paper deals with efficiency problems confronting global as well as nonglobal treaties on global pollution. A stringent global treaty is hardly a realistic objective for the near future. However, if such a treaty is considered to be the relevant long-term objective, it is important to keep this objective in mind when analyzing proposals for initial nonglobal treaties, so that policy design is made compatible with the ultimate goal of an effective and efficient global treaty. In the final section, we focus on an efficiency problem specific to nonglobal treaties.

We discuss five efficiency aspects of policy design:

- The problem of indirect, instead of direct, control of emissions (section 2).
- Potential conflicts between international distribution targets and efficiency (3).

- Efficiency implications of making national emission quotas internationally tradeable (3 and 4).
- Efficiency implications of internationally harmonized carbon (or $CO_2$) taxes vs. tradeable carbon emission quotas (5).
- Efficiency implications of including supply-side measures in an agreement among a small group of countries (6).

The paper draws on and develops some of the author's previous studies of these topics. One study that features prominently in the paper/chapter concerns an evaluation of the efficiency properties of the first treaty on global pollution, the Montreal Protocol on Substances that Deplete the Ozone Layer. This treaty has been considered a possible blueprint for a treaty on GHG emissions.

## 2. Direct vs indirect control of emissions

It sounds wise to let practical aspects, which seem to have innocuous side-effects at the time of a policy decision, determine the design of policy. However, doing so distorts incentives which eventually may turn out to have unwanted consequences. There are many examples of "realistic" policies of indirect control ultimately producing detrimental side-effects or no apparent beneficial market responses.

As a case in point, the Montreal Protocol did not attempt to fully control CFC emissions, the main culprit in ozone depletion. Instead, the treaty targeted CFC production and CFC use. This means that (with some exceptions provided by the Protocol) it is not worthwhile for signatory countries to test suggested, but still economically uncertain, techniques for the recovery and destruction of CFCs in existing products at the scrapping stage. Probably more important, it does not make it worthwhile (again with some exceptions) to devote resources to developing new techniques for such activities. Given that the stock of CFCs in existing products were considerable at the time the Protocol entered into force and were yet to be emitted into the atmosphere (and still are), this is not a trivial matter. Moreover, if efficient techniques to recover these substances from scrapped products were to appear more or less accidentally, it would not be worthwhile under the present policy to design future products so that CFC recovery would be facilitated (Bohm, 1990).

Some CFC-destruction technologies for cooling equipment (the CFC refrigerant as well as the CFC in insulation) have actually been developed and put to some practical use in at least Germany and Switzerland. However, it is unlikely that the scale of these operations is globally efficient. It is also unlikely that the number and efficiency of existing technologies are close to those for technologies which would have been available and used if the appropriate incentives to develop and employ new technologies had been present.

Could this efficiency problem have been avoided by a realistic redesign of policy? Although CFC emissions cannot be measured directly, proof of destruction of recovered CFCs could be agreed to be accepted and quantities destroyed to be subtracted from the CFC use of a country,[2] allowing it to use more new CFCs (or other ozone depleting substances) as long as it does not increase CFC emissions. To encourage actions to reduce CFC emissions from the remaining stock of CFC-containing products after the use of new CFCs or other ozone-depleting products have been phased out, the appropriate incentive would have to eventually take the form of a pure subsidy of CFC destruction.

For similar reasons, consideration should be given to the possibility of allowing credit for (i) proof of carbon removal (e.g., from stack gases) and (ii) use of fossil fuel as feedstock or other verifiable measures, which sequester emissions of $CO_2$, rather than basing a GHG-controlling treaty solely on the attractively simple measurement of the carbon contents of fossil fuel use (production plus net imports). The general principle of this policy approach is that of a deposit-refund system (Bohm, 1981), where proven nonpollution is subtracted from potential pollution (here: carbon contents of fossil fuel consumed) and returned for a tax refund or a quota credit.

To sum up, when direct monitoring of pollution is too costly and when simple indirect approaches, such as monitoring CFC use or the carbon contents of fossil fuel use, have

2 For certain types of CFC destruction relevant at the CFC *production* stage, such proof is in fact accepted by Article 1.5 in the Protocol. Here, it is stated that '"production" means the amount of controlled substances produced minus the amount of controlled substances destroyed by technologies to be approved by the Parties' (see e.g. Benedick, 1991, Appendix B).

distortionary long-term effects, an optimal feasible solution may be to add a corrective subsidy component in money or quota credits to the indirect tax or quota approach, thus neutralizing its distortionary effects.

## 3. Conflicts between efficiency and specific forms of distributional considerations

Treaties on global pollution face formidable problems when it comes to determining a "fair" distribution of the burden of pollution control among countries. One approach is to settle the international distribution of emissions (or use of the potential pollutants, CFC or fossil fuel) as well as a set of "sufficient" international *side-payments*. This is how the issue seems to have appeared to the founding fathers of the Montreal Protocol (for a presentation of the discussions preceding the Protocol, see Benedick, 1991). These problems are likely to be particularly difficult for a treaty on $CO_2$ emissions, given the paramount importance of energy use for economic development.

Attempts to find a fair initial distribution of $CO_2$ emissions among countries are likely to be based on some simple criteria such as current emission levels, GDP, population, etc. or some combination thereof (see Grubb, 1989, and Barrett, 1992). In the final analysis, adherence to a simple formula may not prove feasible. Individual countries or groups of countries are likely to request special treatment, which, if allowed, will either affect the allocation of emission quotas or the design and extent of compensatory side-payments. The outcome may therefore be a patchwork.

Regardless of the pattern of the final solution to the distribution issue, the outcome may be globally inefficient. For example, the result of the negotiations may well be that developing countries, whose participation requires that they be given favorable treatment, could have obtained a still better deal, or that there could have been a larger cut in global emissions, without any party being hurt.

The Montreal Protocol is an illustration of such a Pareto-inefficient outcome. For this treaty -- as is likely to be the case for a global warming treaty -- the primary problem of burden distribution was between industrial countries (ICs) and developing countries (DCs). Essentially, the DCs were allowed to increase their CFC consumption for a considerable period of time and to reach the reduction targets with a ten-year lag. In addition, the DCs were offered certain side-payments, made more specific in the London revisions of the Protocol (see e.g. Benedick, 1991).

Since the national CFC use quotas were not tradeable (except within the European Community), the result of the Protocol was to stimulate the growth of CFC consumption in signatory DCs, where the quota reservation price at the margin is likely to have been, and continue to be, zero for all or most of the transition period. By contrast, the marginal quota reservation price was positive and probably quite high in many ICs. Given the CFC reduction targets for the signatories to the Protocol, both DCs and ICs would have gained from allowing quota trade. Specifically, DCs would have gained from reducing their CFC use as long as their marginal abatement costs fall short of the prices offered by countries that want to increase their use of CFC (Bohm, 1990). Quota trade implies that marginal abatement costs increase in countries (DCs) that increase their abatement and are diminished in countries (ICs) that, due to quota trade, can reduce their abatement. Formally, for any given cap on aggregate emissions, efficiency requires that emissions be allocated in such a way that marginal abatement costs are equalized across countries. When countries buy and sell quota units to the extent they benefit from doing so, at an equilibrium quota price established on a perfect international market for quota trade, the resulting international emission allocation will in fact be efficient.

Thus, solving the distribution problem partly by arranging a nontradeable transfer "in kind" to the DCs, ensuring them what may be viewed as a fair share of global CFC or fossil fuel consumption, is not compatible with efficiency. The feasibility of a tradeable quota (TQ) scheme, which is compatible with efficiency (barring certain problems of imperfect competition on the quota market, see below), indicates that it typically is not efficient to divide the distribution issue into two parts: actual CFC or fossil fuel consumption (or $CO_2$ emissions) on the one hand and financial transfers on the other. Under the TQ scheme, the distribution issue will be one of *initial quota distribution* only. If the initial quota of a country exceeds the country's actual quota demand at the equilibrium quota price, sales of the resulting surplus *minus* abatement costs incurred determine the implicit net transfers received, and *vice versa*. Accordingly, if future quota prices in a TQ system can be reasonably well predicted, the selection of the initial quota distribution will determine the international distribution of the real costs of the treaty. With significant uncertainty about future quota prices, transfer recipients (DCs) are likely to require insurance in the form of additional transfers contingent upon quota prices falling below a certain level, or the quota would simply have to be set at a level sufficiently high to compensate for the risk-bearing incurred.

## 4. Distributional implications of making quotas tradeable

In addition to allowing an efficient allocation of emission abatement as shown above, making quotas tradeable has another important effect. If quotas for $CO_2$ emissions or fossil fuel use are allocated between DCs and ICs in a fashion similar to the allocation of ozone-depleting substances in the Montreal Protocol, the marginal reservation prices for $CO_2$ quotas will be smaller for DCs than for ICs. If nontradeable quota systems are used for $CO_2$ emissions as well, this relation between the marginal reservation prices of $CO_2$ emissions is certain to hold. Accordingly, the opening up of quota trade would mean, as indicated in the preceding section, that the value of marginal quota units would increase in DCs and decrease in ICs to become equal at the international equilibrium quota price.

Now, assume that negotiations about quota allocation are conducted with the precondition that quotas will be tradeable. With the reasonable assumption that, *ceteris paribus*, lower marginal reservation prices will make a negotiating party more willing to make concessions and accept a smaller initial quota and *vice versa*, we will have the following situation: ICs will now be less reluctant to accept lower quota limits than they would with nontradeable quotas, whereas DCs, which now can sell (time-limited, we assume) quota units to other countries at a price exceeding their marginal reservation price for domestic quota use, will be more reluctant to accept a given quota than in the nontradeable quota case. If these changes in the costs of concessions made by the various parties are reflected in negotiated results, the effect of tradeability will be that the (initial) quota distribution moves toward a larger share for DC signatories and a smaller share for IC signatories (Bohm, 1992).

This effect of making quotas tradeable can hardly be overrated. One reason is that the difficulty of getting DCs to participate, which was considerable in the case of the Montreal Protocol, is likely to be even greater for an international global warming treaty. The distributional implications of tradeable quotas now highlighted could significantly increase the incentive for DCs to sign the treaty, which in turn would have efficiency implications. In addition to reducing the costs for a given global limit to $CO_2$ emissions as shown in the preceding section, quota tradeability could make the treaty agreement more efficient by increasing the number of signatories and/or reducing the time needed to reach a global agreement. Alternatively, this additional benefit from making quotas tradeable could be used to lower the treaty limit to $CO_2$ emissions and thus increase the global public good of GHG pollution abatement even further. This latter option would most likely allow greater global efficiency, since under any realistic treaty, aggregate marginal global abatement benefits would still exceed marginal costs (in terms of fossil-fuel use benefits foregone), whereby GHG pollution abatement would still remain suboptimal.[3]

To sum up, we have seen that quota tradeability not only tends to make the international allocation of emission abatement more efficient, but also tends to make initial allocations more attractive for DCs, which in turn boasts efficiency by increasing treaty participation and/or suboptimal emission reduction targets.

## 5. Tradeable quotas vs harmonized domestic taxes

The efficiency implications of making quotas tradeable that were highlighted in the preceding two sections would seem to make a system of tradeable $CO_2$ emission quotas a promising international policy instrument for a future global-warming treaty. There is, however, another potentially efficient policy option, an internationally uniform $CO_2$ tax. Both options have been in focus for quite some time now. Well ahead of international negotiations towards a stringent climate change treaty, the option most suited for meeting the relevant policy criteria needs to be identified. The reason for the hurry is that, once an international agreement has been reached on what policy instrument to use, shifting to another one that later on is found to be superior would be politically cumbersome or even impossible. The Montreal Protocol is, as

---

3 We recall here the rule of optimal provision of public goods, i.e., that the aggregate marginal rate of substitution of private goods for the public good (or aggregate marginal willingness to pay) should equal the corresponding marginal rate of transformation (or marginal costs of the public good).

we have seen, a case in point: it contains deficiencies in policy design, which had not been made obvious prior to the initial accord, nor were they removed later on.[4] The likely costs of similar deficiencies in a climate change treaty would most certainly be many orders of magnitude higher.

An attempt is made here to list what seems to be major relative merits of an international tax and a tradeable quota system, as an illustration of the type of evaluation that is needed. First, it should be noted that we do not envisage a realistic global tax alternative as one where the tax revenue goes *in toto* to an international organization (in some of the scenarios which use a global tax to attain certain oft-stated reduction goals for $CO_2$ emissions, taxes could amount to as much as ten per cent of global GNP [Whalley and Wigle, 1990]). Hence, we assume that most, if not all, of the carbon tax revenue will be kept by the government in the country where it is raised.

Before proceeding to identifying the major differences, we should list major characteristics that in fact are shared by (ideal versions of) systems of TQs and harmonized domestic carbon taxes:

1) Marginal abatement costs tend to be equalized across countries, thus making the allocation of a given global carbon emission abatement internationally efficient.
2) The same set of *de facto* transfers between countries, and hence the same international income distribution, can be achieved (under certainty) by side-payments in the tax case and by adjusting initial allocations in the TQ case.
3) In case of a nonglobal treaty, problems of 'carbon leakage' to nonsignatories would be the same.[5]
4) Both instruments can produce the same potential increase in government revenue, and hence the same benefits from reduced pre-existing tax distortions. The reason is that

---

4 However, taken as a whole, the Protocol seems to have been an undisputed diplomatic success, especially with regard to the speed in which an agreement was reached.

5 Carbon leakage means that increased costs per unit of carbon emissions in signatory countries lead to increases in fossil fuel use in nonsignatory countries through trade and migrating firms (in the absence of carbon tariffs).

countries can choose to use domestic $CO_2$ taxes also in the TQ case or use domestic tradeable permits that are auctioned off by government (Bohm, 1994).

Major relative merits of the TQ approach are:

- *Domestic policy flexibility*: Any domestic policy -- tax, tradeable permits or regulations -- can be used to meet the commitments implied by an (*ex post* trade) quota, whereas harmonized domestic $CO_2$ taxes obviously mean that only taxes can be used in that case.
- *Certainty of the resulting pollution level*: A global (nonglobal) TQ system has a known effect on global (signatory) emissions (barring monitoring fraud), whereas the effect of a given tax rate, other things being equal, is uncertain (regardless of tax fraud).[6]
- *No domestic policy options for 'cheating'*: Since the international agreement in the tax case is in terms of a commitment to introduce a specific carbon tax, possibilities to increase carbon emissions given this tax will exist as long as there are domestic policy options available which indirectly influence emissions (e.g. subsidies of complements to fossil fuel use). In the TQ case, signatories are committed to an (*ex post* quota trade) emission level.
- *Irrelevance of nonconvertible currencies*: Harmonizing domestic tax rates (quoted in, e.g., US dollars) are likely to be complicated by the existence (or possible creation) of non-convertible currencies.
- *One vehicle only for implementing fair distributions*: Dividing an agreed aggregate emission level into initial quotas for individual countries determines the net cost distribution among these countries (in advance only under certainty). In the tax case, the cost distribution is determined (in advance only under certainty) in negotiations concerning (a) the tax level, (b) constraints on domestic policy and (c) what side-payments each country should make/receive. Aside from the added complications, the tax case presupposes large explicit financial transfers whose political acceptability and

---

6 This point is relevant when negotiations are focused on quantitative (moving) targets in terms of global emission levels, rather than the optimal level of marginal net damage values of emissions, which strictly speaking is the basis for the tax approach. If a treaty can meaningfully be negotiated in terms of such marginal net damage values or minimum acceptable marginal abatement costs, carbon taxes would be the relevant approach, at least if its other drawbacks could be neutralized.

sustainability remain to be seen (although the real transfers, as argued in point (2) above, would be the same under certainty in the two cases).

Major relative advantages of harmonized domestic carbon taxes include:[7]

- *the familiarity of the excise tax principle*: The principles of tradeable quotas/permits have not been used outside the United States and a few other countries. Experience from international applications of these principles are strictly limited.
- *the relative certainty of the future costs imposed by the policy*: The harmonized tax rate to be levied on fossil fuel is known for the agreed future period, whereas the quota price, also to be added to fossil fuel prices, will remain uncertain.
- *the relative certainty of actual transfer payments*: Significantly uncertain quota prices make the magnitude of transfers uncertain in the TQ case (in the absence of an insurance policy, see section 3), a problem which is avoided by agreed real side-payments in the tax case.

Given this multitude of dimensions in which the two policy options may differ to a significant extent, future evaluations may well find that one of them is clearly superior to the other, given the relevant policy targets. Hence, there is a large potential loss of efficiency if such evaluations are not made available and, if possible, agreed on, prior to finalizing a first stringent climate change treaty.

## 6. Efficient and sustainable nonglobal treaties

Since it will take time for any treaty on global pollution to get close to being fully global, it is important to discuss efficiency issues specific to nonglobal treaties. One such issue will be raised here.

---

[7] Since there are ways to limit, or even eliminate, the (quota) market power of rich countries (Bohm, 1995) and since no TQ system is likely to be generally accepted without any appropriate such measure being undertaken, we abstain from listing this as a drawback of the TQ system, in contrast to what is elsewhere often done.

Unilateral actions taken by a small country or a group of countries with a small share of global emissions have an insignificant effect on the global environment. Thus, the direct environmental effect of such actions can hardly be the reason for undertaking them. Instead, the reason would probably be (as in the case of the Swedish carbon tax [Bohm, 1994]) to state an example, to show a willingness to make sacrifices and to interest other countries in following suit.

Actions taken by a large group of countries, covering, say, 75 per cent or more of global emissions, would be able through trade policy to penalize nonsignatory countries for not joining. If so, the fringe of reluctant parties could eventually be forced to join.

Here, we will discuss the middle case, an agreement among a group of countries covering some 10 to 50 per cent of global emissions to reduce their fossil fuel use. Assuming that the world fossil fuel market is essentially competitive, at least initially, such actions would be significant enough to reduce equilibrium world-market fossil fuel prices below what they would otherwise be.

A reduction in fossil fuel prices would have three potentially important side effects, the first of which would reduce the net effect of the action taken by such a group of countries:

1) Use of fossil fuel and hence, $CO_2$ emissions, would tend to increase in nonsignatory countries. Thus, the net effect on global fossil fuel use would be less than the reduction made by the signatory group of countries.
2) Fuel exporting countries, many of them poor, would suffer terms-of-trade losses in addition to reduced sales, and such effects might eventually be unpalatable to the industrial countries taking the action.
3) Incentives to form new producer cartels or invigorate existing ones (such as OPEC) would be strengthened.

All three effects obviously contribute to jeopardizing the sustainability of the agreement. Could complementary arrangements be made to avoid these effects? On the demand side, actions could be taken to persuade nonsignatories to enter an associated treaty, whose members would agree, in a practicable and well-defined fashion, to avoid actions that would imply taking advantage of the lower fossil fuel prices. Alternatively, countries initially nonsignatory might be made interested in joining the original treaty in response to offers of resource transfers to support energy conservation or substitution (e.g. to hydropower). If such measures prove only moderately successful, some action on the supply side of the market may be required.

The basis for the supply-side action to be discussed here is provided by a proposal to meet the US strategic petroleum reserve requirement, according to which the US Government would lease on a long-term basis oil stored in oil-exporting countries. A number of countries have shown an interest in participating in an arrangement of this type (USDOE, 1990). The arrangement relevant in the present case is that where the treaty signatories, or an international organization like the World Bank acting in their place, would lease on a long-term basis oil, gas, or coal in their original deposits, deposits that otherwise would have been used for fossil fuel production during, say, a ten-year period. The leasing country would thus avoid extraction costs as well as some processing and transportation costs.

The original option of a reduction of fossil fuel demand can only now be compared with the alternative action, which is designed to have the same effect on global fossil fuel use, but with a smaller reduction in signatory demand coupled with an equally large reduction in global supply. In the latter case, there would also have to be a credible commitment not to release extra fossil fuel in the amount of the deposits leased. If so, the effect on fossil fuel prices would be eliminated in this case. The sign of the net costs to the signatories of replacing the first option with the second one can then be seen to turn on the values of a set of parameters (Bohm, 1993).

Assume that some compensation, primarily to the poorest fossil fuel-producing countries (because of effects (2) and (3) noted above), would be required in the case where only demand is reduced by, say, an amount approximately equal to the gain that the signatories would make from the reduced fossil fuel prices. Then, the main determinants of the outcome would be the size of price elasticities of demand and supply and the share of marginal world fossil fuel supply that would be available for long-term leases of deposits.

It can be shown by way of an example (see Bohm, *op.cit.*), that if global as well as signatory demand price elasticity is about -1, if global price elasticity of supply is below 2, and if the signatory share of global fossil fuel use is around 25 per cent, the combined supply-

and-demand adjustment policy would be the least expensive option, as long as the (absolute) price sensitivity of supply on the global fossil fuel market is less than eight times the price sensitivity on the long-term leasing market. Under such circumstances, the gain from requiring a smaller reduction of signatory fossil fuel use with this policy would exceed the cost of the required long-term leases of fossil fuel deposits.

It is obviously difficult to predict the outcome of a policy combining a reduction of fossil fuel use with long-term leasing deposits for non-use. But the fact that this policy, which at first sight seems prohibitively expensive, is far from ruled out under plausible assumptions, would make it worthwhile to investigate further what the real-world parameter values could be and what institutional options exist for supply-side actions. More importantly, it should be recalled that the alternative, the 'traditional' type of treaty reducing fossil fuel demand only, has the potential drawback of not being sustainable because of the three effects of such treaties highlighted above.

## 7. Concluding remarks

Efficient cooperation is important, possibly crucial, for an international policy to succeed in combating emissions of global pollutants. This is particularly likely to be the case for $CO_2$ emissions, a vital "input" to industrial production and an important factor for material growth.

A number of efficiency issues arise when treaties on global pollution are to be negotiated. For some of these issues a simple solution may not exist. We have seen that long-term efficiency may not materialize if policy amounts to controlling only what is easy to monitor and seems to be well correlated with pollution (e.g., the carbon content of fossil fuel), while disregarding incentives for R&D in pollution removal. We also observed the potentially significant costs of settling issues of international distribution by choosing simple, but globally inefficient, policy solutions such as straightforward (nontradeable) emission quotas. The Montreal Protocol, although an impressive achievement by the negotiating parties, could be seen to provide an illustration of both of these problems. It was shown that using a deposit-refund design and making quotas tradeable would provide solutions to these problems. It was also pointed out that making quotas tradeable would have the additional advantage of increasing the attractiveness of the proposed treaty to developing countries.

Against this background, we revisited the long-standing issue of the major pros and cons of international tradeable quotas and internationally harmonized domestic taxes, the two main contenders for an efficient design of international climate change policy. The main conclusion was that there is no clear picture as to what instrument is the most efficient, partly because the relevant set of policy targets (likely to include more that international distribution and efficiency) is important for the outcome and because it is still far from clear what these targets are. Bearing in mind the efficiency problems of the Montreal Protocol -- problems which are likely to be very costly in the context of a stringent climate change treaty -- it seems important to step up the preparations for policy design and provide more detailed information about the overall efficiency of the two principal international policy alternatives.

In the final section we saw an example of how the design of an efficient treaty among a large, but far from complete, set of countries may require rather complicated solutions. As features prominently among the arguments from special interest groups who advocate a 'do-nothing' policy, simply reducing fossil fuel demand when international competitors do not follow suit may not constitute a worthwhile policy. But a worthwhile option may be for consumer countries to institute a reduction in fossil fuel supply as well, rather than doing nothing now and waiting until something closer to a global treaty can be negotiated. More research is needed, however, in order to implement such a combined supply-and-demand policy and to meet its information requirements.

## References

**Barrett, S.** (1992) 'Acceptable' Allocations of Tradeable Carbon Emission Entitlements in a Global Warming Treaty. *In*: Combating Global Warming, UNCTAD, New York.

**Benedick, R. E.** (1991) Ozone Diplomacy. Harvard University Press, Cambridge, MA.

**Bohm, P.** (1981) Deposit-Refund Systems : Theory and Applications to Environmental, Conservation and Consumer Policy. Johns Hopkins U P/RFF, Baltimore, MD.

**Bohm, P.** (1990) Efficiency Issues and the Montreal Protocol on CFCs Environment. Working Paper No.40, World Bank, Washington, DC, (forthcoming in The Environment and Emerging Development Issues (ed), P. Dasgupta - K.G. Mäler, WIDER and World Bank).

**Bohm, P.** (1992) Distributional Implications of Allowing International Trade in $CO_2$ Emission Quotas. *The World Economy*, 15 (1).

**Bohm, P.** (1993) Incomplete International Cooperation to Reduce $CO_2$ Emissions : Alternative Policies. *Journal of Environmental Economics and Management*, 24 : 258-271.

**Bohm, P.** (1994) Government Revenue Implications of Carbon Taxes and Tradeable Carbon Permits : Efficiency Aspects. Paper presented at the International Institute of Public Finance 50th Congress, Cambridge, MA, August 22-25.

**Bohm, P.** (1995) An Analytical Approach to Evaluating the National Net Costs of a Global System of Tradeable Carbon Emission Entitlements. UNCTAD/GID/9, Geneva, Switzerland.

**Grubb, M.** (1989) The Greenhouse Effect : Negotiating Targets. Royal Institute of International Affairs, London.

**Whalley, J. and Wigle R.** (1990) The International Incidence of Carbon Taxes. Working Paper, Department of Economics, University of Western Ontario, London, Canada.

**U.S. Department of Energy.** (1990) Report to Congress on Alternative Financing Methods for the Strategic Petroleum Reserve. Washington, D.C.

Environmental Policy Between Regulation and Market
ed. by C. Jeanrenaud


# 4.2 Designing a Trading Programme for Emissions of Nitrogen Oxides in the Northeastern United States

**Albert L. Nichols**
**National Economic Research Associates, Cambridge, Massachusetts, USA**

## 1. Introduction

Emission trading offers the potential for substantial cost savings in achieving environmental goals. Although economists have long argued for the use of trading and other incentive-based mechanisms, it is only within the last decade that interest has become substantially more widespread in the United States. The national $SO_2$ allowance-trading programme established by the Clean Air Act Amendments of 1990 (CAAA) is the most significant trading programme to date. In Southern California, the South Coast Air Quality Management District (SCAQMD) has established an ambitious trading programme ("RECLAIM") to help it address that region's air-pollution problems, that are the most severe and intractable in the United States. The U.S. Environmental Protection Agency (EPA) is attempting to spur the development of new trading programmes in other areas as well and has recently issued a proposal to simplify the process of establishing new trading programmes (60 *Fed. Reg* 39668).

*Trading for ozone precursors*

Much of the current interest in trading programmes relates to emission reductions needed to achieve the National Ambient Air Quality Standard (NAAQS) for ground-level ozone, the primary component of urban "smog." Twenty-five years after passage of the Clean Air Act, large sections of the United States still fail to attain that standard, despite cumulative expenditures in the hundreds of billions of dollars. Outside of California, the most widespread problems are found along the eastern seaboard, where ozone is a regional problem involving a dozen or more states. To address the regional nature of the problem, the Clean Air Act Amendments of 1990 established the Ozone Transport Region (OTR), which includes twelve states (from Maine in the north to the northern part of Virginia at its southern end) and the District of Columbia. The Ozone Transport Commission (OTC), which includes representatives from each state, has proposed various measures to reduce emissions of both volatile organic compounds (VOCs) and nitrogen oxides ($NO_x$), the two major precursors of ozone (ground-level ozone is the product of complex photochemical reactions in the atmosphere involving both $NO_x$ and VOCs). Individual states also are developing strategies to attain the ozone NAAQS.

Reductions in $NO_x$ emissions from electric generating stations and other large boilers are a key element of the attainment strategy. Electric generating stations account for about 30 per cent of $NO_x$ emissions in the OTR (U.S. EPA, 1993). Under the terms of a Memorandum of Understanding (MOU) signed by most of the states in the OTR, those sources will have to reduce their emissions significantly by 1999, with additional reductions to follow, if needed, in 2003 (OTC 1994). The states plan to implement the MOU through an emission trading plan that will establish annual "budgets" or "caps" that will be tradeable through a system of allowances, similar to that under the sulfur dioxide ($SO_2$) trading programme established as part of the acid rain portion of the 1990 Clean Air Act Amendments. Two regional organizations of state air-quality officials, NESCAUM and MARAMA, are currently developing model rules for the trading programme. These efforts are building on the work of an earlier task force that produced a series of drafts of a *Feasibility Study* in 1994 (Tech Environmental, 1994).

*EPRI Study*

In 1994, the Electric Power Research Institute (EPRI) contracted with National Economic Research Associates, Inc. (NERA) to study the feasibility of emission trading for $NO_X$. The purpose of this project was to build on previous experience with emission trading, including the $SO_2$ programme, by providing information and analyses for use by utilities and regulators working on the development of new $NO_X$ emission trading programmes. The project's first report focuses on relatively broad design issues (Nichols and Harrison, 1994). The project's second report presents empirical analyses of issues raised by the NESCAUM/MARAMA feasibility studies (Nichols and Farr, 1995). This paper briefly summarizes the key findings of the two reports and their implications for the design of trading programmes more generally.

The remainder of the paper is divided into three sections. The first provides an overview of basic trading concepts and key lessons from earlier experiences with trading. The second addresses two specific issues that have arisen in the design of the $NO_X$ trading programme for the Northeast. The final section provides some concluding remarks.

## 2. Basic concepts and experience

The basic concept behind emission trading is straightforward. By allowing firms the flexibility to reallocate (trade) emission limitations among themselves, trading can reduce costs while achieving environmental results equivalent to those under more traditional emission standards. These gains are possible because the costs of controlling different sources usually vary widely, depending on such factors as a source's size, age, fuel, capacity utilization and technology. Emission standards generally can take into account only a few of these factors, and then only crudely. As a result, standards typically result in situations in which the marginal costs of control vary widely across sources; i.e., some sources could increase control at relatively low cost, while other sources could achieve large cost savings if allowed to reduce their control efforts. By allowing high-cost sources in effect to pay low-cost sources to control more, trading can achieve significant cost savings without sacrificing overall emission-reduction goals. Studies typically find potential cost savings of 40 per cent or more; i.e., trading can almost cut costs in half compared to traditional emission standards.[1] Depending on the details of the design and its implementation, trading can also yield environmental benefits, providing more control over the total quantity emitted.

The concept of emission trading has been known to economists for almost three decades (Dales 1968), and the basic concepts behind cost-effective control have an even longer history, dating back to Pigou (1920). Limited forms of trading have been in use in the United States since 1976, when the offset programme was introduced. Unfortunately, the potential gains from trading are not always realized in practice. A review of experience with trading programmes in practice suggests which elements are most critical to success.

*ET: bubbles, offsets, netting, and banking*

During the late 1970s and early 1980s, the U.S. Environmental Protection Agency (EPA) developed a series of trading programmes -- offsets, netting, bubbles and banking -- that were codified in EPA's *Final Trading Policy* in 1986 (51 *Fed. Reg.* 43814). These trading mechanisms are often collectively referred to as the "ET" (Emission Trading) programme. They have had limited success (see Hahn and Hester 1989b). They all rely upon the concept of "Emission Reduction Credits" (ERCs), which are defined as the amount by which a source exceeds the controls required under existing standards.

Netting programmes were used most widely, but they bear little resemblance to the trading programmes envisioned by economists; they allowed new sources to be exempt from burdensome requirements if other sources within the same facility reduced their emissions enough that the net emissions increase from the new source fell below the limit for New Source Review. Offset programmes have also been used fairly broadly, but only because they are mandatory for major new sources in areas that do not attain the relevant NAAQS. In addition to meeting very stringent emission limits, such sources must show that emissions

---

1 Several such studies are cited later, in the context of specific programs. In addition, see Nichols et al (1992), who report potential cost savings of 42 per cent for $NO_x$ trading in southern Ontario, Canada and ICF (Braine, 1995), which estimates cost savings of 33 to 57 per cent (depending on the year) for $NO_x$ trading in the OTR.

have been reduced sufficiently ("offset") at other sources so that there is no net increase in emissions in the area.

The bubble programmes come closest to economic conceptions of trading. They allow firms to combine standards for several different sources into one overall limit; the name evokes the image of placing a "bubble" over several individual stacks and then measuring the sum of their emissions, rather than each one individually. Because of limitations on their use and extensive administrative requirements (including the need for prior approval), less than 150 bubbles had been approved through 1986. Banking programmes have been the least used ET elements. They allow firms to gain credits for reductions and then "bank" them until they are needed for new activities or they can be sold to another source.

The performance of bubbles and ET has generally fallen far short of the expectations of their early advocates. On a proportional basis, the savings have been very modest, probably much less than 1 per cent of overall control costs. The primary problem appears to be that ET has had many restrictions on allowable trades and many procedural requirements that raise the transactions costs of trading and create uncertainty and delay (e.g., see Dudek and Palmisano 1988). As a result of these problems, relatively few firms have bothered to try to trade, and virtually all trading has been internal (e.g., a bubble within a plant or across two nearby plants owned by the same firm), rather than between companies.

The restrictions on trading in turn reflect strong concerns on the part of many environmentalists and some government officials that many trades are not "legitimate." To some extent, these concerns represent a philosophical opposition to trading: if a firm *can* control emissions more, it should do so, and should not be rewarded for it. Such objections to trading ignore the fact that standards are far from perfect at identifying the feasibility and cost-effectiveness of controls, particularly at the level of individual plants. There is also a more practical element to the objections, however, which is based on uncertainty about the starting point from which ERCs should be given. Many sources routinely emit less than allowed by their permits or the relevant standards. Because of variability in the level of control actually achieved, firms must aim to do better than the standard if they are to be in compliance all (or virtually all) of the time. As a result, average performance is typically better than that required by the standard. Moreover, permits often set overall limits based on peak production levels, which are rarely achieved. Opponents of trading argue -- with some justification -- that giving firms "paper credits" for reductions below the levels allowed by standards or permits will result in increased overall emissions.

Similarly difficult issues arise in trying to assign credits for closing older, relatively dirty plants. From an efficiency perspective, closures of older plants may well be the most cost-effective "control" method in some cases, and should not be ruled out. Closures (or less drastic reductions in production) result in real reductions in emissions. On the other hand, environmentalists typically argue that most closures have nothing to do with emission trading, and would occur even if firms were not given ERCs.

The highly detailed 1986 trading regulations largely eliminated the risk of granting "paper credits," but at the price of a cumbersome system that makes it too expensive and uncertain to pursue most potentially beneficial trades. The impact of the 1986 tightening of the rules can be seen in the fact that of the 40 bubble proposals reviewed under the new rules up through early 1989, 31 -- more than three-fourths -- had been disapproved or were going to be disapproved (Creekmore 1989). Almost half of the disapprovals were based on the use of an "improper baseline", while the second most common reason was "enforceability." Thus, it appears that bubbles will continue to be no more than tiny adjuncts to the system of standards. Offsets, however, continue to be required -- and used -- in nonattainment areas.

*Lead in gasoline*

The approach used to regulate lead in gasoline during the mid 1980s offers a much more successful example of the use of trading. EPA regulated the lead content of gasoline starting in the early 1970s. Up through 1982, lead limits were enforced on a refiner-by-refiner basis, though each refiner was allowed to average concentrations across gallons, in the language of ET, intra-, but not inter-, plant bubbles were allowed). In 1982, the rules were changed to allow trading across refineries. Under the new rules, a refinery could use lead in gasoline above the limit if it purchased a sufficient number of rights from refineries that produced gasoline with less lead than allowed. From mid-1983 (when the new rules took full effect) until early 1985 (when a further phasedown began), an increasingly vigorous market in lead rights developed. In a typical quarter, over half of all refineries participated in the market, and up to one-fifth of the lead rights were traded (Hahn and Hester 1989a). No estimates were made of the cost savings, but refiners were enthusiastic about the programme; indeed, it is clear that many small refiners would have been forced out of business without it.

In 1985, EPA promulgated a rule to reduce the lead limit more than ten-fold, in two steps, from 1.1 grams per leaded gallon (gplg) to 0.5 gplg in mid 1985, and then to 0.1 gplg on January 1, 1986. As part of that rule, EPA allowed refiners to "bank" lead; if they reduced ahead of schedule during 1985, they could save the excess rights for use or sale during 1986 and 1987.[2] EPA estimated that refiners would bank about 7 to 9 billion grams of lead, resulting in savings of up to $ 226 million (in 1985 dollars, discounted at a 10 per cent real rate) over two and one-half years, or about 20 per cent of the estimated cost of the rule over that period (U.S. Environmental Protection Agency 1985). Although no new estimates of the cost savings were made after banking occurred, the level of banking was even higher than predicted -- 10.6 billion grams, almost 17 per cent higher than the upper end of the predicted range (Hahn and Hester 1989a). Thus, it seems likely that the actual savings were higher than the estimate.

Trading and banking were very successful in the case of lead in gasoline for several reasons:

1) There were no constraints on trading among refiners; any refiner could trade with any other.
2) Administrative requirements were minimal, with no preapproval required -- refiners reported trades in their normal quarterly reports to EPA.
3) The implicit allocation of allowable lead each quarter was clear and easy to determine (based on that quarter's production of leaded gasoline), which eliminated problems associated with insecure property rights and problems with paper credits.
4) In both 1982 and 1985, the trading/banking provisions were enacted as parts of other changes that tightened the law. As a result, trading and banking were widely viewed as

---

2 EPA ended all trading and banking on January 1, 1988, because the new standard of 0.1 gplg was set to allow a minimum amount of lead to protect certain older engines against premature valve wear; EPA was concerned that with continued trading after the bank had been exhausted, refineries might reduce lead levels too far to provide that protection. Market forces have now essentially eliminated leaded gasoline from the market as few of the older vehicles allowed to use it are still in use.

important elements in reducing lead quickly (particularly in the case of banking in 1985).[3]

*$SO_2$ allowance trading programme*

Title IV (which deals with acid deposition) of the 1990 Clean Air Act Amendments establishes the most ambitious use of trading to date. Under the law, electric utilities are allocated "allowances" to emit $SO_2$, based primarily on historical production levels and standardized emission rates. The number of allowances is being reduced in two phases, in 1995 and in 2000. During Phase I (1995-1999), only about 110 plants will be covered. In Phase II, however, about 800 plants (virtually all units that emit $SO_2$) will be part of the system. Utilities are allowed to trade allowances freely anywhere in the continental United States, without any requirement for preapproval. Although all allowances are allocated to utilities, other parties can buy them (from the utilities or in annual auctions), either as an investment or to retire them, thereby reducing emissions. Owners will also be able to "bank" unused allowances for use or sale in future years. However, generating stations will continue to be subject to all existing limits set under state implementation plans, new source performance standards and other programmes. Trading has been added to, rather than substituted for, existing regulations.

Despite the limitations imposed by continued application of other standards, the estimated cost savings from allowing trading are substantial. When it proposed the bill, the Bush Administration estimated that the trading programme would reduce costs by about 20 per cent by the year 2010 (Rosenberg 1989 and Schmalensee 1989). An earlier study -- done before the precise proposal was announced -- put the potential cost savings from national trading at 20 to 27 per cent of compliance costs (Temple, Barker & Sloane 1989). More recently, EPA has estimated even larger savings from trading as part of its analysis of the costs of rules implementing the acid-rain allowance programme. In the Regulatory Impact Analysis (ICF 1991, ES-7), EPA estimates that the control costs with trading will have a discounted value of

---

[3] In internal EPA dicussions surrounding the 1985 phasedown, in which the author participated while Director of the Agency's Economic Analysis Division, the fact that banking would allow a faster phasedown was the key argument in securing approval from key officials responsible for implementing the regulations.

$ 9.5 to $ 17.1 billion over 18 years. It also estimates that the discounted cost of those same reductions without allowing trading would be $ 19.1 to $ 30.9 billion, suggesting a savings in control costs of 45 to 50 per cent. The most recent Regulatory Impact Analysis indicates that annual savings from trading provisions will range from $ 1.0 billion in 1997 to $ 2.2 billion in 2010 (ICF 1994). In a similar analysis conducted for EPRI, it is estimated that the allowance trading programme provides savings opportunities of $ 2.9 billion annually, although annual savings of around $ 2.3 billion are more likely to occur (Van Horn Consulting *et al,* 1993).

The actual performance of the $SO_2$ allowance system remains to be seen. Thus far, only a relative handful of trades has occurred; few utilities appear to be relying on purchased allowances to achieve their limits under Phase I. However, the low level of trading activity to date need not mean that trading is working poorly. Most of the predicted savings result from trades *within* utility systems (i.e., between units owned by the same company), which are unlikely to result in public announcements until the exchange is actually made (i.e., during the month-long "true-up" period in the year after the emissions occur). In addition, most early predictions of trading activity did not properly account for the economically attractive option of banking Phase I allowances for use during Phase II, when the limit will be much tighter (Hahn and May 1994). Thus, even if inter-company trading volumes continue to be low, the allowance programme is likely to yield substantial savings through intra-system and intertemporal shifts in emissions.

*Southern California's RECLAIM*

The RECLAIM trading programme in Southern California covers $NO_X$ and $SO_2$. Recently the SCAQMD proposed extending it to VOCs as well. The current programme, adopted in later 1993, replaces 30 previously adopted rules and 12 future rules for the covered sources. Initially, 390 facilities were in the $NO_X$ programme and 41 facilities were in the $SO_2$ trading programme. The $NO_X$ sources represent only 6 per cent of permitted stationary $NO_X$ sources in the District, but account for 65 per cent of stationary source emissions (SCAQMD, 1993).

The RECLAIM programme resulted from a three-year process in which key features were debated in working papers, meetings of Advisory Committees and nine working groups with broad representation of various interests, and public workshops and hearings. The general evolution was toward a programme that was simpler, with trades subject to less government review. Major examples of this evolution include the following:

- *Emissions cap rather than ERC system.* All facilities in RECLAIM receive an emissions cap and an annual rate of reduction. The early proposals were based upon ERCs, which would have required the determination of emission reductions from some more ambiguous starting point. The allowances are described as RECLAIM Trading Credits (RTCs) -- which sound similar to ERCs -- but an RTC actually represents the authority to emit one pound of the pollutant, and thus effectively is an allowance, not a credit.
- *Annual rather than seasonal or daily system.* The programme is based upon annual emissions, rather than daily or quarterly caps as discussed during the deliberations. To provide additional flexibility, facilities are divided into two compliance cycles, one from January 1 to December 31 and one from July 1 to June 30. The overlapping terms of the RTCs increase the liquidity of the market.
- *Pre-approval not required.* The final role of the SCAQMD in the market is limited, with facilities selling RTCs not required to secure pre-approval. The SCAQMD will track RTCs and prices in an official RTC registry.
- *Only two trading zones.* The initial proposal included 38 trading zones, based upon the geographic scheme included in the offset programme. The final programme consists of only two zones, a Coastal Zone and an Inland Zone. Moreover, only some types of trades are limited by the zones.

The SCAQMD emphasizes that RECLAIM will provide benefits to the public in the form of equivalent or better emissions reductions, and benefits to the business community in the form of greater flexibility to achieve emissions reductions and lower compliance costs. The estimated control-cost savings for sources covered by RECLAIM are about 40 per cent (Harrison and Nichols 1992 and SCAQMD 1993).[4]

---

4 Initial trading under RECLAIM has been very light, primarily because the final allocation plan set a relatively lenient starting point. Over time, however, the allocations will substantially decrease, and more robust trading is likely to develop.

*Lessons from experience with trading*

Experience with trading shows mixed results. EPA's long-standing, credit-based ET programme has provided important flexibility, particularly for new sources in non-attainment areas, but for the most part has served as a relatively minor adjunct to traditional standards. In contrast, the lead-in-gasoline trading and banking programmes played important roles both in reducing costs and in achieving environmental goals more rapidly. One of the key factors in determining the success of trading programmes appears to be the clarity with which they define the baseline starting points from which trading can take place. Clear allocations and clear (and simple) trading rules minimize the obstacles to trading while achieving planned emission reductions. Both the $SO_2$ allowance programme and RECLAIM incorporate clear initial allocations and simple trading rules.

*Initial allocations should be clearly defined*

For any market to function well, the initial allocation of the good in question must be clear. If that allocation is not clear, substantial resources will be wasted trying to determine which trades are legitimate and which are not. The ET programme illustrates the problems that can arise when the baselines are not clear. Because of unclear baselines, proposed trades are often controversial, facing charges that they represent "paper credits" -- firms taking credit for actions they would have taken anyway, even if trading were not possible -- rather than genuine reductions in emissions. The 1986 policy statement imposed stringent requirements that helped ensure that paper credits would not be granted, but in the process imposed so many requirements that trading became too uncertain and difficult to undertake except in unusual circumstances or when required, as with offsets for major new sources. In contrast, the lead-in-gasoline trading programme was far more active, in large part because the initial allocations each quarter were clear and did not require case-by-case determinations of the legitimacy of baselines. Similarly, Congress established clear allocations for most of the allowances in the $SO_2$ trading programme.

*Most standards provide ambiguous baselines for measuring credits*

Standards generally provide a poor basis for trading, in large part because they are usually defined in terms of *rates* per unit of input (e.g., pounds of $NO_x$ per million Btu of heat input to a utility boiler), or some other measure related to production levels or technologies. They also often vary by fuel (e.g., different $NO_x$ limits for gas than for coal-fired boilers) or by age, with tighter standards for new plants than for existing facilities. As a result, the level of emissions that a source implicitly is "allocated" varies tremendously depending on those factors.

These dilemmas do not arise if the allocations are defined in terms of *quantities* of emissions (e.g., tons of $NO_x$ emitted per quarter or per year) that are independent of *future* inputs, outputs, technologies, or other factors under the control of individual firms. This is the basic approach taken with the national $SO_2$ allowance programme. Each unit is allocated a certain quantity of emissions in a given year. If it emits less, it frees up allowances to sell or to save for future use, regardless of the reason for the reduction. Conversely, if the unit emits more, it needs to obtain allowances from other sources, or draw down balances from previous years, again regardless of the reason for the increase. EPA does *not* need to inquire as to the reasons for the reduction (or increase) in emissions, or attempt to estimate what emissions would have been without the programme; it only needs reasonably accurate measurements of actual emissions. As a result, firms are free to use any means available to them to reduce emissions. California's RECLAIM programme has also established clear source-by-source allocations that steadily tighten over a ten-year period.

*Trading works best when it is integral to the regulatory design*

To achieve the principles laid out above, it is important that trading be an integral part of the regulatory design. Several of the features important for a smoothly functioning trading programme are less desirable -- if not completely unworkable -- in a command-and-control regime. In particular, as discussed above, caps denominated in tons per year (or a shorter period of time) are desirable for a trading programme to provide well-defined starting points for trading and greater certainty in achieving the NAAQS. However, source-by-source caps would be too rigid under command and control. Fixed, non-tradable caps on individual sources would be particularly inefficient over time, as some firms expanded production and others reduced it; a predetermined cap would be too stringent for a plant with expanded production, too large for a plant that reduced output. With trading, such shifts are accommodated easily by allowance purchases (for those expanding) and sales (for those contracting).

Similarly, with trading it is important to set the caps with the expectation that firms overall will operate right up to their limits; those firms that do better than the limits will sell the excess to others. By contrast, with standards, the limits set for individual sources must often be substantially less stringent than expected performance, to allow for source-to-source and intertemporal variability. As we discuss in more detail later, even with a reasonably well-functioning trading programme, banking is likely to be important in providing the market and individual firms with more intertemporal flexibility.

## 3. The trading system proposed for the OTR

The $NO_x$ trading programme being developed for the OTR reflects these key lessons in its reliance on cap-based allocation systems, rather than one that employs credits measured from ambiguous starting points. Each state now has a fixed $NO_x$ emissions budget for the sources covered by the programme. These caps are based on historical production and emission levels, with reductions applied based on the MOU. Each state in turn will allocate its budget among the sources included in the programme. Thus, the programme will begin with clear allocations based on mass emissions. These budgets will cover the five-month "ozone season," which runs from May 1 through September 30 each year. Based on their budgets, sources will receive allowances each year, which may be used by the source to cover its own emissions, or may be traded to other sources for their use.

In addition to choosing the fundamental structure, establishing a trading programme requires a host of detailed decisions. Some of these decisions -- in particular the specific formulas used to distribute the cap among sources -- are highly charged politically because of their effect on the *distribution* of gains, but have little impact on the efficiency with which the programme will work. Others, however, are more critical to the success of the programme. The most difficult issues in the development of the NESCAUM/MARAMA model rules have revolved around the extent to which the trading programme should attempt to limit *when* emissions occur. This debate has focused on both day-to-day emissions within each season (a "daily consideration") and on the extent to which sources should be able to "bank" unused allowances for use in future seasons.

The questions of whether to impose a daily consideration or to limit banking go well beyond the OTR. Any emission trading programme that deals with ozone or other time-varying pollution problems must address timing issues, at least implicitly. Unfortunately, most of the discussion of timing issues has *not* been informed by empirical evidence. For the most part, concerns about timing have reflected hypothetical situations in which trading could lead to large "spikes" in emissions. The remainder of this section analyzes these issues based on detailed empirical study.

*Daily consideration*

The *Feasibility Study* recommends that sources be subject only to a seasonal cap, without any "daily consideration" to limit maximum emissions on a daily basis. However, many observers are concerned that it could result in excessive emissions on some days, in particular on those hot summer days when demand for electricity is highest and ozone problems are most likely to occur. To address these concerns, the *Feasibility Study* calls on each state to review daily emissions at the end of each season and to reduce the cap in future years if daily emissions are higher than expected. The *Feasibility Study* also recommends that if the U.S. EPA finds this first approach inadequate, states impose a more formal restriction, which would limit each unit's highest daily emissions to 120 per cent of its allowances averaged over the 153-day ozone season.

*Daily emission patterns with standards*

Policy debates about trading often compare *potential* problems with trading to an *idealized* version of emission standards. The more appropriate comparison, however, is between the *likely* outcomes under the two approaches. Thus, our analysis begins by estimating what daily emission patterns would be if the MOU were implemented using traditional emission standards. To conduct that analysis, we obtained historical daily utilization data from two of the three major power pools that exist in the OTR, the New England Power Pool (NEPOOL), which covers six states (Connecticut, Maine, Massachusetts, New Hampshire, Rhode Island, and Vermont), and the New York Power Pool (NYPP), which includes all of the utilities in that state. These data include daily generation levels for each of the several hundred units in these pools. The NEPOOL data cover the five most recent years, 1990-94, while the NYPP data are available for only one year, 1991. For each unit, we apply three different emission rates. The highest rate is the "Baseline" rate. The lowest rate is based on the MOU signed by the states. We also use an intermediate rate, reflecting the rules that states are now implementing based on "Reasonably Available Control Technology" (RACT).[5]

By combining the daily generation data with the estimated emission rates, we can estimate daily emissions for each day in the 1990-1994 seasons in New England and the 1991 season in New York under each of the three emission rates. Figure 1 plots the daily totals for 1994 in New England; the four other years and New York in 1991 show similarly wide variations in daily emissions, reflecting primarily daily fluctuations in demand for electricity and the availability of various types of generating units. In New England and New York, nuclear and hydro units - neither of which emit $NO_x$ typically are "baseload" units that are used whenever they are available. Incremental generation during intermediate and peak demand periods is met by units that emit $NO_x$. As a result, daily emissions vary even more (on a proportional basis) than daily generation.

One of the key issues in discussions of trading has been how states can model daily emissions given the uncertainties involved with a seasonal trading system. At a meeting in early 1995, one state official went so far as to wonder whether states could claim *any* credit for reducing peak-day emissions with a trading plan. However, as Figure 1 makes clear, day-to-day variability is hardly unique to trading, as emissions under conventional emission standards vary greatly from day to day.

5 The data used were obtained both from the pools (NEPOOL, 1995; NYPP, 1995), the inventories prepared from EPA (Pechan & Associates, 1995a,b), and monitoring results maintained by EPA (U.S. EPA, 1995).

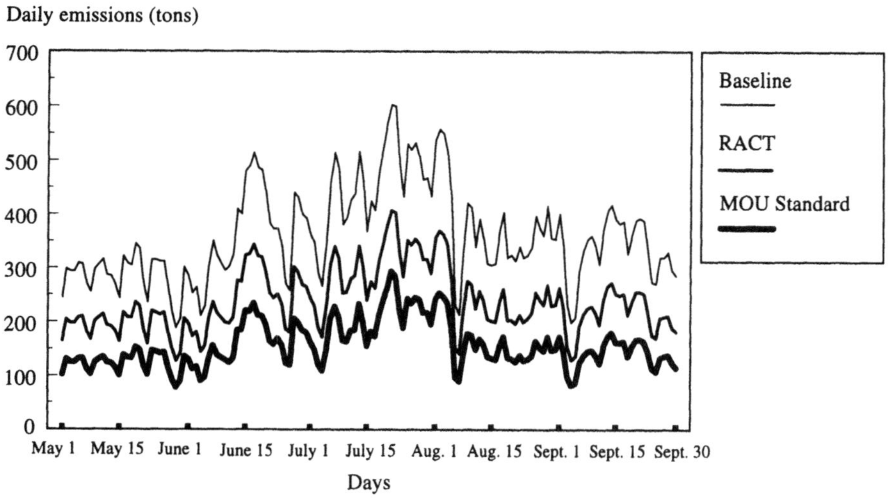


Source: NERA calculations as explained in text.

Figure 1. Daily variation in emissions under standards, NEPOOL 1994 utilization.

Figure 2 presents a range of alternative estimates of "daily" emissions assuming implementation of the MOU through a set of conventional, unit-by-unit emission standards, where all of the estimates are expressed as percentages of average daily emissions. These estimates assume historical unit utilization rates, but individual unit emission rates consistent with the MOU. The first bar in each cluster shows the seasonal average of daily emissions, which is normalized to 100 per cent. The last bar in each cluster shows emissions under the assumption that all units operate at full capacity for all 24 hours of one day, the approach taken in the *Feasibility Study* to estimate daily emissions. This assumption is extreme: it is about three times higher than the actual average. Even the system's actual day of maximum emissions (the third to last bar) is always less than 62 per cent of the hypothetical maximum. On those days when ozone exceedances actually occurred in 1990-1994, estimated emissions are higher than on average days, but only by about 25 per cent.

*Impact of trading*

The ideal approach to estimating the impact of trading would be to use unit-specific control-cost data to estimate how trading would affect emission levels from different units. Unfortunately, such data are not currently available. We have taken a simpler approach, focusing on the key concern raised about trading -- that it will allow utilities to reduce controls on peaking units (relative to what they would be with standards), thus increasing emissions on days with high generation levels, which also tend to be days with ozone problems. Although trading would require that higher-utilization units control more to balance total emissions over the ozone season, the equivalence would be on a seasonal basis, which *could* lead to net increases on days with potential ozone problems.

Emissions/Day (Index: Avg of all days = 100%)

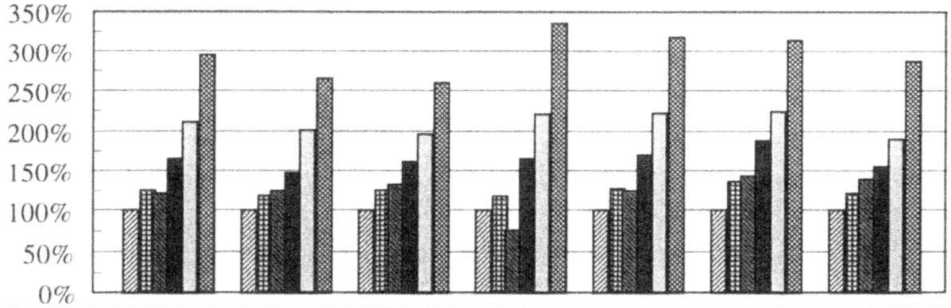


| | NE Avg. | NE '91 | NE '92 | NE '93 | NE '94 | NE '95 | NY '91 |
|---|---|---|---|---|---|---|---|
| Avg of all days | 100% | 100% | 100% | 100% | 100% | 100% | 100% |
| Avg of days with exc | 125% | 118% | 125% | 117% | 127% | 136% | 121% |
| Avg of days>1% MonHr | 121% | 124% | 132% | 76% | 124% | 143% | 139% |
| System peak day | 165% | 147% | 161% | 165% | 170% | 188% | 155% |
| Sum of unit peak day | 211% | 201% | 196% | 221% | 222% | 224% | 190% |
| 100% utilization | 295% | 265% | 260% | 336% | 318% | 314% | 287% |

Source: NERA calculations as explained in text.

Figure 2. Alternative estimates of daily emissions based on historic utilization and MOU emission rates.

Such a shift in control efforts is quite likely with trading. As other studies have shown, trading tends to reduce controls at low-utilization units and increase them at high-utilization units, because $NO_x$ controls tend to be capital intensive, which makes the cost per ton sensitive to utilization. However, the critical issue is the *magnitude* of any *net* resulting changes in emissions on ozone problem days. To investigate that issue, we consider two trading scenarios:

1) *Trading I* assumes that units with utilization rates of 40 per cent or less will emit (on average) at their RACT levels, not their MOU rates (for some units, the RACT and MOU rates are the same). The increase in seasonal emissions from these units is made up by a reduction in emission rates (below MOU levels) by units with utilization rates above 60 per cent.
2) *Trading II* assumes much larger shifts. Units with utilization rates below 20 per cent do not undertake *any* new controls, with emissions staying at their baseline levels. As under Trading I, units in the 20-40 per cent utilization class go to RACT emission rates, and units with greater than 60 per cent utilization make up the difference; however, the units with the highest utilization rates (80-100%) reduce emissions by twice as much (on a proportional basis) as those in the second-highest group (60-80%).

It is important to note that these trading scenarios are quite conservative. They are likely to overestimate the impact of trading both because they assume fairly large changes in emissions levels and that those changes occur solely as a function of utilization -- the characteristic of units that is most likely to be correlated with emissions on exceedance days. To the extent that trading also is influenced by other factors (such as age or size), the shifts toward exceedance days are likely to be less pronounced.

Figure 3 plots the daily emissions estimates for NEPOOL using 1994 utilization rates under the two trading scenarios as well as the MOU Standards. The three lower lines are virtually indistinguishable, and all are much lower than the baseline, which is shown for comparison; i.e., the predicted daily emission patterns under trading are almost the same as those under a more traditional set of emission standards.

Data from the other years and from NYPP show even smaller effects of trading. Figure 4 compares average emissions on days with ozone exceedances. Over the five-year period, exceedance-day emissions in New England under Trading I are 1.9 per cent higher than with the MOU Standards. The difference between Trading II and the MOU Standards is larger, but still only 3.7 per cent. New York exceedance day emissions increase by 0.9 per cent under Trading I and 2.2 per cent under Trading II. In all cases, the differences between the trading scenarios and the MOU Standards are very small compared to the emission reductions achieved relative to the baseline or RACT.

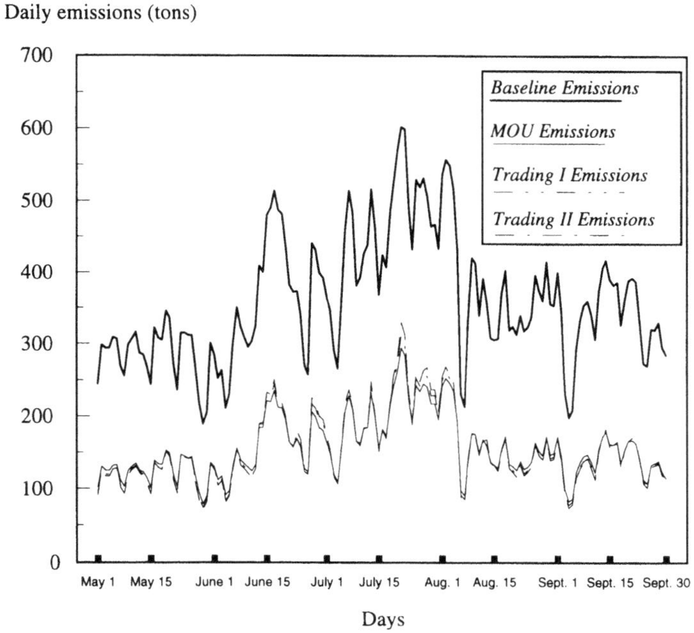


Source: NERA calculations as explained in text.

Figure 3. Impact of trading scenarios on 1994 NEPOOL daily emissions.

We do not have daily dispatch data for the rest of the OTR (New Jersey, Pennsylvania, Delaware, Maryland, the District of Columbia, and Northern Virginia). However, using the 1990 baseline assembled for allocating allowances, we can estimate likely changes in that area. Those calculations suggest that any increases on exceedance days would be modest, perhaps slightly higher than those in New York but lower than those in New England.

These results suggest that implementing the MOU with trading will have virtually the same impact on emissions on days with potential ozone problems as implementing it through standards would have. Thus, there does *not* appear to be any *need* for a "daily consideration."

Subseasonal restrictions on trading are likely to yield little if any benefit, and have the potential to increase control costs significantly by reducing the flexibility afforded by trading.

*Daily considerations proposed in Feasibility Study*

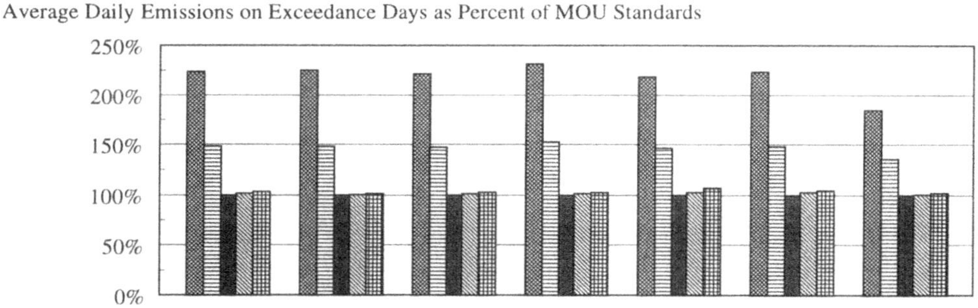


| | NE Avg. | NE '90 | NE '91 | NE '92 | NE '93 | NE '94 | NY '91 |
|---|---|---|---|---|---|---|---|
| Base | 223.8% | 225.1% | 221.6% | 231.7% | 218.6% | 223.5% | 185.0% |
| RACT | 148.3% | 148.1% | 147.3% | 152.6% | 146.4% | 148.1% | 135.8% |
| MOU std | 100.0% | 100.0% | 100.0% | 100.0% | 100.0% | 100.0% | 100.0% |
| Trading I | 101.9% | 100.6% | 101.5% | 101.5% | 103.0% | 102.8% | 100.9% |
| Trading II | 103.7% | 101.5% | 103.0% | 102.7% | 106.9% | 104.8% | 102.2% |

Source: NERA calculations as explained in text.

Figure 4. Summary of estimated average daily emissions on exceedance days.

As noted earlier, the *Feasibility Study* recommends that states supplement a seasonal cap with an annual review of daily emissions. If daily emissions (by which the authors appear to mean *highest* daily emissions) are higher than those modeled for the SIP, state's cap will be reduced proportionately. Analysis of the NEPOOL data indicates that if states wish to avoid having to make such downward adjustments, they will have to make very conservative (high) estimates of emissions for SIP modeling. This is true because maximum daily emissions (with standards or trading) are much higher than average daily emissions. For the five-year period and the six New England states, the ratios of maximum to average daily emissions range from a low of 1.4 (Massachusetts in 1990) to a high of 11.2 (Vermont in 1994). During those five years, all states except Massachusetts had at least one year with a ratio above two. The value for New York in 1991 is 1.5.

The proposed approach has several additional problems. It effectively precludes interstate trading, because the review and adjustment process would take place on a state-by-state basis. Regulators probably would not permit trades that increased emissions within their state. In addition, because the adjustment process is a one-way ratchet that can only get tighter, never looser, it would lead to an increasingly stringent cap on seasonal emissions over time. The process also is independent of ozone problems; it reduces the cap automatically even if there are no ozone exceedances on the maximum emissions day in a given season.

The backup approach proposed in the *Feasibility Study* is even more restrictive. Under it, a source would have to have enough allowances to cover its seasonal emissions *and* to ensure that its maximum daily emissions would not be more than 120 per cent of its number of allowances divided by the 153 days in the ozone season. It appears that the authors of the *Feasibility Study* expected the daily portion of this two-part test to be binding for only a few sources. Over the five-year period, however, we find that over 90 per cent of the units in New England, accounting for over 80 per cent of emissions, would be bound by the daily rather than the seasonal limit. These units would require additional allowances beyond their seasonal limits to accommodate their maximum daily emissions. Overall, the number of allowances issued would have to be roughly 1.8 times higher than the desired level of seasonal emissions.

The severe problems associated with the backup provision do not appear to be offset by significant benefits. The 120 per cent limit focuses only on each unit's day of maximum emissions, but the vast majority of emissions on ozone exceedance days have historically come from units that are not experiencing their highest daily emissions.

*Banking*

The NESCAUM/MARAMA committee and the states also are grappling with whether to allow sources to "bank" unused allowances for use in future seasons. In a budget -- or cap-based trading system, "banking" allows sources to save unused allowances for future use. Banking gives sources additional flexibility, thus reducing compliance costs. Environmentally, banking encourages early reductions, but it also leads to the possibility that *future* emission levels will be higher than they might otherwise be. The final draft of the *Feasibility Study* recommends that the use of banked allowances be subject to unspecified restrictions imposed by the states. Moreover, unless practicable mechanisms can be developed, it recommends that banking be prohibited.

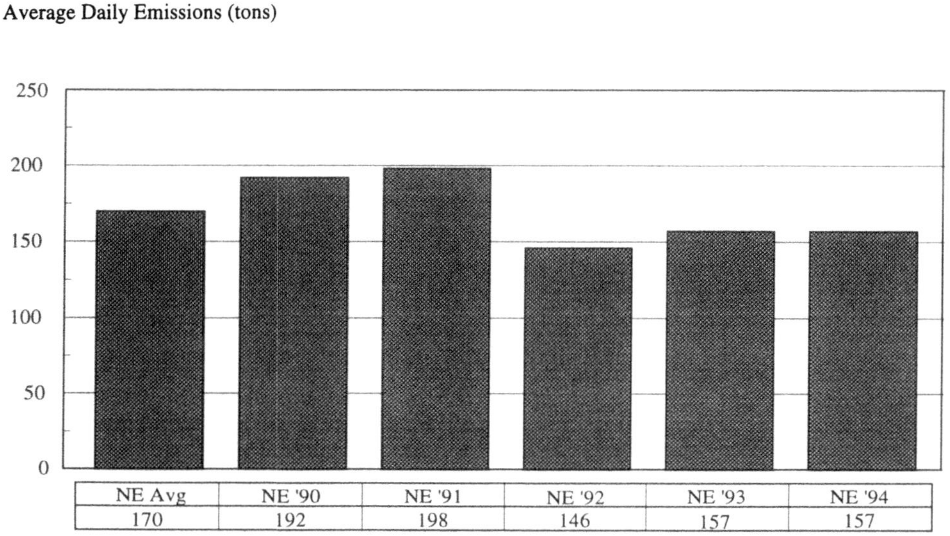


Source: NERA calculations as explained in text.

Figure 5. Year-to-year variations in emissions in average daily emissions under MOU standards.

The major concern about banking is that it will lead to variability in emissions from year to year. Opponents of banking often appear to assume that such variability in annual emissions is a new problem, one not present with traditional standards. However, it is critical to understand that traditional emission standards allow substantial variation because of year-to-year changes in electricity demand and in the mix of units available; such standards automatically increase the implicit overall emissions limit to accommodate higher generation (in response to higher electricity demand) and greater utilization of units with higher $NO_x$ limits (e.g., more intense use of fossil-fired units if nuclear units are out of commission). Figure 5 plots our estimates of average daily emissions from NEPOOL under MOU Standards, by year. Over the five-year period, they range from a low of 146 tons/day in 1992 to a high of 198 tons/day in 1991, a difference of 35 per cent.

In contrast, cap-based trading without banking can be expected to result in virtually no year-to-year variability. Banking obviously allows some year-to-year variability under a cap, but there are at least two reasons to think that it will be less than what is observed with traditional standards: (1) the cap places a firm limit on the long-term, multi-year average, which standards do not, and (2) banking entails an opportunity cost (banked allowances earn no interest), whereas standards impose no cost on year-to-year swings in activity levels and their associated variations in emissions.

Without banking, the value of allowances falls to zero at the end of each season, so that sources will be willing to sell any surplus allowances near the end of the season at very low prices. Those low prices in turn will encourage high consumption of allowances, with consequent increases in emissions, either by the original allowance holder or buyers. In contrast, if banking is allowed, sources will save unused allowances rather than sell them off at very low prices. Thus, allowing banking will also reduce the risk of an emission "spike" at the end of each season.

As experience with other trading programmes (e.g., lead-in-gasoline and the $SO_2$ programme) has shown, banking is a valuable source of flexibility and can encourage early reductions. Even the Southern California RECLAIM programme, which nominally does not allow banking, implicitly permits it by having overlapping terms for allowances. If states in the OTR continue to be concerned that banking may cause problems, the simplest and most effective solution is likely to be a contingency plan to restrict the outflow of banked allowances. For example, overall yearly net withdrawals from the bank might be limited to 15 per cent of the annual allocation. Compared to not allowing any banking, such an approach avoids the problem of precipitous drops in the value of allowances at the end of each season. It also gives sources substantial flexibility to deal with uncertainty about various factors that affect the demand for allowances.

## 4. Concluding remarks

Emission trading offers major potential gains over traditional emission standards and other "command-and-control" mechanisms. The most important and widely recognized gains are from reductions in the cost of achieving environmental targets. In many cases, trading can reduce costs by roughly a factor of two compared to typical standards. However, trading can also provide environmental benefits in the form of tighter or more quickly achieved targets and by providing greater assurance that reduction targets will be achieved.

Achieving these cost-savings and environmental benefits requires careful design of the trading programme. The critical element is defining the initial allocations clearly, so that there is no ambiguity about what sources' obligations are and what they are allowed to trade. Clarity in allocations can generally be achieved most effectively in a system that sets overall caps on emissions and distributes those caps among sources in a fixed manner that does not vary with future production levels or choices of fuel or technology. That is the approach taken in both the $SO_2$ allowance programme and in Southern California's RECLAIM programme. It is also the basic structure for the $NO_X$ trading system being developed for the Northeastern United States. In contrast, credit-based systems -- such as the bubble and other programmes using ERCs -- have inherent problems because they attempt to define allocations based on emission standards that allow differing levels of emissions depending on a host of factors.

Trading systems also work best if they are kept relatively simple, allowing the broadest possible trading across sources. When trading programmes are being developed, many observers raise a wide range of *potential* problems that trading may cause, such as: sources will use trading to concentrate their emissions on ozone exceedance days, or they will use banking to store up allowances for simultaneous use in the future, thereby causing violations of ambient standards. It is critical that such concerns be addressed realistically and empirically, rather than through the construction of hypothetical worst-case scenarios. It is also essential to compare the performance of trading to the performance of standards as they actually are implemented, rather than to some abstract ideal. Many of the issues raised with respect to trading apply with equal force to standards, but have not been considered under extreme assumptions.

Our analysis of the $NO_x$ trading programme for the Northeast focuses on timing issues -- daily emissions and banking. Similar concerns arise in connection with a host of other issues. For example, in the Northeast, one of the concerns has been that trading could lead to shifts in the geographic distribution of emissions. However, there is nothing "optimal" about the distribution of emissions under most types of standards; typically the standards are based at least as much on technology as on air-quality, with uniform rates applying across large areas. Thus, shifts may have positive as well as negative effects, and there is no reason to assume the worst. It also is important to note that uniform emission standards do not lead to uniform emissions; emissions will be high in areas with many sources, low in those with few. In any event, most trading programmes are unlikely to cause large changes in the location of emissions. For example, a recent study suggests that $NO_x$ trading in the OTR is unlikely to lead to significant interstate shifts in emissions (Braine, 1995).

In addressing most of these issues, the best approach is to start with a simple and flexible system. For example, in dealing with concerns about daily emissions, it makes sense to start with a simple seasonal (or annual) cap, without trying to fine-tune daily emissions. Similarly, the economic and environmental advantages of banking are large enough to justify offering it at least initially without restrictions. After that simple system has been in operation for several years, it *may* make sense to introduce refinements. For example, rather than setting rigid daily caps, controls can be targeted on ozone problem days by increasing the number of allowances required per ton emitted on days with weather conditions conducive to ozone formation. Similarly, in the future states may wish to enact limits on the rates at which withdrawals can be made from banked allowances. In general, however, such refinements represent *improvements* in environmental performance relative to traditional standards, not safeguards needed to avoid problems.

*Acknowledgements*
*This paper is based on two reports prepared by National Economic Research Associates (NERA) for the Electric Power Research Institute (EPRI) (Nichols and Harrison, 1994; Nichols and Farr, 1995). In addition to the coauthors of those reports (John Farr and David Harrison), the author thanks Benjamin Ligon, who performed many of the calculations; David Rowland, who provided research assistance; Christiana Olsen, who prepared the manuscripts; and Gordon Hester, the EPRI project officer, who provided valuable guidance*

*on both reports. The views expressed in this paper are the author's, and do not necessarily reflect the views of either EPRI or NERA.*

## References


**Braine, B.** (1995) Preliminary Findings of Northeast Electric Utility $NO_X$ Study. Memorandum to Chris O'Brien, U.S. EPA, John Bachman and Dina Kruger. Fairfax, VA: ICF Kaiser, Inc., July 11.

**Creekmore, A. T.** (1989) Emission Trades for Sources of Volatile Organic Compounds. Presented at the 82nd Annual Meeting & Exhibition of the Air & Waste Management Association, June.

**Dales, J.** (1968) Pollution, Property and Prices. University of Toronto Press, Toronto.

**Dudek, D. J. and Palmisano, J.** (1988) Emissions Trading : Why is This Thoroughbred Hobbled? *Columbia Journal of Environmental Law,* 13(2) : 217-256.

**Hahn, R. W. and Hester, G. L.** (1989a) Marketable Permits : Lessons for Theory and Practice. *Ecology Law Quarterly,* 16 (2) : 361-406.

**Hahn, R. W. and Hester, G. L.** (1989b) Where Did All the Markets Go? *Yale Journal on Regulation,* 6 (1) : 109-153.

**Hahn, R. W. and May, C. A.** (1994) The Behavior of the Allowance Market : Theory and Evidence. American Enterprise Institute, Washington, DC, February.

**Harrison, D. Jr. and Nichols, A. L.** (1992 March) An Economic Analysis of the RECLAIM Trading Program for the South Coast Air Basin. National Economic Research Associates, Inc., Cambridge, MA.

**ICF, Inc.** (1991 September) Regulatory Impact Analysis of the Proposed Acid Rain Implementation Regulations. ICF, Washington, D.C.

**ICF, Inc.** (1994 February) Economic Analysis of the Title IV Requirements of the 1990 Clean Air Act Amendments. ICF, Washington, D.C.

**New England Power Pool** (1995) Detailed 1994-1994 data on unit daily generation, emission, and other characteristics provided on diskette.

**New York Power Pool** (1995) Detailed 1991 data on unit daily generation, emission, and other characteristics provided on diskette.

**Nichols, A. L. et al** (1992 October) Emission Trading Program for Stationary Sources of $NO_X$ in Ontario. National Economic Research Associates, Inc., Cambridge, MA.

**Nichols, A. L. and Farr, J. G.** (1995) The Timing of $NO_X$ Emissions and Emission Trading in the Ozone Transport Region. Electric Power Research Institute, forthcoming as Technical Report, Palo Alto, CA.

**Nichols, A. L. and David Harrison, Jr.** (1994) Key Issues in the Design of $NO_X$ Emission Trading Programs to Reduce Ground-Level Ozone. Document n° TR-104245, Electric Power Research Institute, Palo Alto, CA..

**Ozone Transport Commission (OTC)** (1994 September) Memorandum of Understanding Among the States of the Ozone Transport Commission on Development of a Regional Strategy Concerning the Control of Stationary Source Nitrogen Oxide Emissions. OTC, Washington, DC.

**Pechan, E. H. and Associates** (1995a) Ozone Transport Region Baseline Inventory, version Inv.4. Produced and maintained by Pechan, Springfield, VA. Available on EPA's CHIEFS electronic bulletin board, February 21.

**Pechan, E. H. and Associates** (1995b) Ozone Transport Region Baseline Inventory, version Inv.5. Produced and maintained by Pechan, Springfield, VA. Available on EPA's CHIEFS electronic bulletin board, March 22.

**Pigou, A.C.** (1920) The Economics of Welfare. Macmillan, London.

**Rosenberg, W. G.** (1989) Testimony before the Subcommittee on Energy and Power of the Committee on Energy and Commerce, United States House of Representatives. Congressional Record, Washington, DC.

**Schmalensee, R. L.** (1989) Testimony before the Subcommittee on Energy and Power of the Committee on Energy and Commerce, United States House of Representatives. Congressional Record, Washington, DC..

**South Coast Air Quality Management District (SCAQMD)** (1993 October) RECLAIM: Final Report. SCAQMD, Los Angeles, CA.

**Tech Environmental** (1994 September) Feasibility of a Regional Market-Based $NO_X$ Budget System for the Ozone Transport Region. Final report prepared for NESCAUM/MARAMA. Tech Environmental, Waltham, MA.

**Temple, Barker & Sloane, Inc.** (1989) Market-Based Emissions Trading as an Approach to Acid Rain Control. Temple, Barker & Sloane, Lexington, MA.

**U.S. Environmental Protection Agency** (1985 February) Costs and Benefits of Reducing Lead in Gasoline : Final Regulatory Impact Analysis. EPA 230-05-85-006. Economic Analysis Division, Washington, DC.

**U.S. Environmental Protection Agency** (1993) Regional Interim Emission Inventories (1987-1991), Volume II: Emission Summaries. Office of Air Quality Planning and Standards, Durham, NC.

**U.S. Environmental Protection Agency** (1995) Detailed 1990-1994 daily ozone monitoring readings in New England, provided by Wendy McDougall. EPA, Lexington, MA, U.S.

**Van Horn Consulting, Inc. et al** (1993) Integrated Analysis of Fuel, Technology and Emission Allowance Markets: Electric Utility Responses to the Clean Air Act Amendments of 1990. EPRI, Document n° TR-102510, Palo Alto, CA.

Environmental Policy Between Regulation and Market
ed. by C. Jeanrenaud


# 4.3 Emission Trading: The Basle Experience

**Elke Staehelin-Witt**
**B,S,S., Volkswirtschaftliche Beratung, Basel, Switzerland**

## 1. Introduction

In early 1992, Basle became the first region in Europe to introduce emission trading. Emission credits[1] are intended to help industry cut environmental protection costs and to better achieve the goal of environmental protection. Until today, however, emission trading has existed mainly on paper. The theoretical possibilities of cost-savings for industry have not been taken advantage of in practice.

When the reasons for the lack of interest were analysed more closely, it soon became clear that the Basle emission trading concept suffers from the shortcomings tied to a market-based approach which is too restricted by the traditional command and control policy. Based on this analysis, a feasibility study for the Basle region now suggests emission trading that no longer relies on emission standards but works with mass emissions and an allocation of these emissions through tradeable permits. In the following months, the suggestions of the feasibility study will be discussed by different interest groups.

---

1 Put more precisely, they introduced emission credits and 'bubbles'. For reasons of simplicity, the following contribution deals only with emission credits. However, the analysis why the instruments were not successful holds true for both instruments.

The present paper shows the reasons for the lack of success of emission credits within the command and control policy, and addresses the main questions that arise in the context of a system of tradeable permits with specific regard to the case of a small region where trading has to be kept simple and control costs low.

## 2. The Basle emission trading concept: A critical analysis

*The idea of emission trading within the command and control policy*

Similar to policy in other countries, the Swiss Clean Air Act sets national ambient air quality standards and emissions standards that determine the concentration of harmful substances that a single source is authorized to emit. The Clean Air Act sets out two types of emission standards. The first is the federal emission standard, which is the same for every source in each of the 26 cantons.[2] However, since the amount and type of sources vary widely from one canton to another, the federal emission standard is not sufficient to reach the federal air quality standards especially in the industrialized cantons. In this case, the individual cantons have to set more stringent standards.

This was the case for the two cantons of Basle, which were forced to tighten their emission standards for NOx (nitrogen oxides) and VOC (volatile organic compounds). In order to keep the costs for attaining these standards low, they followed the US pattern (cf. Hahn und Hester, 1989; or Tietenberg, 1989) and introduced emission credits: A plant is given emission credits when it reduces its emissions below the cantonal emission standard. The credits can be purchased by another plant which can not fulfil the stringent standards. Trading is possible within the range of the cantonal and the federal threshold values, the latter forming a limit which may by no means be exceeded. Figure 1 shows this graphically.

---

2 The 26 individual states that make up Switzerland.

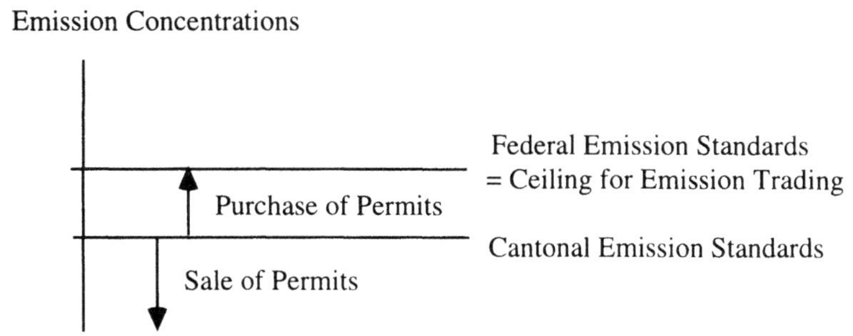


Figure 1. The range for emission trading within the command and control policy in the Basle region.

*Reasons for the lack of success*

It soon became clear that the idea of combining elements of the traditional command and control policy with a market-oriented approach did not work. First of all, in 1992 the federal government revised the Clean Air Act and tightened the federal NOx emission standard to the level of the amount for the Basle region. Since the federal standards form the upper limit, the potential for trading with NOx emission credits had already vanished before emission trading could start.

As far as VOC are concerned, a trading of emissions is still theoretically possible because the state emission value is below the federal value. However, the trading of emission credits has not seemed very attractive, on either the demand or the supply side, due to the following reasons (cf. Staehelin-Witt and Spillmann, 1992):

*Small range for emission trading*

The federal standard is quite close to the cantonal one. The federal standard may by no means be exceeded, which means that facilities which are over the limit have to invest in abatement measures even though it would be cheaper for them to buy credits. In recent years, the federal standard has required in particular the chemical industry, the dominant branch in the Basle region, to adapt itself to newer technologies. Those firms that were required to make large-scale investments in emission reductions in order to reach the federal standard generally also meet the cantonal standard. Therefore, the most important facilities would be, if at all, suppliers of credits. This is not due to the fact that they can reduce emissions at lower cost but rather that they were forced to undertake large-scale investments within the command and control policy.

*No internal compensation*

Since the federal emission standard applies to each individual source, it is not possible to use internal compensation (netting) within a firm. Although a firm may have several sources operating below the stringent cantonal standard, no one source is allowed to exceed the federal standard. When analysing the reason for the lack of success, the large industries in the Basle region stated that, especially in the case of a small market volume, they would be interested in internal compensation.

*Lack of smaller emitters*

Emission standards apply only to large stationary sources. Accordingly, emission trading located within the traditional command and control policy does not encompass the bulk of smaller emitters, and does not include larger sources that emit VOC through the use of solvents. Although the individual output of these sources may be smaller, their total contribution to VOC emissions and therefore to cost-saving within an emission trading programme could be considerable.

*Lack of new emitters*

New emitters are not automatically included in emission trading. As long as a new source complies with emission standards, it does not have to buy emission credits. Since there is no emission ceiling, the total amount of emissions in a region can rise unaccompanied by any demand for credits.

*No full credits for emission savings*

Besides the restrictions on the federal level, there were some additional restrictions in the cantonal Environmental Protection Law. The 80% and 10% clause states that an emission credit will only be issued when a facility lowers its emissions to more than 10% below the state emission standard and that for every further lowering only 80% of the emission reduction will be credited. Furthermore, the regulation states that an emission credit will be devalued by 20% every single year it has not been used. These regulations additionally restricted the supply of credits.

*Pre-approval by the Environmental Agency*

Emission credits are issued by the Basle Environmental Agency and can be used in another location with the consent of the Agency. Applying for emission credits and pre-approval of their use involves a great deal of paperwork. Moreover, the Environmental Agency has adopted a cautious attitude when it came to deciding whether certain emission reduction measures were actually measures that qualified for an emission credit.

The reasons that prevented industry from demanding and/or supplying emission credits are summarized in Table 1.

Table 1. Limiting factors for the demand and supply of emission credits.

| Limiting factors for | Demand | Supply |
|---|---|---|
| NOx emission standards | x | |
| Federal standards as upper limit | x | x |
| No incorporation of small emitters | x | x |
| 10% and 80% clause | | x |
| Devaluation of credits not used | | x |
| Pre-approval by the Agency | x | x |

*Conclusions*

A market-incentive approach that is established within the existing command and control policy can only be a half-hearted venture to cut environmental protection costs with the help of market forces. As long as emission trading can only take place within the range of emission standards, its possibilities are very limited. On the other hand, as long as the environmental quality within a region can only be regulated by means of these emission standards, it is difficult to provide greater flexibility for emission trading.

Consequently, in order to achieve greater eco-efficiency, it is necessary to change the approach from emission concentrations to mass emission limits. The advantage is twofold: first, mass emission limits allow better control of the total amount of pollutants emitted and hence greater certainty in achieving the ambient emission target. Second, firms can reallocate allowable emissions among themselves and thus find the least costly emission reduction measures.

## 3. The design of VOC emission trading in the Basle region: critical questions to be addressed

Based on the aforementioned experiences, the Basle Environmental Agency has decided to go one step further and to design an emission trading concept that comes as close as possible to theoretical suggestions. A working group has prepared a feasibility study that will now be discussed in public. The reactions of the different interest groups and of the Federal Environmental Agency will show whether VOC emission trading will also be feasible in the real (political) world.

While the functioning of tradeable permits has been discussed at length on the theoretical level, until now only a few experiences exist on the practical level. The following section will address some key issues in the design of a concept of tradeable permits. The feasibility study has focused on VOC, notwithstanding, most of the points also hold for NOx tradeable permits.

*The basic design of the programme*

The basic design of the programme follows very closely the idea of the Californian RECLAIM programme (cf. South Coast Air Quality Management District, 1993):

- The Environmental Agency defines an emission target, expressed in tons of VOC per year, for every source category that will be included in the trading programme. It also defines the time span (at least 10 years) within which this aim should be reached.
- In order to keep control and transaction costs low, only emitters of a certain size are included. The RECLAIM programme considers 4 tons as the appropriate threshold. In

the Basle region, the threshold will be somewhere between 2 and 4 tons of VOC per year. Certain VOCs, e.g. those that are carcinogenic, will not be included.

- For each of these sources, the so-called 'baseline', the year-one mass emission limit, is defined. This mass emission limit (also expressed in tons of VOC per year) is the initial allocation of emission permits for each source before the programme starts. In order to avoid economic hardship, the baseline is calculated on the basis of the actual emission output (so-called 'grandfathering'). Following initial allocation, the permits will be devalued within a certain time span (every five years) by a certain percentage. The percentage of devaluation is the same for every facility, and reflects the percentage of reduction of overall emissions necessary in order to reach the emission target.
- The permits are freely tradeable between all facilities that are involved in the programme. Each facility is allowed to emit to the extent for which it holds emission permits. Free allocation between the sources gives each facility the individual flexibility to choose the most cost-efficient way and thus reduces the overall costs of reaching the environmental goal.

While in its basic design the concept for the trading programme follows the Californian RECLAIM programme, there is at least one important difference. The South Coast Air Basin for which RECLAIM was developed encompasses some 14 million people. This is more than twice the population of all Switzerland. The Basle region has no more than half a million inhabitants. Accordingly, the Basle staff may be only a fraction of the Californian staff. A key precondition for the concept is therefore, that trading and control can be kept simple.

The following section takes a closer look at some key issues that have to be addressed in the context of emission trading in the Basle region.

*Key issues to be addressed in the context of an emission trading programme*

*Baseline*

The initial allocation of permits is one of the key components of successful implementation of emission trading. As already mentioned, the baseline is calculated according to the so-called 'grandfathering', i.e. on the basis of the present emission output. This allows an allocation of the permits that takes the current situation of each firm into account and thus avoids placing an additional burden on sources built under prior rules.

The task, however, is to define what 'present' emission can mean. The problem to be addressed is quite clear: grandfathering cannot mean that those firms that have not undertaken emission reduction measures in the last years are rewarded by a large number of permits. This would undermine the environmental policy effort of recent years and endanger the entire programme. Another critical factor is that allocation should not create incentives to emit more in order to qualify for a larger amount of permits.

There are several possibilities for initial allocation. Within the design of the Californian RECLAIM programme more than twenty forms of baselines were discussed at length, and the result is quite a complex system that represents a mixture of these proposals. With specific regard to the situation in Basle, the feasibility study has discussed two forms of allocations more thoroughly.

*Historical measure of activity level:* It is commonly acknowledged that the most appropriate base for allocation is some level of 'historical' activity. For example, the US acid-rain model based allocation for industrial and utility boilers on average heat input over several preceding years. For process emissions, the appropriate measure might be units of throughput (e.g. barrels of crude oil feedstock for a refining process) or output (e.g. tons of steel for a steel-making process). The historical period chosen as a basis for allocation should precede the time the debate about the programme began, and should cover several years in order to help smooth out temporary anomalies in production levels (Nichols, 1990).

For the Basle region, however, allocation based on historical level of activity does not seem feasible. Information about the activity and hence the emissions of smaller and even medium-sized emitters in the past is still incomplete; the Environmental Agency is now gradually gaining knowledge about the size and the emission structure of a larger share of the sources located in the region. Even information about the activity of those sources that come under the environmental legislation is not complete. In general, it does not seem feasible to evaluate for each individual source its input and/or its output and its technology of the previous years, at least for a small agency like the Basle one.

*Maximum emissions allowed under current regulations:* Another possible baseline, at least for the stationary sources which come under the environmental regulation, are the maximum mass emissions under current regulations. Since emission standards are defined as concentration limits per hour, this would imply multiplying the hourly mass emissions that are allowable under the concentration standard by the hours of operation during a year.

However, defining mass emission targets on the basis of concentration targets gives rise to at least one major problem. Figure 2 shows this graphically. The two sources feature a special emission structure during the day. Source A meets the emissions standards of the environmental regulation, while Source B exceeds the standard during some hours and therefore still has to undertake emissions reduction measures. Since the total amount of emissions of Source A, although in compliance with the standard, is much higher than that of source B, the figure already shows the problem that emission targets environmental policy faces today. However, it also shows the problem that arises by basing mass emissions on concentration standards. Actual mass emissions over a period of time, i.e. the area under the curve, do not correspond to the maximal mass emissions a source is allowed to emit. Consequently, if the allocation of permits is calculated on the basis of maximum mass emissions under current regulations, this will probably result in a considerable rise of emissions, at least on paper.

Accordingly, the Basle feasibility study recommends a permit allocation process that is based on the effective amount of VOC emissions a source produces when the programme is scheduled to start. For those sources that do not comply with environmental legislation, permits will be devalued to a certain amount. On the other hand, those firms which are below the legal emissions standard will perhaps receive a bonus for their mass emissions. However, such a bonus is a political decision. It is neither necessary from the efficiency aspect nor from the legal aspect.

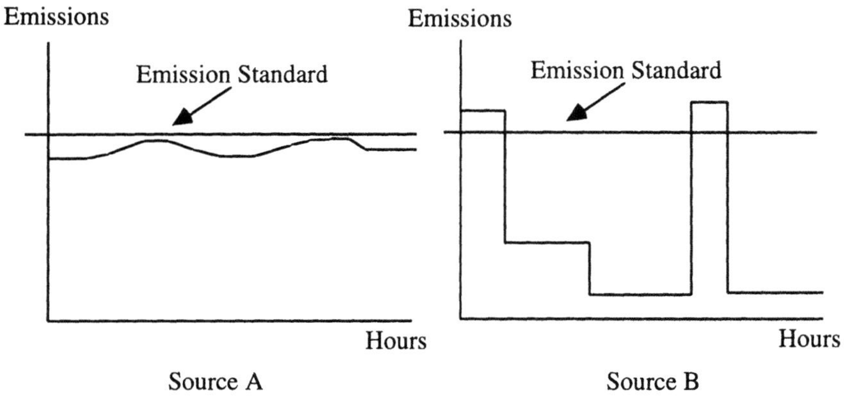


Figure 2. Actual mass emissions compared with maximal mass emissions under current environmental regulations.

*The problem of newcomers*

If the initial allocation of permits follows the grandfathering scheme, only already existing sources can be taken into account. Consequently, the important question arises as to how to treat new sources entering the market at a later stage of emission trading: Shall they buy permits for each single emission or shall they enjoy free allocation of permits in the beginning, like existing sources have done?

The feasibility study has examined several possibilities of how to treat the so-called 'newcomers'. If newcomers experience a free initial allocation of permits the total emission threshold for the region is exceeded. Thus, in the case of free initial allocation, all permits in the market have to be devalued over and above the regular devaluation in order to keep total emissions constant. This solution would come close to present treatment of new sources. Today, a source can be established if it does not exceed the concentration standards. If, e.g due to economic growth, the amount of sources in a region rises and environmental quality worsens, the concentration standards for each individual source have to be tightened. All sources must therefore bear the cost of the stricter standards.

As long as the total amount of emissions in a region remains constant, free initial allocation for newcomers is primarily a distributional and political question. However, there is also a further problem: when is a source a newcomer? Does an existing firm, when building a new plant, create a new source that has the right to free initial allocation? If this is the case, almost every expansion of emissions will have to do with newcomers in one way or another. If on the other hand only 'real new' sources enjoy free allocation of permits, the question arises as to why an existing firm has to buy permits when expanding emissions while another firm receives them free of charge.

Accordingly, an emission system can work best when, once the initial allocation is made, the sources have to buy permits for every additional emission, regardless of whether it is an expansion of an existing source or a completely new source. In recent years, many have regarded this kind of allocation as different treatment for newcomers as against existing sources, and thus as a major obstacle for emission trading under a grandfathering scheme. The treatment would violate existing law, especially Art. 4 of the Swiss Constitution as far as equality of rights is concerned.

When the question of equal treatment is addressed, a distinction has to be made between the economic and legal points of view:

- In economic terms, the answer is quite clear. The costs which new and existing sources face are the same: in the former case, these are direct expenses for buying permits, while in the latter they are so-called opportunity costs in the form of lost profits. Consequently, the grandfathering approach does not result in different treatment of new and existing sources and does not distort competition.
- From the legal point of view, the answer was less unanimous, at least in former years. However, a new legal study regards grandfathering as compatible with existing law (c.f. Trüeb, 1995). According to this study, the grandfathering approach does not violate the postulate of equal treatment. If new sources have to buy tradeable permits, they are treated in the same way as existing sources intending to expand emissions. Moreover, it would be a violation of equality if new sources received emission rights at no charge while existing sources in an expansion phase had to buy them. Another legal aspect is that existing sources built under prior rules have the right to an appropriate adjustment when the law changes (the so-called 'Vertrauensschutz'). Grandfathering is seen as an ideal way to fulfill this condition, because it can take the existing situation into consideration.

Finally, the following reflection may make the aspect of the different treatment clearer: Emission permit, or put more precisely the clean air that is used for the production process, is an input factor like any other , e.g. like the site on which a plant is built. If a firm decides to build a new plant and already owns the site while another firm still has to buy it, nobody will claim that this results in a different treatment of the two firms and necessitates a free distribution of the site for the new firm. In principle, the same holds true for emission permits, although this idea may still be less familiar.

*Trading*

In order to transmit the permit from a seller to a buyer, the RECLAIM feasibility study lists four types of market structure that are possible (South Coast Air Quality Management District, Appendix G, 1992):

- Direct search market: potential buyers and sellers seek out among all possible partners the most appropriate one with which they wish to carry out the transaction.
- Brokered market: potential buyers and sellers use brokers as their agents in seeking suitable partners because of the economies of scale a broker can provide in the search process.
- Dealer market: dealers act on their own as sellers and buyers. Markets with a high volume of trading enable dealers to cover their costs by always being available to take the opposite position of the transaction as a buyer or seller.
- Auction markets: auction markets provide centralized procedures for communicating bid and offer terms to all market participants simultaneously.

In many cases, direct search can result in large transaction costs which can reduce or even eliminate the surplus of the trading system. On the other hand, institutional arrangements like a dealer or auction market require a large number of participants, a precondition which is not met in the Basle region (indeed even the RECLAIM programme does not rely on these types of market structure). However, other institutional arrangements that support the market with the necessary information and do not necessitate a large market volume are quite feasible. For a small region like the Basle area, the following assistance seems to be appropriate:

- Provision of information about suppliers and searchers through the Environmental Agency or, even better, on a private basis through a business association;
- An auction, e.g. conducted through the Environmental Agency or, once again, through a business association within set time intervals in order to bring buyers and sellers together;
- Options and futures are also feasible on a rather small market. In order to establish a solid calculation base for future investments, buyers and sellers may agree on an option or future contract;
- Leasing is another possibility for an enterprise to secure future positions. Leasing may provide more market liquidity, especially in the case of a small market where firms might otherwise be reluctant to sell permits they would need again at a later date.

Thus the institution of trading itself does not seem to provide larger problems. Furthermore, on a smaller market some institutional arrangements may be made without creating additional costs.

Another question to be addressed is whether transactions should be pre-approved by the Environmental Agency. Pre-approval provides the certainty that the emission reduction measures on which the emission credits are based have indeed been implemented. This information is important to both the firm buying a credit and the Environmental Agency. However, pre-approval slows down the trading process and thus imposes additional transaction costs.

On the other hand, some kind of control has to be established. Consequently, each transaction should be reported, e.g. 30 days prior to transaction. Within this time span, the Environmental Agency is empowered to check whether emission reduction measures have been implemented at the source reporting the sale and can, if necessary, bar the transaction. This reporting system provides the same control security as pre-approval (the Environmental Agency is informed about the intended transaction and can carry out checks if it suspects any kind of irregularity) while at the same time giving buyer and seller the certainty that the transaction can be carried out after a fixed time span.

*Market problems*

The question whether emission trading leads to market power by single firms has been discussed in literature at length (e.g. Bonus, 1981). In general, emission permits are only one production factor among many others and buying permits on a large scale in order to keep other competitors out of the market causes large additional costs (again, the comparison with the real estate market may make this clearer: no firm buys additional sites in order to keep its competitors out of the market).

In a small market, some kind of market power is at least conceivable, but only in a special market situation. Not only does the market have to consist of a few firms, but the firms must also belong to the same branch. If, and only if, these firms form some kind of cartel will they be able to keep new competitors out of the market by not selling them the permits they need in order to establish a new plant.

Indeed, the Basle region shows a very specific industrial structure. Three of the world's largest chemical industries have their headquarters in the region, in addition to a couple of smaller chemical companies. Yet despite this concentration of a single branch, the danger of forming a cartel in order to keep other competitors out of the market is virtually non-existent. The chemical industry is competitive on an international or even worldwide level. As a result it hardly makes sense to keep a competitor out of the Basle region by buying additional permits.

Another concern that is often expressed in the context of market problems is that there may not be enough permits available when a firm intends to expand. Of course, from an economic point of view, the availability of permits depends on price, and even this problem can be overcome with the help of futures, options and leasing. Consequently, lack of liquidity is a psychological rather than an economical fear. However, since it has been mentioned by a couple of firms and interest groups, it must nevertheless to be taken seriously because in the final analysis the success of an emission trading programme is determined by the political process. In order to give firms as much security as possible, the Environmental Agency can provide a certain stock of permits for such 'emergency' cases, where regardless of clear willingness to pay, no permits are available. Furthermore, possible lack of liquidity in a small market can be overcome by the industry itself, e.g. through business associations that provide permits for their members.

Finally, in any discussion of market imperfections, one should always keep in mind the fact, that in the real world, markets seldom achieve perfection. Consequently, the market for tradeable permits does not have to meet special requirements with regard to the perfection of its functioning.

*Control*

The question how mass emissions can be measured and controlled is a question that is seldom addressed in literature but one which is regarded as a crucial point in practice. The RECLAIM report on emission trading of VOC (South Coast Air Quality Management District, 1992) lists a couple of rather technically ambitious control methods. However, the control elements cannot be transferred across the board to the Basle model because they are linked to the specific structure of US environmental legislation and because they are simply too expensive for a small region like Basle.

In order to keep control simple, annual mass emissions are to be calculated and controlled on the basis of emission declarations made by the firms themselves. The firm must declare all inputs used that are likely to cause VOC emissions, either directly or through the use in the production process. This declaration has to be made in addition to the annual accounts. In conjunction with other informations about the production process (technology, output etc.), the Environmental Agency is able to caculate the mass emissions. The Environmental Agency is empowered to check the accuracy of the emission declaration. In doing so, it can require further information about product output or sales and it can ask to see the normal accounts.

## 4. Conclusions and prospects

The failure of emission credits in the Basle region is one more in the row of market-based approaches which are supposed to work within a restrictive command and control policy. Consequently, if the market approach is to be pursued further, an emission trading programme seems the logical consequence.

Most emission-trading programmes developed to date have been designed for large stationary sources of NOx. It will become clear in the near future whether or not a VOC emission trading programme within the Basle region is feasible. The advantage with VOC is that it stems mainly from industrial activity, while traffic can account for up to 60% of NOx emissions in the Basle region. However, traffic cannot be included in an emission trading programme on the cantonal or regional level; rather, it must be addressed on the national level, e.g. through the oil importers. On the other hand, small sources and solvents are responsible for a substantial share of VOC emissions. The survey about the structure of the VOC emitters in the Basle region is not yet finished. It will show just how many emitters could be included in a trading concept.

Of course, VOC emissions can also be regulated through an environmental tax. However, compared with environmental taxes, the strongest argument for emission trading can be seen in the following two points:

- The allocation of permits through a grandfathering scheme imposes no additional taxes, hence no additional costs on industry, and does not lead to additional revenues for the State. Thus, emission trading allows a clear separation of environmental and fiscal policy.
- Emission trading makes it possible to control the quantity of emissions itself, in other words, it allows direct control of the aim of clean air policy.

Accordingly, environmental taxes can be useful for small emitters where emission trading imposes high control and transaction costs compared with the gains of the concept. But as soon as a facility emits substantial shares of VOC, through either stationary sources or solvents, the financial burden of an environmental tax can be considerable and emission trading can offer a promising alternative means of achieving environmental targets at the lowest possible costs. The problem of including solvents used in the industrial process can probably be overcome. Thus, whether or not emission trading in the Basle region can be implemented in practice will depend on the final information on the emitters structure. However, the reactions and the support of industry, environmentalists and the Federal Environmental Agency will be just as important in this connection.

## References

**Bonus, H.** (1991) Emissionsrechte als Mittel der Privatisierung öffentlicher Ressourcen aus der Umwelt. *In*: Marktwirtschaft und Umwelt, Lothar Wegehenkel (ed.), J.C.B. Mohr, Tübingen.

**B,S,S.** (1995) Von der Konzentration zur Fracht. Handelbare Emissionslizenzen in der Luftreinhaltung. Machbarkeitsstudie im Auftrag des Lufthygieneamtes beider Basel, Basel.

**Environmental Protection Agency (EPA)** (1993) EPA Programs to Control Sulfur Dioxide in the Atmosphere. Washington, D.C.

**EPRI (Electric Power Research Institute)** (July 1994) Key Issues in the Design of NOx Emission Trading Programs to Reduce Ground-Level Ozone. Prepared by National Economic Research Associates, Inc., Cambridge, MA.

**Hahn, R. H. and Hesters, G.** (1989) Where Did All the Markets Go? An Analysis of EPA's Emission Trading Program. *Yale Journal on Regulation*, 6 : 109-152.

**Hahn, R. W.** (1984) Market Power and Transferable Property Rights. *The Quarterly Journal of Economics,* November 1994 : 752-765.

**Nichols, A. L. and Harrison, D. Jr.** Using Emissions Trading to Reduce Ground-Level Ozone in Canada: A Feasibility Analysis. Final Report prepared for Environment Canada. National Economic Research Associates, Inc., Cambridge, MA.

**South Coast Air Quality Management District** (October 1993) RECLAIM - The Regional Clean Air Incentives Market, Final Volume I. SCAQMD, Diamond Bar, CA.

**South Coast Air Quality Management District** (Spring 1992) RECLAIM, Diamond Bar, CA.

**Staehelin-Witt, E. und Spillmann, A.** (1992) Emissionshandel - Ein marktwirtschaftlicher Weg für die schweizerische Umweltpolitik. WWZ-Studie n° 40, Basel, WWZ.

**Tietenberg, T.H.** (1991) Reductions in Emissions: Command and Control or Market-Based Mechanisms? Paper prepared for the Conference on "Economy and Environment in the 1990s", Neuchâtel, August 26-27.

**Tietenberg, T. H.** (1989) Marketable Permits in the U.S.: A Decade of Experience. *In*: Public Finance and the Performance of Enterprises, MI Wayne State University, Press Detroit.

**Trüeb, H. R.** (1993) Economics and the Environment : The Case for Tradeable Emission Permits. *In*: Biology, Culture, and Environmental Law, McGuire, M. T. and Rehbinder, M. (eds.), Berlin.

**Trüeb, H. R.** (April 1995) Handelbare Emissionszertifikate - Grandfathering, die Behandlung von Newcomers und weitere ausgewählte Fragen. Rechtsgutachten im Auftrag der Bau- und Umweltschutzdirektion des Kantons Basel-Landschaft.

Environmental Policy Between Regulation and Market
ed. by C. Jeanrenaud


# 4.4 A Tradeable Permit Market for $NO_x$: An Application to the Chablais Region

**Marc-Alain Stritt**
**University of Neuchâtel, Neuchâtel, Switzerland**

## 1. Introduction

The interest shown in economic instruments (incentive charges and tradeable permits) is due in particular to the need to seek the least costly options for achieving environmental policy objectives. It is true that command-and-control regulation, consisting mainly of directives and prohibitions, has brought about an improvement in the quality of the environment in several areas, but it has revealed its limits as far as cost-effectiveness is concerned. This is primarily due to the fact that emission standards are imposed without taking account of the widely varying marginal costs of cleaning up the different sources of pollution.

The specific characteristic of economic instruments is that they allow emitters themselves to choose the sources that they intend to clean up, as a function of their costs. Compared with regulation, a permit system should thus allow a marked reduction in the cost of reducing emissions, but without sacrificing the quality of the environment: the objective to be achieved is identical in both cases. The increase in effectiveness depends on several factors, in particular the pollutant being controlled, market operating rules, number of participants, etc. As theory is not able to indicate the scale of the saving that can be achieved, the latter must be measured on a case-by-case basis. The purpose of the present paper is to determine the benefit that would be achieved by implementing a tradeable permit system for controlling nitrogen oxides in a small region of Switzerland, the Chablais.

## 2. Characteristics of the Chablais region

The Chablais is the region that extends from St. Maurice to Lake Geneva. It is a plain with a surface area of some 130 $km^2$ (50 sq. mi.) which is almost entirely surrounded by mountains. Windless situations are frequent there, as are temperature inversions, so that there are few atmospheric exchanges with the neighbouring regions. The Chablais is thus a genuine *air shed* : pollutants emitted there tend to stagnate. The primary consequence of this is that the environmental damage is due mainly to local emissions: it is thus easier than in other regions to establish a causal link, which is a favourable precondition for the introduction of an instrument based on the "polluter pays" principle.

Moreover, $NO_x$ per capita emissions are proportionally greater in the Chablais than in the country as a whole. During 1990 more than 3,000 tons of $NO_x$ were emitted there, all sources included (Table 1). The share of emissions originating from industry is higher in the Chablais (virtually half of the total emitted) than at the national level, where it reaches only 27% (SFOEFL 1987, p. 45).

Current forecasts show that overall $NO_x$ emissions should decrease by about one-third between 1990 and the year 2000 (Ecoscan 1994, p. 25). The reduction will be more marked in the transport sector (spread of catalytic converters) than in the industry one. Consequently, the proportion of emissions stemming from industry will be predominant and might account for more than half of the total by the year 2000.

Air quality is greatly influenced by stagnation of emissions. On an annual average, $NO_2$ ambient concentration exceeds the limit value of 30 $\mu g/m^3$, and the daily limit value (80 $\mu/m^3$) is also exceeded, even frequently, whereas it should under no circumstances be overstepped more than once a year according to the Air Protection Ordinance (OPair). The same applies to ozone, of which $NO_x$ are some of the precursors (Ecoscan 1994, pp. 43-46). Thus, it can be said that in the Chablais region, $NO_2$ and $O_3$ ambient concentrations are excessive as defined by article 2, paragraph 5 of the OPair. Under the EPA classification, the Chablais would thus be a "nonattainment" region. However, no special measure is envisaged: the authorities consider that the clean-ups required within the framework of OPair will be enough to eliminate the excessive ambient concentrations (Rollier 1995, Schnydrig 1995).

Table 1. Annual $NO_x$ emissions in the Chablais, by category of polluter, in 1990 and 2000.

| | 1990 | | 2000 | | |
|---|---|---|---|---|---|
| Sources | Tons per year | Relative shares | Tons per year | Relative shares | Variations over 1990 |
| Households | 50.3 | 1.6% | 50.3 | 2.4% | 0% |
| Transport | 1507.2 | 48.9% | 918.6 | 44.8% | -39.1% |
| Industry* | 1525.3 | 49.4% | 1082.3 | 52.8% | -29.0% |
| TOTAL | 3082.8 | 100.0% | 2051.2 | 100.0% | -33.5% |

* including heating

Source: adapted from ECOSCAN 1994, p. 18 and p. 21

## 3. Design of the market

It should be stressed first of all that, as Noll (1983, p. 214) has pointed out, "selecting a design for a system of marketable emissions permits is not a purely technical, scientific matter". Indeed, a permit market can take several distinct forms, and the possible options are numerous. However, the options are always *conditioned* since the system to be implemented must form part of an existing context. As a result, it is necessary to take account of the legal bases in force, the environmental policy implemented to date and the pollution control action already undertaken, the political considerations, the structure of companies, the presumed acceptability of the instrument, etc. The characteristics of the market proposed for the Chablais are defined below.

*Type of market*

The literature traditionally describes two types of permit markets, depending on whether one reasons in terms of emissions or of air quality (*emission permit market* or *ambient air quality permit market*, see Atkinson and Tietenberg 1982). In the case of the Chablais, the hypothesis that air quality is homogeneous in the entire region is accepted, so that any reduction of emissions, wherever it may take place, helps to reduce ambient concentration. In this case, permits are defined in terms of emissions. They grant the right to emit one ton of $NO_x$ per year, and the obligation to respect the emission concentration standards (in $mg/m^3$) is waived.

*Participants*

Although most $NO_x$ emissions come from transport, the permit market is aimed primarily at industry. The system proposed for the Chablais would encompass all companies whose manufacturing process generated at least one ton of $NO_x$ during 1994, thus yielding a market comprising fifteen participants. Several industries are represented, such as chemicals, oil refining, electricity generation, refuse incineration, machine construction, metallic construction, cement production, etc. It should be pointed out that the diversity of the sectors involved makes it unlikely that strategic behaviour will emerge on the permit market.

The companies are distributed over the entire region: their location is such that it does not cause a concentration of the emissions. On the other hand, the structure of the emissions is not very homogeneous in that five companies emit more than 100 tons of $NO_x$ each year, whereas seven others only emit less than 3 tons. For the permit market to function properly, no participant should be large enough to be able to influence the price of the permits. This condition is only imperfectly fulfilled in the case of the Chablais.

*Supervisory authority*

The supervisory authority is the agent responsible for setting up the market and making it function. In the literature, this body is usually assigned one or more of the following tasks:

- handling the initial allocation of the permits;
- putting the potential permit sellers and buyers in touch with one another;

- studying the proposed trade transactions and authorizing (or prohibiting) them.

While the first task is clearly the responsibility of the supervisory authority, the situation is different for the other two. Companies offering and requiring permits can be put in touch with each other through brokers, or the process can take place at an existing exchange centre. It may also be left to the companies' initiative, i.e. without intermediaries. In the case of the Chablais, the contacts between potential buyers and sellers would be made by the supervisory authority. It is notified of permits available for sale or required for purchase, which are kept up to date in a special database that may be consulted at any time by the companies. The database could also be made available to other interested parties (the public, ecological organizations, etc.).

The problem of prior approval of trading by the supervisory body is more complex. In theory, any obstacle to free permit trading reduces the gain that can be expected of the system. However, from an ecological viewpoint, not all emission reductions have the same value, depending on local ambient concentrations, winds, etc. Consequently, the authority may wish to retain a degree of control of emission reductions, and thus of permit trading. The decision as to prior approval could be described as a trade-off between the ecological effectiveness and economic effectiveness of the permits. Prior approval prevents the appearance of hot spots, since the authority can limit, or even prohibit, the purchase of permits by a company located at a place where the ambient concentration measured exceeds the limit value. This supervision increases ecological effectiveness and -- probably -- acceptability. On the other hand, by limiting trading possibilities it reduces economic effectiveness and increases administrative costs.

Naturally, the supervisory authority's staffing requirements and operating costs depend on the tasks it is assigned. In the case of the Chablais, it would have to perform all the above-mentioned tasks. However, since the number of participants is limited, putting the players in touch with each other and studying the proposals for permit trades does not constitute a particularly large volume of work. It would represent on an annual average a maximum of one half-day per week, which corresponds to the creation of one-tenth of a post. The resultant cost would thus amount to one-tenth of a salary (CHF 10,000 per annum).[1] The necessary equipment

---

1 All the amounts quoted in the text are expressed in 1995 Swiss francs (CHF).

and premises could be provided by the existing administrations at no additional expense. Furthermore, the cost of establishing the market can be evaluated at about CHF 100,000 for the first year.

*Initial distribution*

The number of permits put on the market "materializes" the maximum volume of the emissions authorized each year in the region, in tons per year. In the initial allocation phase, this quota must be distributed among the companies concerned. If the market works well, the permit-allocation process will not affect the gain that can be achieved over regulation. Tietenberg (1991, p. 9) notes that whatever the mode of distribution that is selected, "[it] will ultimately result in a cost-effective allocation of the control responsibility among the various polluters as long as they are all price-takers, transaction costs are low, and the permits are freely transferable". This remains true even if the initial allocation is done randomly (*ibid.*). Under these conditions the supervisory authority could thus, during the initial allocation, pursue distributive goals without this reducing the effectiveness of the proposed system (*ibid.*). However, the mode of initial allocation conditions the way in which the gain is subsequently distributed among the market participants. For the proposed market to enjoy good acceptability among companies, it is important that the mode of allocation selected be equitable, transparent and simple.

There are numerous possible ways of allocating permits to the different participants. As the OECD has pointed out (1993, p. 34), "no consensus has emerged for the moment about the most acceptable allocation formula (...) Economic theory cannot furnish any turnkey solution". The literature traditionally describes two modes of initial allocation: auction and free-of-charge distribution by means of a formula based on past emissions (*grandfathering*). However, these two solutions pose such large practical problems that they cannot be used in the case of the Chablais. The first certainly suffers from too low a level of acceptability from companies and poses the problem of allocating the sums collected, while the second penalizes companies that have already carried out clean-ups within the framework of command-and-control regulation.

The procedure selected for the Chablais involves carrying out allocation on the basis of the standards currently in force (OPair 92). With this mode of allocation, companies are allocated a number of permits corresponding to the volumes that they would emit if they observed the regulations. Taking company A as an example, let us suppose that OPair authorizes it to have an emission concentration of 250 mg/m$^3$ maximum. Let us also assume that this company's manufacturing process entails an hourly discharge of 12,500 m$^3$ of polluted air, by one or several chimneys, and that the company operates eight hours a day and 320 days a year. In this case, company A would be allocated eight permits (number obtained by multiplying authorized concentration by volumes emitted annually).

It should be recalled that companies are no longer obliged to respect the emission concentrations. However, with the procedure described above, these standards do not disappear completely since they are used for initial permit allocation. Since the standards are generally accepted (in Switzerland, they have been through the consultation procedure), it is conceivable that there will be good acceptability for their use as a formula for free-of-charge initial distribution. Moreover, the companies which, at the end of this operation, have more permits than they use are those that already comply today with environmental legislation. Thus, this type of allocation offers the advantage of "rewarding" those companies that have undertaken clean-up programmes within the framework of OPair. Finally, basing allocation on the standards prevents any discrimination against the new entrants: once their facilities come on line, the new entrants would obtain permits in the same way and on the same terms as existing companies (equal treatment).

However, this method of allocating permits also poses some problems. First of all, it is based on the hypothesis that standards have been set efficiently, which has not been demonstrated. Next, the procedure becomes more complicated when flows are not constant. Finally, the procedure grants a kind of "filling entitlement" which does not conform to the legislator's spirit and which involves a volume of reductions that is *lower* than that required by the regulation (Figure 1). Some authors talk in this respect about the disappearance of the safety margin contained in the standards. Yet this problem can easily be solved by reducing somewhat the number of permits that the authority puts on the market. It should be stressed that this mode of allocation has already been used in practice: in Basle, Switzerland, as in Los Angeles (RECLAIM), initial permit allocation was based on the standards of the *Basle Environmental Protection Law* and the *California Clean Air Act.*

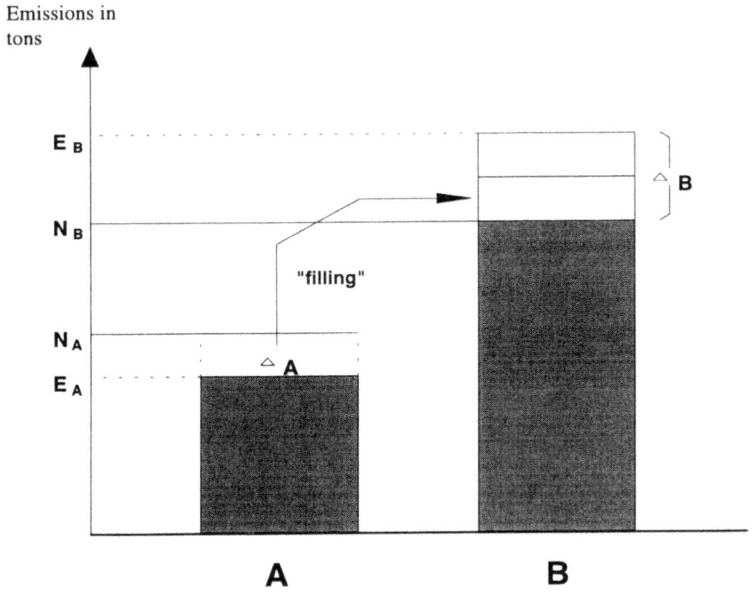


Figure 1. "Filling" effect due to an initial distribution based on standards.

Let A and B be two companies; let $E_A$ and $E_B$ be the volumes of their annual emissions in tons. Let $N_A$ and $N_B$ be the maximum authorized emission levels (initial entitlements), in tons per year, calculated by multiplying the cubic metres emitted by the concentrations laid down by OPair. Company B emits ΔB more than is authorized, while company A ΔA less. In the case of command-and-control regulation, company B is called on to reduce its emissions by ΔB (clean-up), whereas company A, which complies with the standards, does not take any particular action. With a permit system, company A has surplus permits worth the value of ΔA. It can sell them to B, which allows B to reduce its emissions by a limited amount (ΔB - ΔA). Part of the reduction required in the case of regulation has thus been purchased from company A. By virtue of the "filling" effect implicitly contained in an initial distribution based on the standards, the total reduction of emissions is *lower* than in the case of regulation. The supervisory authority can correct this phenomenon by adjusting the total number of permits which it puts on the market.

*Permit trading modalities*

After the initial allocation, companies compare the number of permits they hold with their actual emissions. The differences determine the quantities of permits that are offered and sought. In a first stage, these quantities therefore depend only on the initial allocation. Indeed, until a sufficient amount of trading has taken place, the companies lack information about the market price of permits. As this prevents comparison with the marginal pollution control costs, it is difficult to make a decision on whether to clean up or purchase permits. This phenomenon could delay the launch of the market.

Permit trading possibilities are devised on the basis of the supply-and-demand database. Companies wishing to purchase permits get in touch directly with the companies selling. The transaction is subject to prior approval by the supervisory authority, which evaluates its environmental impact. If the plan does not involve any ecological disadvantage (a local increase in ambient concentration), the supervisory authority gives its approval and the transaction may go ahead. If the ambient concentration recorded near the purchaser durably exceeds the air quality standard, the authority may restrict the number of permits traded or even prohibit the transaction. It should be recalled that the purpose of prior approval is to reduce the risk of hot spots appearing, hence to increase the system's acceptability. Once the first transactions have been completed, new trading possibilities appear only when emission reductions have been achieved, giving the companies concerned access to surplus permits.

*Trading ratio*

The term "trading ratio" denotes the ratio between the quantity of permits traded and the volume of emissions that this transaction authorizes for the permit purchaser. This ratio can be 1:1 if each permit purchased entitles the holder to emit one ton. However, if the supervisory authority aims to enhance the ecological effectiveness of the permit market, it may reduce the ratio in a proportion to be determined, for all the transactions or for only some of them. In this case, a permit is available if it corresponds to a reduction of one ton, but it only allows the purchaser to emit an additional *fraction* of a ton. Altering the trading ratio makes it possible to take fuller account of the differences in air quality if the latter is not homogeneous in the region under consideration. This can happen even with "uniformly mixed pollutants" such as

$NO_x$, for example if there is a concentration of emission sources or if the prevailing winds are such that they result in the creation of hot spots.

It must be stressed that by authorizing each emission through a greater reduction, altering the trading ratio restores the safety margin that the standards-based allocation had caused to vanish. A permit market with this characteristic allows an ecological gain to be made in addition to the economic gain, which probably enhances its acceptability for ecological organizations and the public. Conversely, a trading ratio of 1:1 maximizes the economic benefit that can be derived from the instrument, making the market more attractive to companies.

Several authors have expressed reservations about the effectiveness of a reduction in the trading ratio based on the distance between the buyer and the seller of permits. Dwyer (1992, p. 51) notes that "the rule effectively creates numerous small markets, rather than a single district-wide market"; Hahn (1993, p. 119) thinks that "such rules are likely to put a damper on trading (...). While such approaches have been used in the past, it is generally more desirable to develop a system that does not discriminate against parties that wish to trade". Finally, Roberts (1983, p. 1032) emphasizes that "to set an exchange rate adequate to prevent hot spots, a complicated model of prevailing winds and topography would be necessary. (...) the cost of administering such a market would be high."

In conclusion, it should be pointed out that the choice of trading ratio consists of a trade-off between economic effectiveness and the ecological effectiveness of the permit system. In this respect, it is a *political*, not an economic choice. Consequently, two evaluations of the gain will be performed in the case of the Chablais: one with a ratio of 1:1 and the other with one of 1:0.8 (the value used in the Basle market). This double evaluation will enable us to quantify the reduction of the economic gain that results from keeping the safety margin laid down in the standards.

*Validity and banking of permits*

The options are numerous as far as the life of the permits is concerned. Noll (1982, p. 121) notes that the latter can have a limited validity (for example, one year, at the end of which a new initial allocation would take place) or, on the contrary, perpetual validity. Lifetime may also be unspecified: "emission permits could be valid until a formal regulatory procedure declared them invalid or changed the amount of emissions allowed by a single permit" (*ibid.*). When setting up the market, the supervisory authority must clearly explain the rules governing the term of validity of the permits, the banking conditions and the events that might lead to a reduction of their value. The effect of a short period of validity is to increase market liquidity, as this reduces the benefit of banking. On the other hand, this would make the system more complex and the administrative costs higher. The same applies to an annual reduction of the permits' value.

As far as the Chablais market is concerned, there would be no periodic devaluation: the number of permits put on the market in the initial allocation already represents the objective to be achieved in terms of tons emitted.[2] Banking of permits would, on the other hand, be allowed: the companies would be able to store their permits for subsequent use for as long as they wish, without the latter losing any of their value. Thus, the permits, just like the currency to which some authors sometimes compare them (see Pasek and Beckerman, 1994, p. 30), would be perpetual, as advocated by certain authors (Heister *et al,* 1990, p. 8).

## 4. $NO_x$ emission reduction technologies and costs

Permit supply and demand depend mainly on the pollution-control costs encountered by the market participants. It is therefore important to be familiar with the technologies that enable $NO_x$ emissions to be reduced and to be aware of their implementation costs. Fundamentally, action may be taken at three levels: (i) fuels, (ii) combustion processes or (iii) emanations (Figure 2). Action concerning fuels implies either resorting to a substitution product (for example, switching to natural gas instead of fuel oil) or using a combination of fuels. Alterations to combustion installations are so-called "primary" measures because they are taken upstream of the actual formation of $NO_x$. Various technologies make it possible to obtain so-called "Low-NOx" combustion, that is, emitting quantities of nitrogen oxides that are markedly reduced compared with conventional installations. Although this is not the rule,

---

2 Periodic devaluation is used when the number of permits put onto the market authorizes more emissions than the volume fixed by the objective to be achieved. In this case, a date is set by which air quality must be satisfactory, and the value of the permits is discounted accordingly.

these technologies sometimes allow savings to be made, for example, following a drop in consumption. In contrast, action concerning emanations is called "secondary" measures. One then speaks of "add-on" or "end-of-pipe" technologies: a *black box* is added to the combustion process without altering it. It is an unproductive investment, that is, one that cannot be made profitable.

What companies compare with the market price of the permits is the value of the pollution-control investments expressed as the total annual cost per ton of reduction (TAC/t). Figure 3 shows the components of this cost, which is calculated by means of the following formula:

$$TAC/t = \frac{\frac{TCI}{n} + AOC}{\text{tons reduced}}$$

| | | |
|---|---|---|
| where | TAC/t : | total annual cost per ton of reduction; |
| | TCI : | total capital investment; |
| | n : | service life of pollution control equipment; |
| | AOC : | annual operating cost; |
| | tons reduced: | emission reduction, in tons per year, achievable by the investment in question. |

Figure 2. $NO_x$ emission reduction technologies.

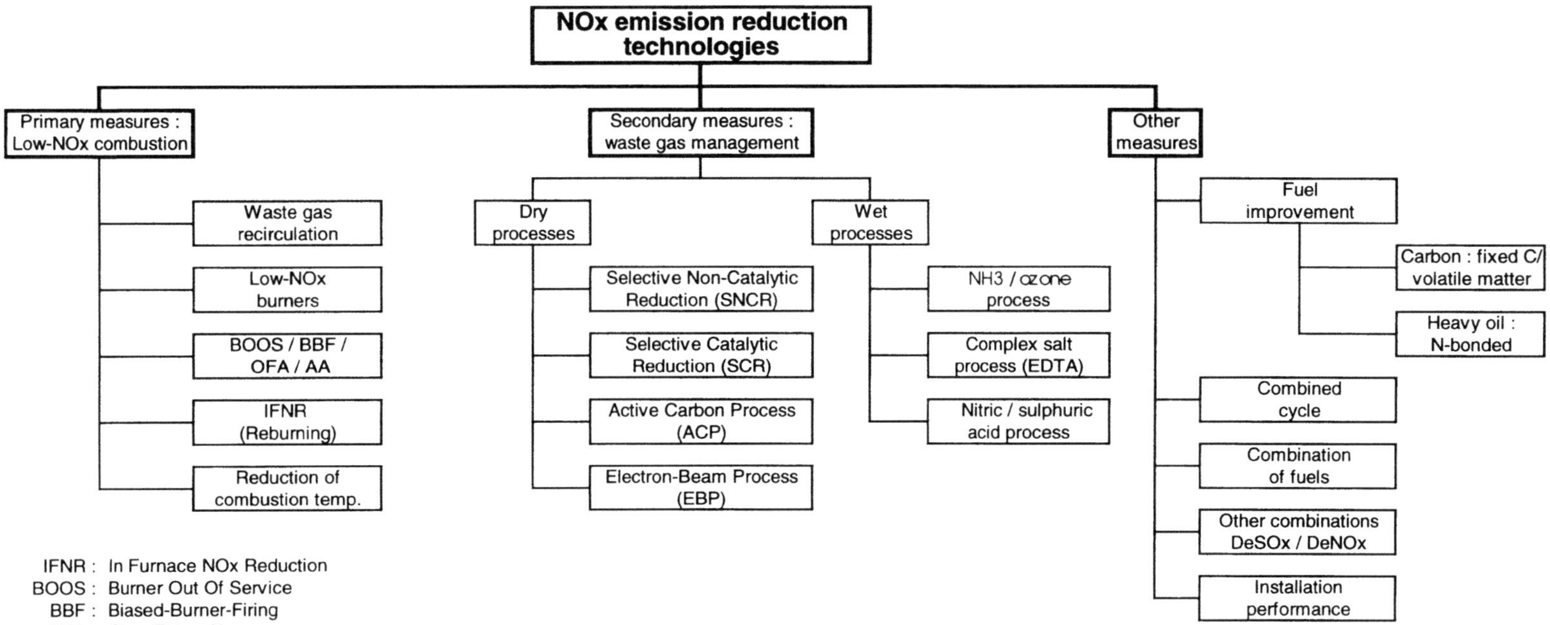


IFNR : In Furnace NOx Reduction
BOOS : Burner Out Of Service
BBF : Biased-Burner-Firing
OFA : Over Firing Air
AA : Additional Air
SCR : Selective Catalytic Reduction
SNCR : Selective Non-Catalytic Reduction
ACP : Active Carbon Process
EBP : Electron-Beam Process

Source: Release by *Energie Ouest Suisse* (EOS), September 1995

Figure 3. Types of pollution-control costs.

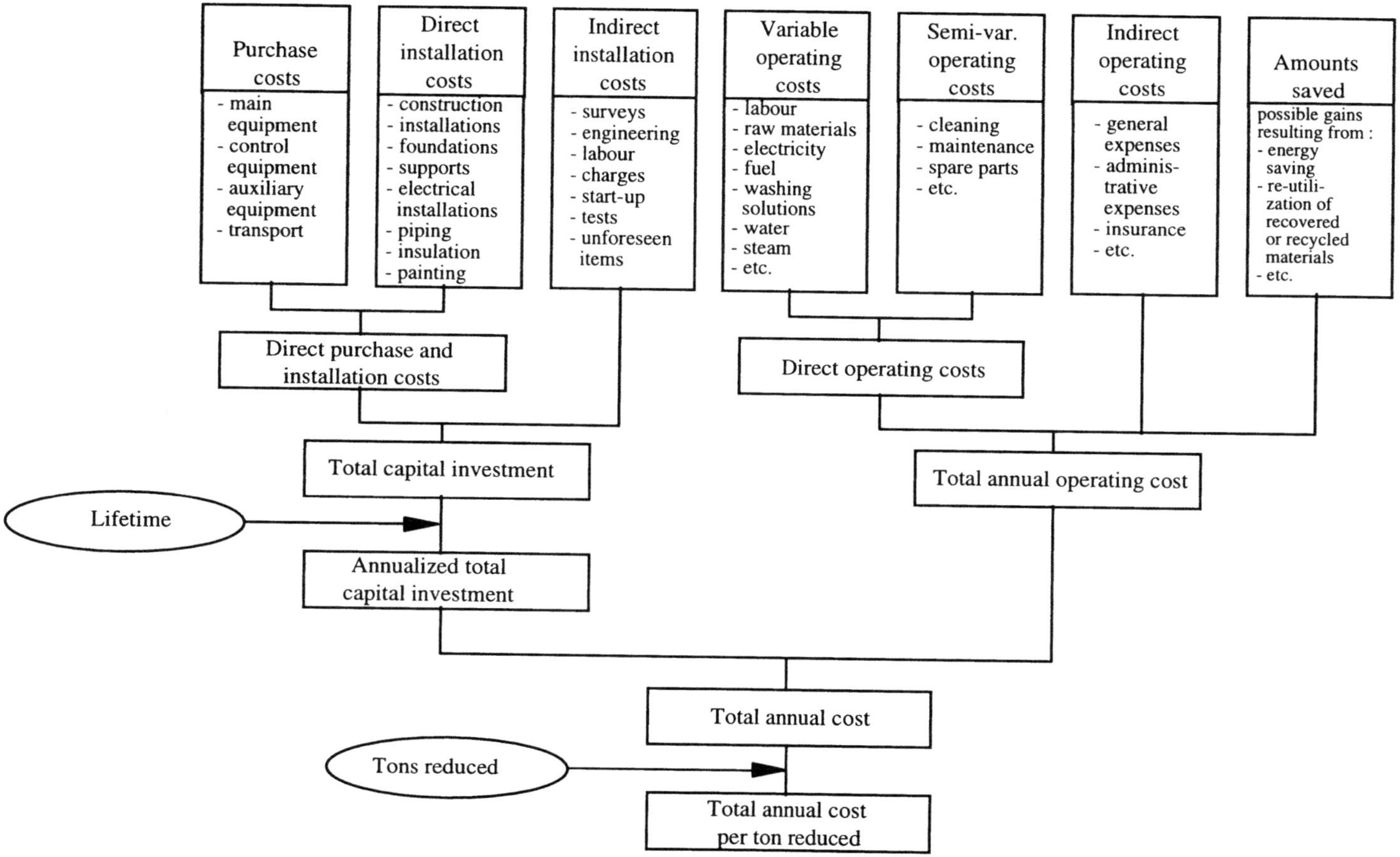


Source: adapted from NERA 1992, pp. 2-1 to 2-14.

In the Chablais the costs of the completed and ongoing clean-ups were obtained directly from the fifteen companies by means of a questionnaire. Only clean-ups required for $NO_x$ since the introduction of the OPair (1986) were considered. The questionnaire is structured around a relatively complete grid to determine costs, with a breakdown based on the above figure. In some cases data collection was supplemented by an interview with a company manager. The amounts obtained for the completed clean-ups were updated on the basis of the production cost index of the Swiss Contractors' Association. Amortization was assumed to be straight-line and taken over the foreseeable lifetime of the installations. When this datum was not known, an economic lifetime of 10 years was assumed, which corresponds to the lower limit of the range estimated by the EPA (EPA 1990, p. 2-6).

Possibilities for future reductions were also researched. For the large companies the values come from feasibility studies currently in progress, whereas for the small companies it proved necessary to resort to another source of information: suppliers of pollution-control technology. Starting from the technical characteristics of the installations (boiler type, burner type, year of implementation, power, fuel used and annual consumption), it was possible to determine what new installation would allow the emissions to be further reduced, in what proportion and at what cost.

A total of 37 observations was collected in this way. The existence of a large scatter in marginal emission reduction costs was again verified (for a first study on the subject, see Stritt and Jeanrenaud, 1992). The costs, as calculated above, are contained within a range extending from CHF 11 to more than CHF 130,000 per ton reduced (Figure 4).[3]

---

3 The lower value is an exceptional case caused by the adoption of a markedly cheaper fuel.

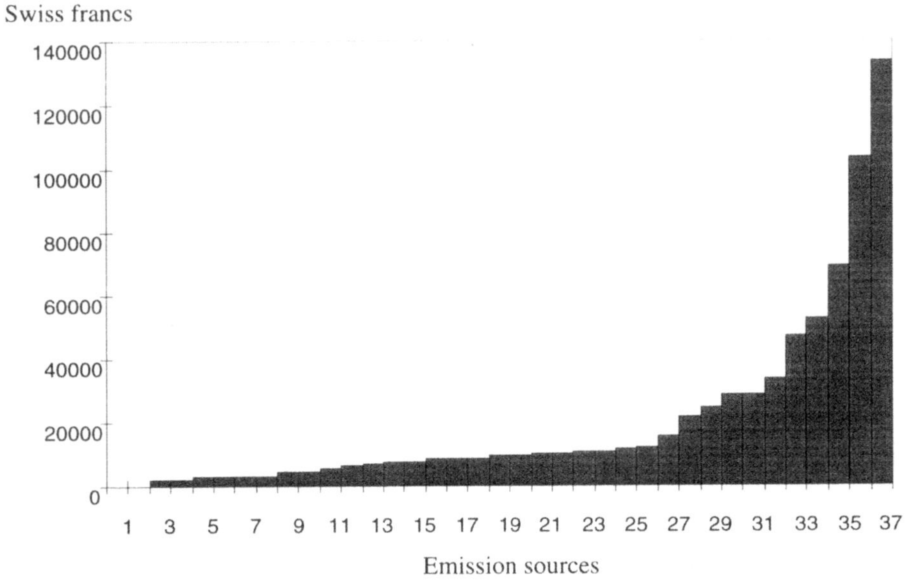


| | |
|---|---|
| Minimum value : | 11 |
| First quartile ($Q_1$) : | 4,557 |
| Median : | 9,259 |
| Mean : | 19,570 |
| Third quartile ($Q_3$) : | 21,600 |
| Maximum value : | 133,578 |

Note: Values calculated using the TAC/t formula described above.

Source: Survey by questionnaire, IRER, March 1995.

Figure 4. Marginal costs of reducing industrial NOx emissions in the Chablais region, in Swiss francs per ton reduced per annum.

## 5. Evaluation of cost-saving

*Introductory remark*

"Cost-saving" is taken to mean the reduction in the cost of the resources necessary to achieve the environmental objective. Cost-saving derives from the fact that part of the pollution-control effort is shifted from high-marginal-cost sources (on the right in Figure 4) to sources with a low marginal cost. In a permit system, emission reductions are actually made in ascending order of marginal costs.

Measurement of the increase in effectiveness is in fact a comparison of costs. The cost of implementing a permit market is compared with that of command-and-control regulation when the two instruments must allow an identical environmental objective to be achieved (for example, a given annual volume of emissions). As far as regulation is concerned, "implementation cost" is taken to mean the sum of the cost of inspections carried out by the regulatory authority (emission measurements) and of the cost of emission reductions, the latter encompassing the initial investment and the operation and maintenance costs of the installations (Luken 1990, p. 98). In the case of permits there is an *administrative* cost over and above these elements, that is, the sum of the permit management cost borne by the supervisory authority, and of the transaction cost borne by the permit purchasers and sellers (Table 2).

Table 2. Cost elements of command-and-control regulation and permit market.

| Cost of implementing CAC regulation | Cost of implementing the permit market |
|---|---|
| Cost of emission measurement | Cost of emission measurement |
| + | + |
| Cost of emission reductions | Cost of emission reductions |
| | + |
| | Administrative cost |

The literature proposes several methods of evaluating the increase in effectiveness (for example, comparison with the Least Cost policy, or use of a general equilibrium model, etc.). The approach selected for the Chablais is that which consists of establishing the functions of permit supply and demand. From their point of intersection, it is possible to calculate the cost of emission reductions made in the context of a permit market and to compare it with that of the clean-ups required by regulation.

The supply function is relatively simple to establish. The supervisory authority fixes the annual volume of emissions allowable in the region as a function of the air quality it wishes to obtain. The emission "ceiling" determines the number of permits to be put on the market. Supply is thus quite inelastic; graphically, it is represented by a vertical straight line (Hahn and Noll 1982, p. 128; NERA 1992, p. 66; Kula 1994, p. 195).

Determining the demand function is more complicated. Hypothetically, the price of a permit is equal to the price of a ton that is not emitted, that is, to the cost necessary to reduce emissions by one ton (Bergman 1991, p. 54). As a result, we are looking for the total annual cost per ton reduced for every possible operation at every source (Figure 4). The demand curve can be plotted from these values. The volume of emissions, in tons per year, increasing since the origin, is shown on the abscissa. The emission level in 1986 (date when the OPair came into force) serves as a reference. It is assumed that if a permit system had been introduced at that date, the clean-ups would have been made in ascending order of pollution-control costs. Accordingly, the marginal costs of Figure 4 are shown in inverted order, with the maximum value on the left and the minimum value on the right. Subsequently, for each of the 37 observations, the reduction (number of tons) brought about by the clean-up is associated with the marginal cost. As a result, the horizontal segments of the bars correspond to the reductions made in tons. Figure 5 shows the demand curve obtained in this way. On account of the capacity limits of pollution-control technologies, the curve takes the form of a staircase. Otherwise, it has a conventional appearance: the lower the number of permits put on the market, the higher their price. Conversely, the greater the number of permits, the lower their price.

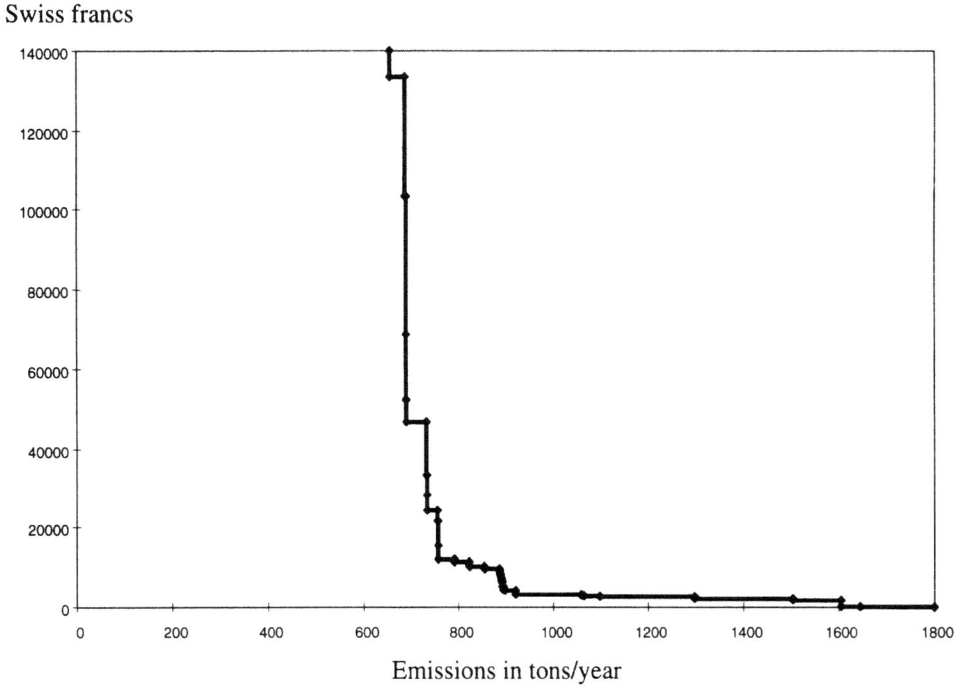


Figure 5. Permit demand curve in the Chablais region.

The intersection between the supply straight line and the demand curve provides the means of determining the market price of the permits. Above all, this also makes it possible to identify the companies requiring and offering permits, depending on their marginal costs. On the graph, the sources situated to the right of supply are those in which emission reductions must be effected, since the pollution-control cost is lower than the permit selling price. Conversely, the sources situated on the left of supply have marginal costs greater than the price of permits: they will therefore not be reduced, and their owners will have to purchase permits. The cost of the emission reductions to be made in the context of the permit market is given by the surface area delimited by the demand curve, the supply curve and the x-axis.

Several separate evaluations of cost-effectiveness will be made. First of all, two periods will be considered: a first one extending from 1986 to 1994 and concerning the clean-ups completed under the OPair and a second one for the emission reductions currently being studied which will be carried out in coming years. Next, the estimate will be performed in a first stage with a trading ratio of 1:1, and then with 1:0.8. The difference will enable us to quantify the cost of maintaining the safety margin contained in the OPair. Lastly, hypotheses will be made as to the divisible or indivisible nature of the emission reduction programmes.

*Cost-effectiveness between 1986 and 1994*

The aim is to determine the gain that could have been achieved by a permit system *had it been introduced in 1986 in place of the OPair*. Between this date and 1994, the fifteen companies under consideration reduced their $NO_x$ emissions by almost 400 tons, cutting their total from nearly 1,800 tons per year to about 1,400 tons per year. For the purposes of the evaluation, it is thus assumed that the supervisory authority would have put 1,400 permits on the market in 1986. Accordingly, permit supply is thus represented by a vertical straight line crossing the abscissa at the 1,400 mark (Figure 6).[4] The intersection between the two curves determines the market price of the permits, which in this case amounts to just over CHF 2,000 (CHF 2,187).

The total annual cost of the clean-up programmes completed between 1986 and 1994 is known; it amounts to CHF 1.2 million (CHF 1,199,999) (source: questionnaire to companies). With a permit system, emission reductions are made in the order of marginal costs. Thus, the 400-ton reduction would imply a total annual cost which can be calculated by adding up the emission reduction costs whose marginal cost is lower than or equal to the price of permits. To calculate the increase in effectiveness, it is therefore important to distinguish between two scenarios:

- Clean-ups are perfectly divisible, i.e. it is possible to reduce the emissions by exactly the desired quantity. The cost of the last clean-up programme is not entirely accounted for: only the share which allows emissions to be reduced to 1,400 tons/year is considered;
- Clean-ups are not divisible and the emission reductions can only be achieved in stages; in this case, the cost of the last clean-up must be accounted for *in toto*.

---

4 Figure 6 shows only part of the demand curve, which is why it appears to be different from the one shown in Figure 5.

Whatever the hypothesis used, the ratio between the cost of the OPair and that of the permit market is much higher than one: it amounts to 2.7 if it is assumed that clean-ups are divisible, and otherwise to 1.8 (Table 3).

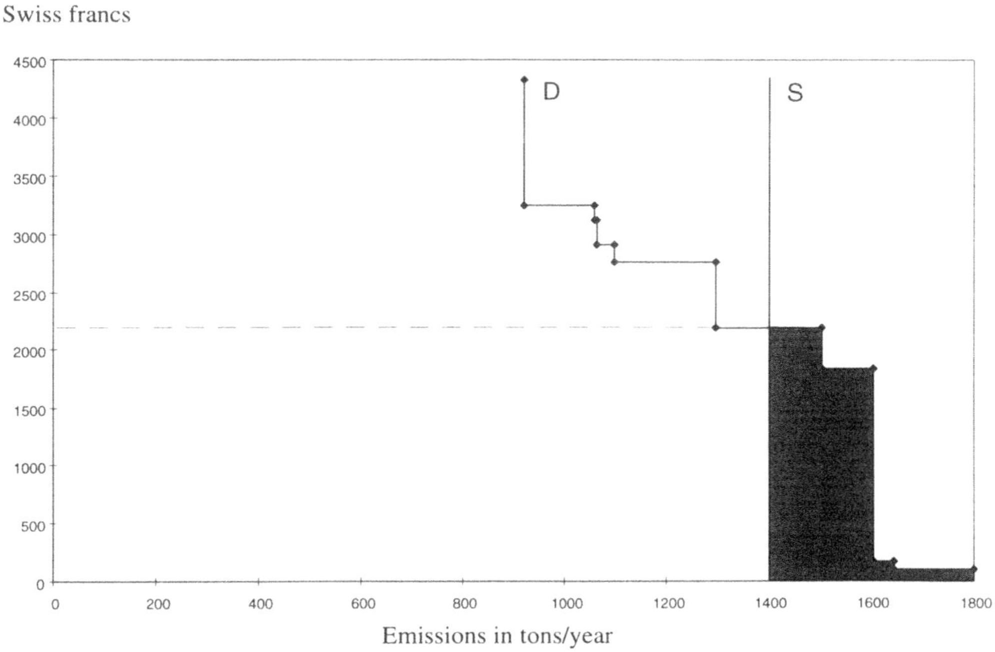


Figure 6. Equilibrium on the permit market with an emission level of 1,400 tons/year (trading ratio 1:1).

If the trading ratio is reduced to 1:0.8, this means that each permit traded authorizes an emission of 800 kilograms instead of one ton. Consequently, the permits available following the 400-ton reduction allow only an emission volume of 320 tons. The 80 permits lacking have to be obtained by further emission reductions amounting to 100 tons. The alteration of the trading ratio therefore implies a greater volume of emission reduction, viz. 500 tons instead of 400 previously. In fact, this situation corresponds to the instance in which there would be 1,300 permits each worth one ton. Graphically, it is as if the supply vertical straight line had shifted to the left. The additional cost compared to the previous case is indicated by the dotted area in

Figure 7. The increase in effectiveness is calculated in the same way as before, on the hypothesis that the alteration of the trading ratio still does not affect the administrative cost. It is worth noting that the price of permits does not increase on account of the steps shown by the demand curve.

Table 3. Estimation of the increase in effectiveness between 1986 and 1994 in 1,000 Swiss francs per year (trading ratio 1:1).

| | OPair | Permit market (divisible clean-ups) | Permit market (indivisible clean-ups) |
|---|---|---|---|
| Total annual cost of reductions | 1,200.0 | 420.9 | 643.5 |
| Administrative cost: | | | |
| - one-tenth of a post | --- | 10.0 | 10.0 |
| - annualized initial expenditure * | --- | 10.0 | 10.0 |
| Total | 1,200.0 | 440.9 | 663.5 |
| Differences in annual costs | --- | 759.1 | 536.5 |
| Ratio of OPair cost / permit cost | --- | 2.7 | 1.8 |

* Note: The expenditure necessary to implement the market (CHF 100,000 in the first year) was annualized over ten years.

Even with the reduction of the trading ratio, the permit system remains superior to regulation, and quite markedly so, since the cost ratio is close to 2. The divisible or indivisible nature of the clean-ups is hardly felt at all in this case, as it is very close to the capacity limit. Shifting to a ratio of 1:0.8 leads to an additional annual cost of almost CHF 220,000. However, this additional cost appears only in the event that the clean-ups are assumed to be perfectly divisible, in which case the cost of the permit system rises from CHF 440,000 to CHF 659,000. In the event that the clean-ups are not divisible, altering the ratio does not cause a switch to a higher step, so the cost remains identical.

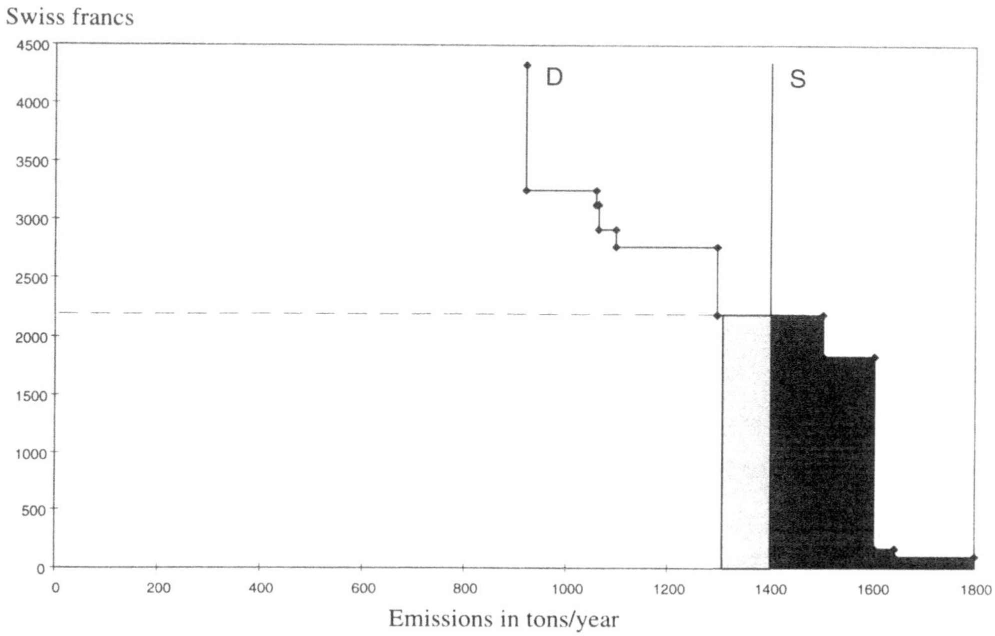


Figure 7. Equilibrium on the permit market with an emission level of 1,400 tons/year (trading ratio 1:0.8).

Table 4. Estimation of the increase in effectiveness between 1986 and 1994 in 1,000 Swiss francs per year (trading ratio 1:0.8).

| | OPair | Permit market (divisible clean-ups) | Permit market (indivisible clean-ups) |
|---|---|---|---|
| Total annual cost of reductions | 1,200.0 | 639.5 | 643.5 |
| Administrative cost: | | | |
| - one-tenth of a post | --- | 10.0 | 10.0 |
| - annualized initial expenditure | --- | 10.0 | 10.0 |
| Total | 1,200.0 | 659.5 | 663.5 |
| Differences in annual costs | --- | 540.6 | 536.5 |
| Ratio of OPair cost / permit cost | --- | 1.8 | 1.8 |

*Cost-effectiveness of future reductions*

The estimates shown above concern the period 1986-1994. Assuming that the permit market is introduced today, let us evaluate the increase in effectiveness that could be achieved between 1995 and the year 2000. Various clean-up possibilities are currently being studied by several of the fifteen market participants. They could be achieved in the coming years and ought to allow emissions to be cut by some 450 tons, bringing the present total of 1,400 tons/year down to about 950 tons/year by the year 2000 (source: questionnaire to companies). For the purposes of the simulation, it will be assumed that the supervisory authority puts 950 permits on the market. The demand curve is re-plotted from the 1,400 point; the new equilibrium determines a permit price of CHF 3,247, that is, an almost 50% increase over the previous situation (Figure 8). The increase in effectiveness is estimated as before (Table 5).

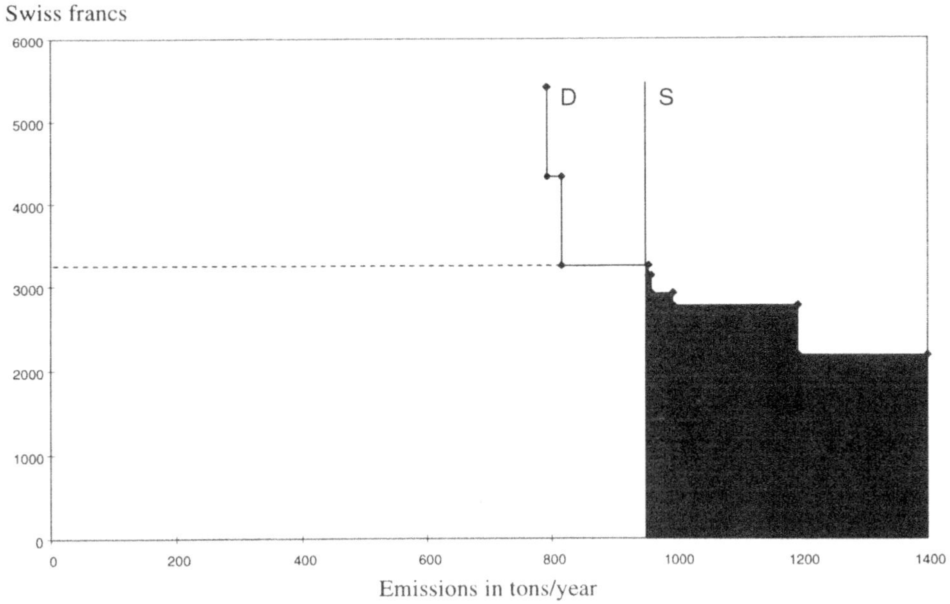


Figure 8. Equilibrium on the permit market with an emission level of 950 tons/year (trading ratio 1:1).

Table 5. Estimation of the increase in effectiveness between 1995 and the year 2000 in 1,000 Swiss francs per year (trading ratio 1:1).

| | OPair | Permit market (divisible clean-ups) | Permit market (indivisible clean-ups) |
|---|---|---|---|
| Total annual cost of reductions | 3,900.0 | 1,136.4 | 1,570.1 |
| Administrative cost: | | | |
| - one-tenth of a post | --- | 10.0 | 10.0 |
| - annualized initial expenditure | --- | 10.0 | 10.0 |
| Total | 3,900.0 | 1,156.4 | 1,590.1 |
| Differences in annual costs | --- | 2,723.6 | 2,289.9 |
| Ratio of OPair cost / permit cost | --- | 3.3 | 2.4 |

The clean-ups which the OPair will impose on companies in the Chablais over the next few years will be particularly costly, owing to the exponential nature of marginal pollution-control costs. A permit system implemented today would allow a substantial increase in effectiveness to be achieved during the next years, with an annual cost saving of between CHF 2 and 3 million. Here, it is important whether or not clean-ups are divisible or indivisible: the last reduction (i.e. the one that allows emissions to be brought down to 950 tons/year) is an important operation which makes it possible to achieve a result well below 950 tons. Graphically, the intersection of supply and demand occurs right at the beginning of a new emission reduction stage. If the cost of this last operation has to be totally accounted for, the cost-effectiveness of the permit system diminishes markedly. Even in this case, however, the cost ratio (2.4) remains clearly favourable to permits.

With a trading ratio of 1:0.8, the 450 permits available following the reductions would only allow an emission volume of 360 tons. The shortfall of 90 has to be obtained by new emission reductions amounting to 112.5 tons. The new trading ratio therefore implies a greater emission reduction, increasing from 450 to 562.5 tons/year. This situation thus corresponds to the case in which 837 permits each worth one ton would be on the market. Graphically, the additional cost is indicated by the dotted area of Figure 9. The increase in effectiveness is calculated as before (Table 6). It is worth noting that the price of the permits does not change (same stage).

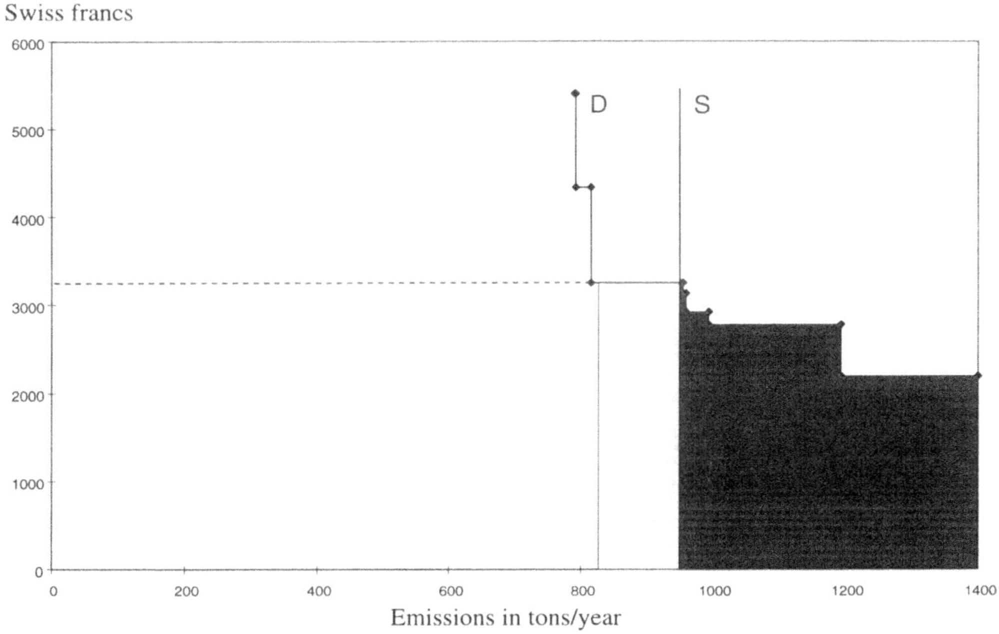


Figure 9. Equilibrium on the permit market with an emission level of 950 tons/year (trading ratio 1:0.8).

Table 6. Estimation of the increase in effectiveness between 1995 and the year 2000 in 1,000 Swiss francs per year (trading ratio 1:0.8).

| | OPair | Permit market (divisible clean-ups) | Permit market (indivisible clean-ups) |
|---|---|---|---|
| Total annual cost of reductions | 3,900.0 | 1,526.0 | 1,570.1 |
| Administrative cost: | | | |
| - one-tenth of a post | --- | 10.0 | 10.0 |
| - annualized initial expenditure | --- | 10.0 | 10.0 |
| Total | 3,900.0 | 1,546.0 | 1,590.1 |
| Differences in annual costs | --- | 2,334.0 | 2,289.9 |
| Ratio of OPair cost / permit cost | --- | 2.5 | 2.4 |

During the period 1995-2000, modifying the trading ratio would entail an additional cost of nearly CHF 400,000 per annum. However, this is true only if the clean-ups are actually perfectly divisible. Should the reverse hold true, the sums remain identical, since one remains within the same capacity limit (one does not move up to a higher step). Whatever hypothesis is made about the divisible character of the clean-ups, the permit system would remain easily superior to command-and-control regulation during the period under review, with a total annual cost of over 50% less.

## 6. Conclusion

The superiority of a permit system over regulation, as far as cost-effectiveness is concerned, has been known for a long time. It has been demonstrated once again. Yet permits offer other advantages, such as a greater freedom of choice for companies and above all a permanent incentive to carry out further emission reductions, thereby possibly stimulating technological progress.

However, the introduction of such a system does pose some practical problems. From the legal point of view, the implementation of the proposed market in the Chablais would not be feasible without some amendments to the Federal Environmental Protection Law and to the Air Protection Ordinance, but the necessary modifications are minor (see paper by R. Jacobs in this volume). As far as acceptability is concerned, there could conceivably be negative reactions from the population and ecological groups. It is also possible that small businesses might be afraid that large companies could acquire, to their own detriment, a position of strength on the permit market, and that they might accordingly prefer regulation to be maintained.

Yet these implementation problems do not appear to be insurmountable and above all do not appear to be greater than the expected benefits of the instrument. It should be recalled that the objective to be achieved is indeed the same as that of command-and-control regulation but by a different, less expensive means. Consequently, it is rather difficult to explain why permits are so rarely used in Europe (whereas they are fairly widely employed in the United States) and in particular it is difficult to understand why this instrument seems to arouse mistrust in Switzerland.

*Acknowledgements*
*The author is grateful to the Swiss National Science Foundation for funding this research within its Priority Programme « Environment » (Project No. 5001-035281).*

## References

**Atkinson, S. E. and Tietenberg, T. H.** (1982) The Empirical Properties of Two Classes of Designs for Transferable Discharge Permit Markets. *Journal of Environmental Economics and Management*, 9 : 101-121.

**Bergman, L.** (1991) General Equilibrium Effects of Environmental Policy : A CGE-Modelling Approach. *Environmental and Resource Economics*, 1 : 43-61.

**Dwyer, J. P.** (1992) California's Tradeable Emissions Policy and its Application to the Control of Greenhouse Gases. *In*: Climate Change : Designing a Tradeable Permit System, OECD (ed.), Paris, 41-77.

**Ecoscan** (1994) Etude d'assainissement de l'air dans le Chablais. Report to the Nuisance Control Department of the Canton of Vaud, Lausanne.

**Environmental Protection Agency (EPA)** (1990) Office of Air Quality Planning and Standards, OAQPS Control Cost Manual, 4th ed., EPA Report 450/3-90-006, Research Triangle Park, NC.

**Hahn, R. W.** (1993) Getting More Environmental Protection for Less Money: A Practitioner's Guide. *Oxford Review of Economic Policy*, 9 (4) : 112-123.

**Hahn, R. W. and Noll, R. G.** (1982) Designing A Market For Tradeable Emissions Permits. *In*: Reform of Environmental Regulation, Magat W.A. (ed.), Ballinger Publishing Company, Cambridge, MA, 119-146.

**Heister, J. ; Michaelis, P. and Mohr, E.** (1990) Praktische Einsatzmöglichkeiten für Zertifikate im Rahmen marktwirtschaftlicher Umweltpolitik in der Bundesrepublik Deutschland und in der EG, Stickoxide (Abridged version). University of Kiel, Institut für Weltwirtschaft, Kiel.

**Kula, E.** (1994) Economics of Natural Resources, the Environment and Policies. Chapman & Hall, London.

**Luken, R. A.** (1990) Efficiency in Environmental Regulation. Kluwer Academic Publishers, Boston.

**National Economic Research Associates Inc. (NERA)** (1992) Emission Trading Program for Stationary Sources of NOx in Ontario. Report prepared for the Toronto Advisory Group on Emission Trading, Cambridge, MA.

**Noll, R. G.** (1982) Implementing Marketable Emissions Permits. *The American Economic Review*, 72 : 120-124.

**Noll, R. G.** (1983) The Feasibility of Marketable Emissions Permits in the United States. *In*: Public Sector Economics, Finsinger, J. (ed.), The Macmillan Press Ltd., London, 189-225.

**Organization for Economic Cooperation and Development (OECD)** (1993) International Economic Instruments and Climatic Change. OECD, Paris.

**Pasek, J. and Beckerman, W.** (1994) International Justice and Environmental Policy: Tradeable Permits for Carbon Emissions. Working Paper GEC 94-24, Center for Social and Economic Research on Global Environment (CSERGE), London.

**Roberts, M. W.** (1983) A Remedy for the Victims of Pollution Permit Markets. *The Yale Law Journal*, 92 (6) : 1022-1040.

**Rollier, H.** (1995) Head of the Nuisance Control Department. Meeting on September 22 and telephone conversation on July 13, Epalinges, Switzerland.

**Schnydrig, J.-P.** (1995) Head of the Environmental Protection Department of the Canton of Valais, Sion, meeting on September 27, 1994 and telephone conversations on January 20 and August 7, 1995.

**Stritt, M.-A. and Jeanrenaud, Cl.** (1992) L'effet des mesures de protection de l'environnement sur les coûts de production dans l'industrie. Report published by the Federal Office of Economic Affairs, Bern.

**Swiss Federal Office of the Environment** (1987) Forests and Landscape (SFOEFL), Emissions polluantes en Suisse dues à l'activité humaine (de 1950 à 2010). Les Cahiers de l'environnement n° 76, Bern.

**Tietenberg, T. H.** (1991) Reductions in Emissions : Command and Control or Market-Based Mechanisms? Economy and Environment in the 1990s, International Conference, IRER, University of Neuchâtel, August 26-27.

Environmental Policy Between Regulation and Market
ed. by C. Jeanrenaud


# 4.5 Tradeable Permits in Switzerland : The Legal Perspective

**Reto Jacobs***
**Hochschule St-Gallen, Switzerland**

## 1. Introduction

In the past, Swiss Environmental Law has relied almost exclusively on command and control type regulations. With this strategy, major improvements have been achieved. However, it has also become obvious that enforcement of the rules causes serious problems. To overcome this shortcoming, the introduction of market-based instruments has been proposed. The currently discussed amendment of the Swiss Environmental Protection Act (Umweltschutzgesetz, USG[1]) will introduce charges on VOC and certain types of fossil fuels.[2] According to Art. 39 Para. 2 of the recently amended Aviation Act (Luftfahrtgesetz, LFG[3]), airport charges vary according to the noise and the air pollution caused by the airplanes. A tax on $CO_2$ was

---

* Attorney at Law, LL.M., Berkeley/Zürich.

1 Bundesgesetz vom 7. Oktober 1983 über den Umweltschutz, SR 814.01.

2 *See* Botschaft des Bundesrates vom 7. Juni 1993 zu einer Änderung des Bundesgesetzes über den Umweltschutz (USG), BBl 1993 II at 1515 et seq.; for a discussion of the proposed charges *see* K.A. VALLENDER, Gedanken zur Ökologisierung des Steuerrechts, 62 (1993/94) ASA, 642-668 at 658 et seq.

3 Bundesgesetz vom 21. Dezember 1948 über die Luftfahrt, SR 748.0.

proposed in 1994.[4] However, this tax will not come into force before the year 2000 and then only if other measures have failed to achieve the reduction goal.[5]

Whereas charges and taxes have got quite a bit of attention, tradeable permits have not been seriously discussed so far on the federal level.[6] Despite this reluctance by the federal legislator, some cantons (states) have introduced tradeable permits or intend to do so in the near future. In 1992, the cantons of Basel-Landschaft and Basel-Stadt enacted an emission trading programme. The cantons of Waadt and Wallis are currently studying a tradeable permit scheme for the Chablais Region.

The scope of this article is to assess the potential for the cantons to introduce tradeable permits. Consequently, the approach used by the USG to combat air pollution is described in section II. It will then be possible to examine the freedom of the cantons to implement tradeable permits. Since the system in Basel has already been studied quite extensively, this article will mainly focus on the proposed system in the Chablais Region. The proposal will be assessed in the light of the USG and the Constitution. The question to be answered is whether the proposed programme can be introduced. It will be shown that the present federal Environmental Law prohibits the implementation of such a permit programme. However, we will also see that only minor amendments of the federal law would be necessary to integrate permit programmes into Swiss environmental policy.

## 2. The concept of the Swiss Environmental Protection Act to combat air pollution

According to the Swiss Constitution (BV),[7] legislative power lies with the cantons unless it is specifically transferred to the federal level by the Constitution itself.[8] Art. 24$^{septies}$ BV

---

4 *See* Vorentwurf vom 1. März 1994 zu einem Bundesgesetz über eine $CO_2$-Abgabe auf fossilen Energieträgern.

5 *See* Pressemitteilung des Bundesrates vom 31. Mai 1995.

6 *See* Botschaft des Bundesrates, *supra* note 2, at 1515 et seq., where tradeable permits were rejected because of difficulties in enforcing them.

7 Bundesverfassung der Schweizerischen Eidgenossenschaft vom 29. Mai 1874, SR 101 (BV).

8 *See* Art. 3 BV.

authorizes federal legislators to enact rules for the protection of the environment. Federal legislators have made use of this power and in 1985, the federal USG, which contains the basic rules with respect to air pollution,[9] entered into force.

Federal legislators decide what level of air quality has to be achieved. For this purpose, they have set national ambient air quality standards (Immissionsgrenzwerte, IGW).[10] In order to achieve the IGW[11], the USG follows a twofold concept: first, irrespective of the air quality, emissions have to be reduced as far as is technologically feasible and economically bearable[12]. For several categories of plants, the federal law describes which measures have to be considered as technologically feasible and economically bearable. Based on these findings, the federal law sets emission standards for various plants (attainment provisions).[13] These emission standards have to be met by all plants throughout the country. If -- despite compliance with federal emission standards -- the IGWs are not met, a second set of measures comes into force (nonattainment provisions). The cantons are then obliged to enact more stringent emission reduction measures.[14] They must draw up an implementation plan for nonattainment areas which contains the measures necessary to achieve the IGW.[15] However, when setting up the implementation plans, the cantons are not entirely free to choose among the possible measures. They are not only bound by their own laws but by the federal law as well. With regard to the federal law, Art. 12 USG is of paramount importance because it

---

9 Art. 11-18 USG.

10 *See* Art. 13-14 USG and Anhang 7 of the Luftreinhalte-Verordnung vom 16. Dezember 1985, SR 814.318.142.1 (LRV).

11 For a discussion of the imperative nature of the IGW *see* R. JACOBS, Lastengleichheit - Ein sinnvolles Prinzip bei der Anordnung verschärfter Emissionsbegrenzungen?, 8 (1994) URP, 341-362 at 345 et seq.

12 Art. 11 Para. 2 USG.

13 In Swiss legal literature the term "Vorsorgebereich" is used.

14 Art. 11 Para. 3 USG.

15 *See* Art. 9 Para. 4 and Art. 31 LRV. For a discussion of the role of cantonal implementation plans *see* U. BRUNNER, Das Vollzugsinstrument des Massnahmenplans gemäss Art. 31 ff. LRV, 2 (1988) URP, 24-42 at 24 et seq.; T. JAAG, Der Massnahmenplan gemäss Art. 31 der Luftreinhalte-Verordnung, 4 (1990) URP, 132-150 at 132 et seq.; K.A. VALLENDER, Ziele und Instrumente des schweizerischen Umweltrechts, in: Festschrift zum 65. Geburtstag von Mario M. Pedrazzini (Eds.: E. BREM *ET AL*), Bern 1990, 197-226 at 231 et seq.

contains an exhaustive list of the emission reduction measures available to the cantons. Furthermore, several other provisions of the USG and the Constitution have to be respected. As a result, the freedom of the cantons to enact various types of measures is limited.[16]

## 3. Opportunities to introduce tradeable permits

*The Programme in the Cantons of Basel-Landschaft and Basel-Stadt*

*Summary of the Programme*

In the cantons of Basel-Landschaft and Basel-Stadt, the IGW for $NO_x$ and VOC were not met. The cantons therefore tightened the federal emission standards for these pollutants. To lower the costs for achieving the stricter standards, an emission trading programme was introduced.[17]

Plants which lower their emissions below the more stringent cantonal emission standards by an average of 10% receive an emission reduction credit (ERC) equalling 80% for each additional reduction.[18] No ERCs are issued for marginal reductions[19] and for complete shutdowns.[20] ERCs which are not used within five years are discounted by 20% each year.[21]

---

16 *See* on this question H.R. TRÜEB, Zertifikate als Instrumente der Luftreinhaltepolitik - Theoretische Modelle und praktische Erfahrungen, 113 I (1994 I) ZSR, 233-262 at 253 et. seq.

17 For a discussion of this programme *see* U. BRUNNER, Zulassungsbeschränkungen für neue Anlagen bei Überschreitung von Luftimmissionsgrenzwerten?, 4 (1990) URP, 213-233 at 212 et seq.; M. KÖLZ, Emissionsgutschriften und Emissionsverbund in den Entwürfen zu den Umweltschutzgesetzen der Kantone Basel-Landschaft und Basel-Stadt, 4 (1990) URP, 194-211 at 195 et seq.; E. STAEHELIN-WITT, A. SPILLMANN, Emissionshandel - Ein marktwirtschaftlicher Weg für die schweizerische Umweltpolitik, Basel 1992 at 61 et seq.; E. STAEHELIN-WITT, A. SPILLMANN, Emissionshandel - Erfahrungen in der Region Basel und neue Ansätze, 17 (1994) ZfU, 207-223 at 207 et seq.; TRÜEB, *supra* note 16, at 257 et seq.; K.A. VALLENDER, Tendenzen im Umweltrecht, 2 (1992) AJP, 1073-1085 at 1084 et seq. An unofficial English translation of the rules (as proposed in 1989) was done by H.R. TRÜEB, Economics and the Environment: The Case For Tradeable Emission Permits, in: Biology, Culture and Environmental Law (Eds.: M.T. MCGUIRE, M. REHBINDER), Berlin 1993, 43-116 at 111 et seq.

18 *See* § 10 Para. 1 USG BL (sGS BL 780) and § 9 Para. 1 USG BS (sGS BS 780.100).

19 Reductions of less than 200 kg $NO_x$ or less than 500 kg VOC are considered as marginal; *see* § 11 Para. 4 USV BL (sGS BL 780.11).

20 *See* § 10 Para. 2 USG BL and § 9 Para. 2 USG BS.

21 *See* § 10 Para. 6 USG BL and § 9 Para. 6 USG BS.

The ERCs are tradeable.[22] They can be used by plants which do not comply with the more stringent cantonal emission standards. However, each transfer has to be approved by the authorities, which have to ensure that no hot spots occur.[23]

Plants which are obliged to meet the more stringent cantonal emission standards are allowed to form bubbles.[24] The authorities approve the bubble if the aggregate emissions are at least 15% less than the sum of all allowable emissions of each plant and no hot spots occur.[25]

*Legal Assessment*

The system in Basel is designed for a nonattainment area. It should help to reduce air pollution below the IGW. The cantons will be obliged by the USG to enact measures which ensure that the IGW are met[26]. In so doing, they will have to respect the guidelines set by the federal law.

*Art. 12 Para. 1 USG* contains an exhaustive list of measures to reduce emissions[27]. Since emission trading is not mentioned in this list, one could argue that the federal law prohibits the introduction of such schemes. However, the emission trading system in Basel is not mandatory. Each plant is free to decide whether it will participate or whether it will simply comply with the more stringent cantonal emission standards. Since such an emission trading

---

22 *See* § 10 Para. 3 USG BL and § 9 Para. 3 USG BS.

23 *See* § 10 Para. 5 USG BL and § 9 Para. 5 USG BS.

24 *See* § 11 Para. 1 USG BL and § 10 Para. 1 USG BS.

25 *See* § 11 Para. 3 USG BL and § 10 Para. 3 USG BS.

26 *See* Art. 11 Para. 3 USG.

27 *See* BGE 119 Ib 480, 491 f.; A. SCHRADE, Kommentar zu Art. 11-18 USG, in: Kommentar zum Umweltschutzgesetz (Eds.: A. KÖLZ, H.-U. MÜLLER), Zürich 1987 at Art. 12 N 10; TRÜEB, *supra* note 16, 255 et seq.; R. WOLF, Führt übermässige Luftverschmutzung zu Baubeschränkungen und Auszonungen?, 5 (1991) URP, 69-95 at 73 et seq.

programme does not restrict the freedom of the polluters, most scholars agree that the programme is in conformity with Art. 12 Para. 1 USG.[28]

The *mandatory character of the federal emission standards* limits the possibilities for emission trading severely. Each and every plant[29] has to comply with the federal emission standards.[30] Even if a plant holds ERCs, it is not allowed to emit more pollutants than authorized by the federal emission standards.[31] Consequently, the range within which the programme can be applied is rather small. Plants which hold ERCs are only allowed to exceed the more stringent cantonal emission standards up to the federal standards. Thus, the system can only work within the narrow zone between the cantonal and the federal emission standards. This restriction is seen as the major reason for the poor performance of the emission trading programme.[32] So far, only four bubbles and three options for ERCs have been granted.[33]

### *The Proposed Programme in the Chablais Region*

#### *Summary of the Programme*[34]

The Chablais Region is located in two different cantons (Vaud and Valais). To improve air quality, a tradeable permit system has been proposed. Participants in the programme are industrial plants which emit $NO_x$. Nowadays, 15 such plants exist. Mobile sources and the

---

[28] *See* J. HOFER, Rechtliche Bedeutung und Wirksamkeit von Massnahmen-Plänen nach der Luftreinhalteverordnung, 4 (1990) URP, 313-333 at 328 et seq.; KÖLZ, *supra* note 17, at 208 et seq.; TRÜEB, *supra* note 16, 256.

[29] A definition of the term "plant" can be found in Art. 7 Para. 7 USG. However, this definition is not very helpful because it does not make a distinction between the terms "plant" (Anlage) and "source" (Emissionsquelle). *See* on this problem TRÜEB, *supra* note 17, at 113.

[30] *See* Art. 3 Para. 1 and Art. 7 LRV.

[31] *See* HOFER, *supra* note 28, at 328; SCHRADE, *supra* note 27, at Art. 16 N 24; TRÜEB, *supra* note 16, at 254 et seq.

[32] *See* on this question the study done by STAEHELIN-WITT/SPILLMANN (1992), *supra* note 17, at 65 et seq.

[33] Telephone interview with ANTON NYFELER, Lufthygieneamt beider Basel, Sept. 1, 1995.

[34] For a full description of the programme *see* the work in this book by M.-A. STRITT, Tradeable Permit System to Control $NO_x$ Emissions in the Chablais Region.

private sector have not been included because of the high administrative costs related to their participation.

Existing plants receive the permits free of charge (grandfathering). The allocation is based on the federal emission standards and the actual time of operation. For new plants, allocations will be calculated in the same manner and they will also be granted the permits free of charge. The permits are tradeable. According to the proposed rules, plants will even be allowed to surpass the federal emission standards as long as their emissions are covered by the permits they hold.

To facilitate transactions, a joint authority has to be set up by the two cantons. This body will provide potential buyers and sellers with the necessary information. Furthermore, it can restrict transactions if hot spots are likely to occur as a result of the trade.

### *Legal Assessment*[35]

In the Chablais Region, the IGW for $NO_x$ are quite often exceeded. It seems that the Chablais Region should be considered a nonattainment area. However, one of the reasons that the IGW are not met is that several plants do not comply with the federal emission standards (attainment provisions). Consequently, it will only be possible to decide whether the Chablais Region is a nonattainment area or not, once all plants have complied with the federal emission standards. The proposed permit programme should help to lower current actual emissions to the legally allowed level. The main objective of the programme is not to ensure that the IGW are actually met; rather, it aims to reduce present emissions to the level which is allowed in attainment areas. If -- despite these reductions -- the IGW are not met in the Chablais Region, the cantons would have to draw up implementation plans which could, *inter alia*, recommend reducing the number of permits. Since the proposed programme is primarily designed for attainment areas, it differs substantially from the one in Basel which is only applicable in nonattainment areas.

---

35 An in-depth legal analysis of the programme is due to be published by R. JACOBS in the December volume of the journal URP (Umweltrecht in der Praxis).

*Conformity with the USG and the LRV*

The *mandatory character of the IGW* obliges the cantons to reduce the emissions below the IGW. The transfer of permits can lead to local concentrations of emissions. There is a possibility that many permits will be transferred into the very same neighborhood. As a consequence, the IGW may be exceeded locally (hot spots), thereby constituting an infringement of the USG. A permit system must have safeguards to prevent hot spots. According to the proposed programme, the authorities may oppose a transfer if there is a risk of hot spots. Consequently, it appears that unlawful local concentrations of emissions can be avoided with the programme.

As already mentioned, the *mandatory character of the federal emission standards* must be respected by the cantons. Each plant has to be run in compliance with the federal emission standards.[36] The cantons may not grant any exemptions.[37] Even the internal trade of emissions (netting) is not allowed if one of the plants exceeds the standards set by the federal law. Nevertheless, according to the proposed permit programme, plants would be allowed to exceed the federal emission standards as long as their emissions are fully covered by permits. This concept is not compatible with the present federal Environmental Law. If the proposed programme is to be implemented, the USG will have to be amended.

The exhaustive list of measures to reduce emissions in *Art. 12 Para. 1 USG* does not constitute an obstacle for the proposed permit programme. The permit programme does not impose any additional restrictions on the operation of plants. On the contrary, it gives the entrepreneurs more freedom because they can choose whether to comply with the emission standards or to buy permits. Since Art. 12 Para. 1 USG aims only to protect the entrepreneurs against restrictions not listed therein, a programme which does not impose any restrictions at all is in conformity with this article.[38]

---

36 *See* Art. 3 Para. 1 and Art. 7 LRV.

37 *See* HOFER, *supra* note 28, at 328; SCHRADE, *supra* note 27, at Art. 16 N 24; TRÜEB, *supra* note 16, at 254 et seq.

38 *See* HOFER, *supra* note 28, at 328 et seq.; KÖLZ, *supra* note 17, at 208 et seq.; TRÜEB, *supra* note 16, at 256.

The *precautionary principle* requires that emissions be reduced even before the IGW are exceeded.[39] Art. 11 Para. 2 USG specifically stipulates that emissions must be reduced as much as is technologically feasible and economically bearable. It has been argued that a permit system which does not require from each and every plant to take all measures which are technologically feasible and economically bearable contradicts the precautionary principle.[40] However, it should be recalled that the present Law does not set an individual standard for each and every plant. Rather, an average plant of the industry is chosen and an effort is made to determine which standard is technologically feasible and economically bearable for the plant.[41] A plant of the industry operates in conformity with the precautionary principle as long as it meets the standard set this way. In other words, a plant has the "right" to pollute the air up to the standard applicable to the industry. In attainment areas, the authorities can not require more stringent measures from an individual plant, even if it would be technologically feasible and economically bearable to achieve them.[42] Accordingly, the precautionary principle -- as laid down in the USG -- does not require that each and every plant take all measures which are technologically feasible and economically bearable. As a consequence, a permit system which does not follow a plant-by-plant approach has to be considered in conformity with the precautionary principle as well.[43]

From this reasoning, another conclusion can be drawn: since every plant is allowed to raise its emissions up to the federal standards, the initial allocation of the permits which is based on the federal standards is in conformity with the USG. As long as every plant has the "right" to pollute up to the federal emission standards, it seems reasonable to grant every plant permits

---

39 *See* Art. 1 Para. 2 USG.

40 *See e.g.* D. KOECHLIN, Das Vorsorgeprinzip im Umweltschutzgesetz (Diss. Bern), Basel/Frankfurt a.M. 1989 at 80.

41 *See* Botschaft des Bundesrates vom 31. Oktober 1979 zu einem Bundesgesetz über den Umweltschutz (USG), BBl 1979 III at 790; Art. 4 Para. 2 and Para. 3 LRV.

42 *See* on this question the statement of BUWAL (federal environmental protection agency), 8 (1994) URP, 179 et seq.

43 *See* on this question HOFER, *supra* note 28, at 329.

which equal these emissions. Otherwise, plants which have lowered their emissions voluntarily in the past would be at a disadvantage.

*Conformity with the Constitution*

The notion of *equal treatment* (Art. 4 BV) requires that no Law make distinctions which are not based on objective reasons and that it does not treat substantially different facts alike.[44] The proposed permit programme is in conformity with Art. 4 BV even though only the industry takes part in it. The exclusion of mobile sources is justified because of the high costs related to their incorporation in the programme. Since new plants will get the same amount of permits as the old ones did, there is no problem in respect of equal treatment.[45]

The *freedom of commerce* clause of the Constitution (Art. 31 BV) is not hampered by the proposed programme because it does not restrict the freedom at all. On the contrary, the owner of the plant can decide whether to reduce the emissions to the federal standards or whether to acquire additional permits.

The *division of powers between the cantons* has to be assessed because the Chablais Region lies within the territory of two cantons. However, the Constitution does not hinder the cantons from setting up joint legislative acts as long as they are in conformity with federal law.[46] The cantons will have to comply with their own rules of lawmaking and it will be necessary to coordinate the legislative process in the two cantons. The final intercantonal agreement must be approved by the federal authorities.[47]

---

44 *See e.g.* BGE 100 Ia 322 at 328.

45 Even if new plants had to buy the permits, the programme would not necessarily conflict with Art. 4 BV. There are good reasons why new plants should be treated differently from existing ones.

46 *See* on this question U. HÄFELIN, W. HALLER, Schweizerisches Bundesstaatsrecht, 3rd ed., Zürich 1993 at 153 et seq.

47 *See* Art. 7 Para. 2 BV.

## 4. Conclusions

The federal Environmental Law contains mandatory emission standards which have to be met by every single plant. The cantons are not allowed to grant any exemptions. Even a permit system which ensures that total emissions do not rise cannot be invoked to justify an exemption. The mandatory character of federal emission standards is the major obstacle to introducing tradeable permit systems. It is the main reason for the poor performance of the emission trading programme in Basel and it hinders the introduction of the proposed permit system in the Chablais Region.

Unless the USG is amended, there is little hope that permit systems will be able to work properly in Switzerland. Only minor amendments of the USG would be necessary to set a more favorable framework for tradeable permits. Art. 12 USG should be amended by a third paragraph which would allow plants to exceed the federal emission standards if there is no rise in total emissions.[48] However, no such provision has been included in the currently discussed amendments of the USG. As a result, it seems rather unlikely that the conditions for tradeable permit programmes will be more favorable in the immediate future.

## References

**Brunner, U.** (1990) Zulassungsbeschränkungen für neue Anlagen bei Überschreitung von Luftimmissionsgrenzwerten ? *Umweltrecht in der Praxis,* 4 : 213-233.

**Brunner, U.** (1988) Das Vollzugsinstrument des Massnahmenplans gemäss Art. 31 ff. LRV. *Umweltrecht in der Praxis,* 2 : 24-42.

**Häfelin, U. and Haller, W.** (1993) Schweizerisches Bundesstaatsrecht, 3rd ed. Zürich.

**Hofer, J.** (1990) Rechtliche Bedeutung und Wirksamkeit von Massnahmen-Plänen nach der Luftreinhalteverordnung. *Umweltrecht in der Praxis,* 4 : 313-333.

**Jaag, T.** (1990) Der Massnahmenplan gemäss Art. 31 der Luftreinhalte-Verordnung. *Umweltrecht in der Praxis,* 4 : 132-150.

**Jacobs, R.** (1994) Lastengleichheit - Ein sinnvolles Prinzip bei der Anordnung verschärfter Emissionsbegrenzungen ? *Umweltrecht in der Praxis,* 8 : 341-362.

**Koechlin, D.** (1989) Das Vorsorgeprinzip im Umweltschutzgesetz (Diss. Bern). Basel/Frankfurt a.M.

**Kölz, M.** (1990) Emissionsgutschriften und Emissionsverbund in den Entwürfen zu den Umweltschutzgesetzen der Kantone Basel-Landschaft und Basel-Stadt. *Umweltrecht in der Praxis,* 4 : 194-211.

**Schrade, A.** (1987) Kommentar zu Art. 11-18 USG. *In*: Kommentar zum Umweltschutzgesetz, Kölz A. and Müller H.-U. (eds), Zürich.

---

48 There is a proposal (Postulat) pending which aims to amend the USG in such a manner (*see* Amtl. Bull. SR 1994 at 489).

**Staehelin-Witt, E. and Spillmann, A.** (1992) Emissionshandel - Ein marktwirtschaftlicher Weg für die schweizerische Umweltpolitik. Basel.

**Staehelin-Witt, E. and Spillmann, A.** (1994) Emissionshandel - Erfahrungen in der Region Basel und neue Ansätze. *Zeitschrift für Umweltpolitik & Umweltrecht,* 17 : 207-223.

**Trüeb, H. R.** (1993) Economics and the Environment : The Case For Tradeable Emission Permits. *In*: Biology, Culture and Environmental Law, McGuire M. T. and Rehbinder M. (eds), Berlin, 43-116.

**Trüeb, H. R.** (1994 I) Zertifikate als Instrumente der Luftreinhaltepolitik - Theoretische Modelle und praktische Erfahrungen. *Zeitschrift für Schweizerisches Recht,* 113 I : 233-262.

**Vallender, K. A.** (1990) Ziele und Instrumente des schweizerischen Umweltrechts. *In*: Festschrift zum 65. Geburtstag von Mario M. Pedrazzini, Brem E. et al (eds), Bern, 197-226.

**Vallender, K. A.** (1992) Tendenzen im Umweltrecht. *Aktuelle Juristische Praxis,* 2 : 1073-1085.

**Vallender, K. A.** (1993/94) Gedanken zur Ökologisierung des Steuerrechts. *Archiv für Schweizerisches Abgaberecht,* 62 : 642-668.

**Wolf, R.** (1991) Führt übermässige Luftverschmutzung zu Baubeschränkungen und Auszonungen ? *Umweltrecht in der Praxis,* 5 : 69-95.

# *5. Covenants*

Environmental Policy Between Regulation and Market
ed. by C. Jeanrenaud


# 5.1 Covenants from Instrument of Environmental Policy to Implementation Tool

**Wim Hafkamp**
**Erasmus University, Rotterdam, Holland**

## 1. Introduction

This contribution focuses on the role of covenants in environmental policy, with particular reference to the Netherlands. It does so on the basis of the observation that the number and qualities of environmental covenants have changed considerably over the past decade. The position of covenants with respect to other instruments of environmental policy will be discussed in section 2, using a familiar typology of instruments which is then expanded. Section 3 discusses briefly the change in configuration of actors involved in environmental policy. This change has an explanatory value for the increased importance and number of covenants. Subsequently, a discussion of experiences with covenants will be given in section 4. In this section the (perceived) advantages and disadvantages will be presented, along with the wide range of applications of this instrument. It will further propose some rules, "10 commandments" for effective covenants. Section 5 concisely presents one particular covenant, the CFC Action Programme for the Netherlands. Some conclusions will be presented in section 6.

## 2. Instruments for environmental policy

The present discussion of instruments for environmental policy, both in literature and in the field of environmental management, presupposes one particular *institutional configuration*. This configuration can be described in political, legal, economic or sociological dimensions, or in its dimensions of public administration.

In each dimension it is the state, or the government, that assumes a role which can be best described by the Dutch word "overheid", or in German by the word "Obrigkeit". In English the closest term would be "overhead", in a literal sense something which is "over our heads" (as humble earthlings). Collectively, as a matter of democracy, we mandate this "overheid" to be responsible to take actions which will improve our present and future welfare. One key area in which this is traditionally the case refers to the natural environment. Managing this collective initial endowment and resource is considered a core task for the "overheid". From here on I will use terms like state, government or authorities instead of "overheid".

A good analogy for this institutional configuration is that of the Copernican universe. In this universe, the sovereign state uses its dominant position to modify the behaviour of lesser bodies, its subjects, in order to achieve greater welfare for all. It has a full arsenal of instruments at its disposal to impose on indicated citizens, consumers, households, travellers, companies, etc. Thus, it forces changes in their behaviour in order to reduce their (negative) environmental impacts; if not by force, then there are other instruments to induce or entice them in doing so.

Various authors give typologies of instruments of environmental policy. Whether they have a background in law, economics or public administration, the typologies by these authors can be shown to be variations of one and the same (Table 1).

It is striking to note that there are no typologies where other actors with *their* instruments (companies, consumers, environmentalists) are included.

Table 1. Typologies of instruments of environmental policy.

| Archetype | Other types | | | |
|---|---|---|---|---|
| Direct regulation | Direct instruments | Regulation | Command & control | Force |
| Economic instruments | Indirect instruments | Stimulation | Carrot & stick | Transaction |
| Social instruments | Moral persuasion | Communication | Tambourine & hoople | Persuasion |
| | Baumol & Oates (1988) | Aalders (1993) | Winsemius (1987) | WRR (1992) |

In order to gain a better understanding of the meaning of social instruments, and covenants in particular, we need to look more closely at this typology and at the institutional setting in which it is defined. WRR (1992) adds a second dimension to the above typology which is very illuminating. It distinguishes between the type of influence on behavioral mechanisms that is exerted through instruments and the legal basis on which the state uses them. Thus, the following typology is obtained:

Table 2. A two-dimensional typology of instruments of environmental policy.

| | Legal Basis | | |
|---|---|---|---|
| Type of influence on behaviour of actors | Public Law | Private Law | |
| Force | Direct regulation | - | - |
| Transaction | Charges<br>Ecotaxes<br>Subsidies | Contracts<br>Liabilities<br>Property | |
| Persuasion | E.I.A.*<br>Right to know<br>Ecolabels<br>EMAS** | Covenants | Gentleman's agreements<br><br>Information<br>Education |

* E.I.A : Environmental Impact Assessment ** EMAS : Eco-Management and Audit-Scheme

This typology, just as the previous uni-dimension typology, still does not provide a strict separation between types of instruments, but is more useful for analytical purposes. Some instruments, for example, have a strict hybrid character, such as tradeable emission permits. The latter are a combination of force and transactions, and have a basis in both public and private law. Other instruments, such as information and education on environment, evidently also have a legal basis.

## 3. A shift in the configuration of actors

Over the last 10 years, two trends can be observed in environmental policy. One is that private law has become more important: individual and corporate liability for environmental damages has had a pervasive effect on the assumption of environmental responsibilities by individual actors in society (especially in the case of soil pollution, disasters/spills, see also Van Dunné, 1995). More recently, covenants have been made explicitly under private law.

The other trend is that social instruments have become relatively more important. Their number and effectiveness have increased. Processes of interaction between the state, other authorities and other actors in society have become increasingly important. The "Copernican universe" no longer holds as an analogy, and is replaced by the analogy of a network.

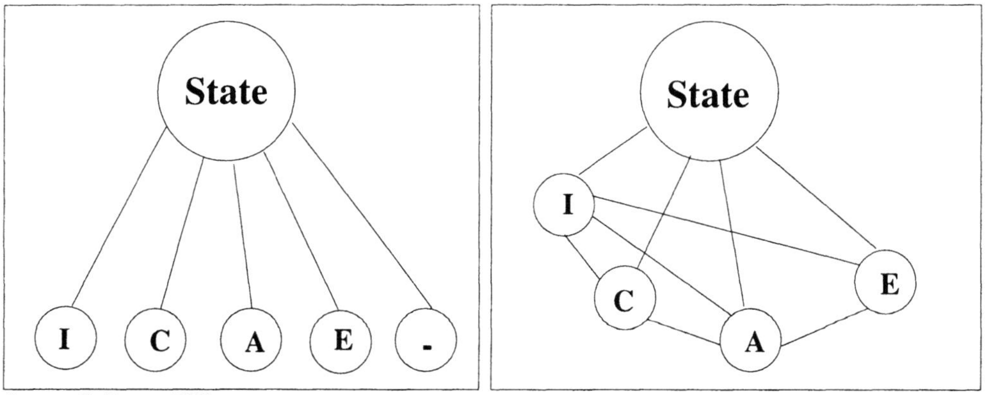


Source: Hafkamp, 1990.

Figure 1. Environmental policies from the welfare state to the network state.

In the "Copernican" model, only the state has instruments for environmental policy. These are the instruments which are usually the focus of analysis, to such a degree that we restrict ourselves almost exclusively to this model. In the network model, all actors have their instruments: industry, consumers, environmentalists, other public interest groups, etc. In the network model the state no longer presides over nor provides for its subjects. Rather, it communicates, participates and negotiates. Environmental policy is not a one-sided issue; rather, it is two-sided or multilateral.

## 4. Covenants in environmental policy

Although any agreement between persons or organizations may be referred to as a covenant, in this contribution I will use a more specific definition (after: Van Acht and Damen, 1991):
*A covenant in environmental policy is a written agreement between the government (typically the national government) and individual companies and/or groups of companies aimed at reaching specific targets in government environmental policy.*

Covenants, even defined by this more restrictive definition, have a long history in environmental policy. Covenants were used as early as the 1960's. At the time, in the total absence of an adequate legal framework, they solved urgent issues of air and water pollution. In the 1970's and early 1980's, a large number of mostly sectoral environmental laws went into effect in the Netherlands. It is only over the last 10 years that covenants have once again become popular. This took place in the societal context of the "retreating government" (a political climate which called for less state interference, less complex and detailed legislation, and for -- in general -- deregulation). More specifically, it took place in a context of increasing evidence that conventional environmental policy and legislation were simply not effective, inefficient due to a lack of acceptance and implementation by both industry and government. Supposedly, the pros in terms of effectiveness and efficiency (through a more expedient process, market-conformist implementation, higher acceptance by parties involved) would be significant. Many argued against this position, saying that covenants were essentially "soft".

Table 3. Perceived advantages and disadvantages of covenants in environmental policy.

| Advantages: | Disadvantages: |
|---|---|
| - effectiveness | - "softness" of covenants |
| - efficiency | - no clear results in targets or in efforts |
| - speed | - no monitoring |
| - market-conformity | - no binding commitment from individual industries |
| - acceptance | - unclear legal status |
| - implementation | - no guarantees for third parties |
| | - lack of democratic control |

This inherent "softness" would render covenants ineffective and preempt any discussion on their efficiency. Covenants would not be specific and clear in terms of their results, targets to be reached, or in efforts to be made by parties. Consequently, it was felt that there would be no monitoring and no binding commitment for all companies represented by the associations or foundations that signed the covenant, not to mention the role of outsiders. The legal status would be unclear. Whereas "second parties" (industry) were given a more prominent role, "third parties" (environmentalists, consumers) were excluded from the process. In fact, they lost rights (of consultation and participation) which they had in current environmental legislation. Last, but not least, covenants would deprive parliaments of their rights.

At the time, both among proponents and opponents of covenants, the idea prevailed that covenants were a full alternative to other (types of) instruments.

Nonetheless, numerous covenants have been made over the last 10 years. The following list of 30 is not exhaustive (the actual number is well over 100), but it contains the major covenants concluded between 1985 and 1994.

Table 4. Covenants in environmental policy since 1985 in the Netherlands.

| | |
|---|---|
| * Mercury oxide batteries | * Tank storage facilities |
| * Packaging of beverages | * Waste water in dentist practice |
| * Alkaline batteries | * Electric utilities air emissions |
| * PET bottles | * Printing industry (2) |
| * Trucks/lorries | * Packaging |
| * Waste transportation (2) | * Water emissions to the Rhine by German chemical industries |
| * Phosphates in detergents | * Metal industry |
| * CFC's in spray cans | * Green label |
| * Cadmium in plastic crates | * Plastic waste and industry |
| * Packaging of pesticides/herbicides | * Plastic waste and agriculture |
| * Asbestos | * Chemical industry |
| * Nickel-cadmium batteries | * Tropical wood |
| * Inland tanker transport waste (2) | |

In 1988 and 1989, the first evaluation studies of covenants were published (Van Rossem, 1988; Klok, 1989). These studies provide a mixed picture. Most of the covenants actually show the flaws expected by the opponents of covenants. Yet their effectiveness was good in the majority of the cases investigated. One of the authors, Klok, questions the causality in these cases: effective covenants, he says, would have reached their targets anyway, while ineffective covenants would have achieved no better results with other instruments. These evaluations and the more recent debate on covenants in both literature and public administration have had a far-reaching effect on covenants in more recent years. In some countries, like Belgium and Denmark, new legislation was introduced on covenants in environmental policy. This legislation aims at a generic solution for most of the problems discussed earlier in this section (Bochen, 1993; and Jorgensen, 1993). The main lines are:

1) The *conditions* for successful covenants have become clear. They include "... (a) a mature environmental policy that underpins mutual respect and understanding, (b) experienced and credible parties, within both government and industry, that are represented by respected negotiators, and (c) a tradition of consensus-seeking and joint problem-solving" (Winsemius, 1993).

2) Selection of policy instruments is no longer seen as the choice of one, dominant government for one instrument of a single type. Selection now involves *the adequate mix of instruments*, see Vos, elsewhere in this volume.
3) Policy formulation, legislation and implementation are seen as a *dynamic process* rather than a static affair. Covenants play their part, not as the one substitute for all other instruments, but as anticipatory and supplementary. In the latter case, they focus on implementation and monitoring.
4) In this dynamic process, one of the vital characteristics of covenants is that government provides *reliability and dependability* for parties in industry as to the government's future decisions (both in adopting new legislation and in interpreting and enforcing existing legislation). Industry trusts that government will act in order to reach a set of given, constant environmental targets.

In other words, covenants are functional in executing environmental policy that has been established in the political process, under democratic control. Some authors go as far as to suggest that those who oppose covenants only do so for opportunistic reasons. These opponents would actually reject the policy (or legislation) whose implementation is the subject of a covenant, rather than covenants *per se*.

Sap (van der Ploeg and Sap, 1995) argues that "... the opinion on the role of the state is changing, and the prerogative is moving to civilians and organizations. Solutions for problems in the public sector require first of all that a check is made on the need for legislation (civilians can do a great deal themselves; self-regulation). Covenants are more adequate to allow for/reflect these freedoms and self-regulation. If well phrased and well implemented covenants strengthen democratic legitimacy, rather than undermine it." Sap does so on the assertion that, along with democratic control, efficiency, influence of civilians and acceptance of government policy are important as well.

Michiels (1994) proposes the following "... 10 commandments for covenants in environmental policy":

Table 5. 10 commandments for covenants in environmental policy.

| | |
|---|---|
| 1. | Precede and implement legislation, rather than replace it. |
| 2. | Targets and Efforts: precisely record the results. |
| 3. | Be consistent with existing legislation and competencies. |
| 4. | Provide for settlement of conflicts. |
| 5. | Provide for major changes in (the understanding of) environmental issues. |
| 6. | Identify parties precisely. In case of representation, mandates should be given. |
| 7. | Be explicit about the legal character. |
| 8. | Provide procedural guarantees for third parties. |
| 9. | Send the final draft to parliament immediately. |
| 10. | If necessary, send the covenant to the European Commission. |

## 5. The Dutch CFC Action Programme

The more recent covenants in the Netherlands, including the Dutch packaging covenant, were drafted, largely and implicitly, on the basis of the above 10 commandments. We conclude with one example to illustrate the "new lines" of a more recent covenant.

The Dutch implementation of the Montreal Protocol on ending the production and use of CFCs and halons and its subsequent revisions, as well as EC-directive (594/91, later 3952/92) are covered by the "CFC Action Programme". This programme fulfils most of the recent requirements on covenants: it contains specific clear measurable targets, as well as an adequate monitoring system. There is a clear time horizon, at which regulation will go into effect and ban the production and use of CFCs and halons. It is binding for all industries that use or produce CFCs and halons (including the army). Third parties, though, are not explicitly party to the covenant or the process.

At present, four years after going into effect, this covenant is ahead of schedule. In the process, without the insertion of specific provisions in the covenant, a "halon bank" has been established by those who have "essential use" of halons (e.g. in army equipment, computer rooms). Given the termination of the production of halons, parties enter their remaining stocks in a "bank", and create a "market" where those who need less of their current stock sell the remainder to those who need more.

The CFC Action Programme is proving to be a fascinating example of a dynamic policy process. Over a 5-year period it constitutes a full mix of diverse instruments, in all classes of the typology in Figure 1:

- direct regulation (legal ban by 1996/1998);
- economic incentives (tradeable emission permits in halon bank);
- social instruments, in two ways:
  - covenant (programme signed by industry and government)
  - self-regulation (halon bank was a private initiative).

## 6. Conclusions

Covenants have become a well-established practice in environmental policy-making. Evaluation studies of the early generation of covenants (pre 1990) have revealed the major weaknesses and practical flaws. These evaluations subsequently provided a basis for the design criteria for covenants (Table 5). Simultaneously, our concept of the covenants evolved. Covenants are no longer seen as "stand-alone" instruments, but rather as essential elements in a policy mix. In particular, the covenant is becoming an implementation tool which serves as a script for governments, authorities and other actors (especially industry) for the process of change from here and now to a future situation where a particular environmental issue is solved.

## References

**Aalders, M. V. C.** (1993) Het milieuconvenant wordt `salonfähig': convenanten `nieuwe stijl' in het milieubeleid. *In*: Beleidsinstrumenten bestuurskundig beschouwd (Policy instruments from a public policy management perspective), Bressers e.a. (ed.), Van Gorcum, Assen/Maasstricht.

**Aalders, M. V. C. and R. J. J. van Acht** Milieuconvenanten (Environmental Covenants). Centrum voor Milieurecht, University of Amsterdam.

**Acht, R. J. J. van and Damen, L. J. A.** (1991) Effectieve toepassing van milieuconvenanten (Effective application of environmental covenants). Centrum voor Milieurecht, University of Amsterdam.

**Baumal, W. J. and Oates, W. E.** (1988) The Theory of Environmental Pollution. Cambridge University Press.

**Bestuurlijke overeenkomst** (1993) Raad voor het Binnenlands Bestuur (Management agreement). The Hague.

**Bochen, H.** (1993) Covenants in Belgium Environmental Law. *In*: Environmental Contracts and Covenants, Dunné J.M. van (ed.), Koninklijke Vermande, 57-72.

**Broek, J. H. G. van den** (1995) Rechter worstelt met betekenis milieuconvenant (Judge has problems with the meaning of environmental covenant). *Staatscourant, 15*, 20 January.

**CFK Aktieprogramma** (1993) Een samenwerkingsprojekt van Overheden en Bedrijfsleven (CFC Action Programme; A Cooperation between Governments and Industry; Annual Report 1991). Jaarraportage 1993. CFK Aktieprogramma, CFK-Commissie, Tilburg.

**Convenanten en vergunningverlening** (Covenants and Licensing) (1993) Commissie voor de Toetsing van Wetgevingsprojecten. Ministry of Justice, The Hague.

**Department of Trade and Industry** (1992) EC Directive on Packaging and Packaging Waste, Compliance Cost Assessment, Final Report. David Perchard Associates/KOMG Management Consulting, The Hague.

**Dunné, J. M. van** (ed.) (1995) Environmental Contracts and Covenants. Koninklijke Vermande.

**EUROPEN** (1995) Facts about Single Use and Reusable Packaging: The Netherlands Experience. Proceedings of a seminar presented by EUROPEN in conjunction with SVM, Brussels, 8 March 1995.

**Hafkamp, W. A.** (1990) Contouren van een schone economie. *In*: Afscheid van een vervuilende samenleving / Contouren van een schone economie ( Farewell to a polluting society / Contours of a clean economy), Nelissen N. and W.A. Hafkamp (eds), Kerckhoff, Zeist.

**Jorgensen, J.** (1993) Legislation on Eco-Contracts in Denmark. *In*: Environmental Contracts and Covenants, Dunné J.M. van (ed.), Koninklijke Vermande, 73-86.

**Klok, P. J.** (1989) Convenanten als instrument voor milieubeleid. Faculteit Bestuurskunde (Covenants as instruments of environmental policy). University of Twente, Enschede.

**Kok, E. and H. Saint, B.** (1994) Issue Profile : Voluntary Environmental Initiatives: What Role for EMAS? *European Environment*, 4 (4) : 14-17, August 1994.

**Lorenzen, K. H.; Pedersen, M. L.; Thomsen, H. and Klemmensen, B.** (1994) Voluntary Strategies in Pollution Prevention: Experience from Denmark and the Netherlands. *European Environment*, 4 (4) : 18-21, August 1994.

**Michiels, F. C. M. A.** (1994) Handhaving van bestuursrecht, Ars Aequi Libri, Nijmegen.

**Ministerie van Volkshuisvesting** (1992) Milieu & Industrie; opzet en achtergrond van het doelgroepenbeleid (Environment & Industry; Structure and Background of the Target Group Policy). Ruimtelijke Ordening en Milieubeheer, VROM 6068/110, The Hague.

**Raad voor het Regeringsbeleid** (1992) Milieubeleid: strategie, instrumenten en handhaafbaarheid. Wetenschappelijke (Environmental policy: strategy, instruments and enforcement). SDU, The Hague.

**Ommeren, F. J. van and de Ru, H. J.** (eds) (1993) Convenanten tussen overheid en maatschappelijke organisaties (Covenants between Government and Societal Organizations). SDU, The Hague.

**Ploeg, T. J. van der and Sap, J. W.** (1995) Rethinking the Balance. VU University Press, Amsterdam.

**Pröpper, I. M. A. M. and Herweijer, M.** (1992) Effecten van plannen en convenanten (Effects of Plans and Covenants). Kluwer.

**Rossum, E. R. C. van** (1998) Milieuconvenanten (Environmental Covenants). *Centrale Raad voor de Milieuhygiëne, 88* (7), The Hague.

**Suurland, J.** (1994) Voluntary Agreements with Industry: The Case of Dutch Covenants. *European Environment*, 4 (4) : 3-7, August 1994.

**Toonen, T. A. J. and ten Heuvelhof, E. F.** (1993) Democratiseren door convenanten (Democratization through Covenants). Vakgroep Bestuurskunde, Erasmus University Rotterdam/RU Leiden.

**Tuininga, E. J.** (1998) Going Dutch in Environmental Policy: A Case of Shared Responsibility. *European Environment*, 4 (4) : 8-13, August 1994.

**Veld, R. J.; de Bruijn, J. A.; ten Heuvelhof, E. F.; Niemeijer, A.; Baas, L.W.; Hommes, R. W. and Huisingh, D.** (1992) Processtandaard voor uitvoering van het Convenant Verpakkingen: advies aan de Stichting Verpakking en Milieu (Process Standard for Implementation of the Packaging Covenant: Advice to the Foundation for Packaging and the Environment). Vakgroep Bestuurskunde/Erasmus Studiecentrum voor Milieukunde, Erasmus University Rotterdam.

**Verpakkingsontwikkelingen** (1992) uitvoering Convenant Verpakkingen in beeld (Packaging Developments 1992: Implementation of the Packaging Covenant, Illustrated). Stichting Verpakking en Milieu (SVM), The Hague.

**Winsemius, P.** (1987) Gast in eigen huis: beschouwingen over milieumanagement (A guest in ones own house: perspectives on environmental management). Samsom H.D. Tjeenk-Willink, Alphen aan den Rijn.
**Winsemius, P.** (1993) Environmental Contracts and Covenants: New Instruments for a Realistic Environmental Policy? *In*: Environmental Contracts and Covenants, Dunné J.M. van (ed.), Koninklijke Vermande, 5-18.
**WRR** (1992) Milieubeleid : strategie, instrumenten an handhaafbaarheld, SDU, The Hague.

Environmental Policy Between Regulation and Market
ed. by C. Jeanrenaud


# 5.2 Covenants as Central Elements in an Effective Environmental Policy Mix

**Ernst. A. Brugger and Barbara Gorsler**
**Beratungen Hanser und Partner, Zurich, Switzerland**

## 1. Introduction

Environmental policy can be implemented through various forms of cooperation, the most elegant way being self-regulation where industry organizes itself on a voluntary basis to achieve a given environmental goal. These voluntary agreements seek to anticipate upcoming regulation by voluntary action. If the preconditions for self-regulation in industry are not met -- in most cases because of free-riders -- there exists a specific form of cooperation called "covenants". These consist of contracts which are concluded between government and industry within the framework of civil law. The Netherlands already have broad experience with this form of cooperation as far as environmental policy is concerned (see Hafkamp in this volume). The government sets goals for specific target groups or sectors in such a way as to ensure that the targets stipulated in the law are met. Industry is entirely free to choose the ways and means for achieving these goals.

It is obvious that covenants open up opportunities for both sides: they reduce uncertainty over the entire period covered by the covenant while minimizing the risk of companies finding themselves subject to inappropriate measures. Companies know what they are required to achieve and within which timeframe; investments can be planned without fear that the rules of the game will be changed. Within the target framework, business is largely free to determine its own priorities and balance sometimes conflicting needs. Environmental targets which apply to an entire sector minimize the risk of distorting competition within it.

Government is spared the need to prepare legislation which is difficult to draft in detail and expensive to enforce. Responsibility with regard to choice of measures and monitoring is by and large shifted to industry. Lastly, covenants provide an opportunity for companies to influence measures and timetables during negotiations with government.

*Are covenants market-oriented?*

At first sight, it may seem contradictory to view covenants as market-oriented instruments. Essentially, markets use price as an incentive to encourage producers to be as efficient and innovative as possible. Thus, the price system plays a central role as an incentive, leaving it entirely up to economic agents to choose the most cost-effective abatement strategies.

Within a framework of agreed objectives and time scales, covenants also give private sector actors a great deal of leeway in choosing how they achieve their goals. Even though the companies participating in the covenant pursue their own interests, they are guided by the "invisible hand" to opt for the most effective pollution control strategy. When viewed in this light, covenants are dearly incentive-based instruments. Frey (1991) classifies covenants under the general heading of "market-oriented environmental protection" tools, arguing that, like the market, voluntary agreements are an effective means of encouraging industry to behave more responsibly.

The Centre for European Studies (CEPS), on the other hand, feels that "strictly speaking, voluntary self-restraint agreements should not be categorized as market-based instruments" (CEPS, 1993). This assertion stems from the view that the main characteristic of market incentives is that they work through prices, whereas voluntary agreements are ultimately induced by government requirements. In view of the heterogeneous nature of covenants, it is difficult to categorize them once and for all as regulatory or market-based instruments. We would thus classify a covenant as market-based if it is designed and implemented in ways that leave maximum leeway for creativity and innovation, and where there is room for innovative ideas to generate the best possible solutions.

Any improvement in the efficiency and effectiveness of environmental policy will require a balanced mix of instruments. Strict command-and-control regulation imposed on emitters will no longer be adequate, and it is precisely in this connection that the cooperative process essential to covenants will make a useful contribution to the environmental policy mix. The process of cooperation could well become an established principle for all other environmental measures.

*The cooperative principle in the Swiss Environmental Law*

The cooperative principle has long been an essential element in the Swiss Environmental Law (USG). It is based on a mutual obligation to provide and the right to receive information, as well as the obligation and the right to cooperate. Although voluntary agreements are already used frequently to implement cantonal legislation, this does not yet apply at national level.

Under proposals to revise the Swiss Environmental Law (USG), the cooperative process would be institutionalized. Cooperation between the authorities and private sector is promoted and described in Article 38 (USG) as follows:

- The government cooperates with industry with a view to implementing the law.
- Before issuing new regulations, the government first examines the effectiveness of voluntary measures taken by the private sector.
- The government may promote voluntary agreements in industry by stipulating quantitative targets and reasonable deadlines and declaring these to be generally binding.

## 2. Evaluating covenants

*Research concept*

The method adopted for the analysis of the voluntary agreements is based on observation of the actors' behaviour, the economic mechanisms which lead actors to conclude or reject specific contracts and the institutional framework in which negotiations take place. This initial analysis led to the formulation of six hypotheses relating to the conditions which determine the success or failure of a voluntary agreement. An analysis of the covenants signed in the Netherlands and in Germany -- two countries with long-standing experience in this field -- provided an opportunity to test these hypotheses. It further made it possible to identify three criteria that are likely to play an important role in the success of covenants: the nature of the environmental problem, the type of participants and the characteristics of the institutional framework.

*Understanding the actors*

A politico-economic approach can be used to identify the motives of the actors involved and hence to deduce the necessary conditions for successful covenants. Depending on their institutional background, the partners will have different negotiating styles as well as varying amounts of room for manoeuvre. The difference in available resources will also play a crucial role: apart from purely material resources, these include negotiating skills. The fact that timescales differ is also important.

As Brugger/Steib (1992) show, private companies generally prefer the cooperative approach because it gives them an opportunity to be actively involved in the decision-making process and a certain freedom of choice as to how the agreed targets are achieved. The authorities are generally thought to prefer direct regulation and economic instruments, particularly when the solutions to a problem are clear-cut. However, they are more willing to accept the cooperative approach when the problems involved are highly complex and where cooperation is likely to facilitate acceptance of the necessary measures and their implementation.

Communication and transparency also play a vital role. This applies both to the parties involved in the covenant process as well as third parties, such as interest groups representing the concerns of various sections of society. Perhaps the most important condition of all for successful cooperative agreements is a genuine will to reach consensus and compromise.

*Understanding the process*

An analysis of the covenant drafting process which examines the motivations of the parties (do the individual interests of both parties generate a win-win situation?) and the legal and institutional environment (rules of the game which are likely to reduce transaction costs) reveals that the scope of application for this type of agreement is particularly broad. Their flexibility also constitutes a major asset, as the covenant can be tailored to meet participants' needs and to fit a given set of circumstances. In any case, the covenant process can be fostered by making it part of the political process, as is planned with the revision of the Swiss Environmental Law.

## 3. Hypotheses explaining the success of a covenant

The above-mentioned analysis provides the basis for the following hypotheses, that illustrate the preconditions for a successful covenant, possible areas of use, efficiency, effectiveness and acceptability to the individuals involved:

1) Covenants are a useful complement to the existing environmental policy mix. Any evaluation of the covenant as an instrument should consider its overall impact in terms of ecological effectiveness and economic efficiency as well as its acceptability to entrepreneurs, politicians and environmental authorities.
2) When the environmental problem is clearly defined, a voluntary agreement is more effective in solving the environmental problem.
3) A covenant's cost-effectiveness is enhanced when the transaction and enforcement costs are low. For this to be the case, emitters must be readily identifiable, potential participants must be well organized and incentives or sanctions must be clearly defined.
4) Parties have the greatest incentive to participate in an agreement if the interests of the potential participants dovetail, existing or planned regulation is deemed quite unsatisfactory and covenants are given a legal status.
5) The chances of persuading producers to adhere to a covenant are greater if they are convinced of the effectiveness of the proposed agreement, they fear the adoption of new,

more stringent and less effective regulations, and companies in the sector to which they belong are well organized.

6) Environmental authorities will be more receptive to a voluntary agreement if direct regulation has not achieved its targets, the search for a solution to the problem requires a great deal of information, and the authorities have accumulated a good amount of experience with voluntary agreements.

These hypotheses were initially tested by taking a closer look at experience in Germany and the Netherlands. This provided an opportunity to verify the hypotheses, that must be amended as follows:

- Covenants and the cooperation upon which they depend are an element of market-oriented environmental policy. The drafting of covenants relies heavily on dialogue, and repetition of the process can help identify solutions to complex problems. In principle, the instrument can be used at any level and for any environmental problem.
- Covenants are more likely to be ecologically effective if solutions are clearly identifiable. The more open the covenant process is to new insights, the greater chance they have of being sustainable. And precisely in situations where there are several -- often conflicting -- views about the solution required, the dialogue principle inherent in covenants can often help generate sustainable solutions.

Keeping transaction costs low is the most effective way to make covenants economically efficient; thus, existing association/communication structures should be used as a basis for their drafting. Although the contracting process takes a certain amount of time, the advantages of a solution found through cooperation with industry are clearly visible. Acceptance of the covenant solution at political, governmental and entrepreneurial levels is most likely where there is a clear win-win situation. Government is most likely to embrace it where existing legislation has failed or proved too difficult to implement, or where the information basis for drawing up new legislation is too weak. Business may see its advantages in the opportunity to discover feasible solutions and prevent broader legislation and restrictions that erode entrepreneurial creativity and innovation.

It should also be noted that it is still relatively early to be conducting such studies because our experience of covenants is rather limited and most of them are still in the drafting stage.

## 4. Key success factors for covenants

We found it would be helpful to establish a list of criteria for evaluating the risks and benefits of a covenant in a given context, as follows:

- *The environmental problem:* a covenants's success depends on the actors' awareness of the problem and the extent to which they agree on how to solve it. Other significant factors are the number and identifiability of polluters.
- *The protagonists*: some of the major decisive factors are market sructure (economic criterion), producers' level of organization within a given branch, the severity of the sanctions applied in case of non-compliance (institutional criterion) and the actors' propensity to seek negotiated solutions (cultural criterion).
- *The framework:* a tradition of dialogue in political institutions, good economic prospects and the perception of inadequacies in the environmental policy are all conducive to the implementation of negotiated solutions.

The following table summarizes the factors affecting the success of covenants:

Table 1. Factors affecting the success of covenants.

| | Chances of success.... ...high | ...medium | ...low |
|---|---|---|---|
| **Environmental Problem** | | | |
| Knowledge/acceptance of the problem | Perception high; highly affected | perception, but only little affected | perception diffuse; low acceptance of the problem |
| Knowledge about the solution | solution clearly indicated | ⟶ | solution can only be found gradually |
| Polluter | polluters clearly identifiable | | many polluters with widely varying degrees of responsibility |
| Extent of the environmental problem | local | ⟶ | global |
| Kind of problem - substitution - energy efficieny - product quality | substitutes are known; technological progress/ innovation is in pipeline | ⟶ | research still needed |
| | **Chances of success.... ...high** | **...medium** | **...small** |
| **Protagonists** | | | |
| Economic criteria | | | |
| Market structure | one competitor ⟶ | some | ⟶ many |
| External influence | low | ⟶ | high |
| Institutional criteria | | | |
| Level of organization | all protagonists organized in a single organization | the most important protagonists are organized | many small protagonists; low level of organization |
| | influence of association high | ⟶ | low influence of association |
| Enforcement of environmental law | sanctions work; can be put through | ⟶ | no sanctions planned |
| Cultural criteria | | | |
| Readiness for dialogue | protagonists experienced in consensus-finding | ⟶ | no experience of dialogue |
| Capacity for dialogue | experience of contracting and strong influence in own organization | ⟶ | no experience of contracting; weak personalities with low influence |
| **FRAMEWORK** | | | |
| Environmental policy | dissatisfaction with existing/upcoming environmental policy (inefficient; deficit in implementation) | ⟶ | existing environmental instruments work well |
| Market conditions; business prospects | strong, expanding economy | ⟶ | recession |
| Political and economic culture | open to dialogue | ⟶ | command-control-approach |
| Institutional framework | covenants institutionalized in legislation | ⟶ | covenants tolerated as exceptions to the rule |

## 5. Empirical studies

The list of criteria defined in the above table was used to analyse four voluntary agreements concluded in Switzerland:

- Country-wide agreement between the graphic arts sector and the authorities to reduce VOC emissions.
- Industry's involvement in the drafting of legislation on drinks packaging design. This includes a system of self-imposed monitoring as a means of achieving specific targets set by the industry's own "Association for Environmentally Sound Drinks Packaging".
- A statement of intent by the paint and varnish industry that it would reduce its use of organic solvents by 20% between 1986 and 1990 (see above).
- A covenant on the recycling and disposal of computers and other electronic office equipment, spearheaded by SWICO (Swiss Industrial Association for Information, Communication and Organization Technology).

The four case studies -- one of which will be presented in detail below -- revealed growing acceptance of the principle of cooperation. This is due partly to fundamental changes taking place in society: a greater willingness to foster dialogue rather than impose -- or accept -- the existing order. The fact that cooperation-based environmental policy is soon to be institutionalized by Art. 38 of the Swiss Environmental Law will give it added weight.

*The covenant in the Swiss paint and varnish industry*

In 1986, Switzerland introduced new legislation designed to reduce air pollution. One of the aims of the new ordinance was to bring VOC emissions down to the 1960 levels by 1995. In the late 1980s, one-third of the 300,000 tons of VOC emitted annually was generated by traffic and the remaining two-thirds by industry. At that time, the paint and varnish sector was identified as a major culprit, accounting for no less than 10% (30,000 tons) of total VOC emissions. In this case, the emissions arise almost exclusively from the way in which paints and varnishes are used. The solvents themselves play a limited role: they are used to dissolve and apply the substances that produce the final coat and then evaporate completely. Contrary to widespread assumptions, households -- that account for about 5-8% of total consumption -- are not heavy users. The vast majority of paints and varnishes -- approx. 95% -- are used for industrial or commercial purposes. The approximately 30,000 users belong to a wide range of industries.

Individually, industrial paint users emit relatively small amounts of pollution. This, combined with the fact that the technology needed to purify emissions (catalytic post-combustion, activated carbon processing, combined adsorption/desorption processes and thermal post-combustion) is complex and expensive, has led to most of them taking a rather passive, "wait-and-see" attitude.

This unsatisfactory situation led the Swiss Paint Makers Association to spearhead the move by the parties involved to find new solutions. In the process, it became clear that the use of solvents could be cut drastically during production of paints and varnishes, i.e. at source. Manufacturers will henceforth be encouraged to develop low-solvent or solvent-free products, and their use will then be guaranteed by a wide and varied range of consumers. Without close cooperation between manufacturers and consumers, such a revolutionary approach to the problem would have been unthinkable.

Prior to this, there had already been a certain amount of unofficial cooperation within the industry on the question of joint market analysis and specific services. This resulted in the formation of a special task force called the "KEL-CH" (Conference on saving solvents during production and coating). The participants were the Swiss Paint Makers Association and ten associations representing users. In 1986, KEL-CH addressed a statement of intent to BUWAL (the Swiss Federal Administration for Environmental Protection): within five years (1986-90), the amount of production-linked VOC emissions would be reduced by 20%. This target was almost achieved: in 1990, there was a reduction of 18.9%. If we compare the 1992 figures with those for 1985, the reduction is 25%. Admittedly, if we consider absolute figures, the result is not as good because overall production during that time increased by 17%. Nevertheless, the covenant played a significant role in increasing awareness of the problem, and efforts are continuing to ensure that this step in the right direction is maintained. Paint and varnish users were given an additional incentive to become involved: they are entitled to use a KEL-CH label on their products as a marketing boost. Two labels are available: one designating solvent-free coatings (up to 2% in weight) and one for low-solvent coatings (2-15% in weight). The label is available to both Swiss and foreign firms.

## 6. The factors that made this covenant a success

*Environmental criteria*

Awareness of the environmental damage caused by VOC emissions was already widespread and there was a relatively high degree of acceptance of the need to find a solution and willingness to do so. The covenant initiated a race between the companies involved to come up with products with lower solvent levels. The direction was clear: the catalytic combustion technology needed to clean air and/or treat emissions was already known; water-based paints were also being developed. Yet another positive factor was that these products were of the same quality and not more expensive, so the only additional cost involved was that of changing from one type to the other.

*The protagonist*

As mentioned earlier, the market was highly diversified in terms of both manufacturers and users. However, the producers' efficient organizational structure, combined with an awareness of changing conditions and an environmentally driven increase in the demand for solvent-free coatings, helped the parties involved to reach the covenant's objectives. As the label was also available for foreign firms, conditions were the same for all competitors. Another decisive component was the lead taken by some of the more innovative members of the Swiss paint and varnish industry, who developed a culture of "environmental performance".

*The framework*

The Air Protection Ordinance of 1986 gave a clear indication of the restrictions that would apply to industry. Thus, in a very real sense, the statement of intent was a means of pre-empting the demands that government would place on the industrial sector. The progress and the innovations that have been made in the use of low-solvent coatings will make it much easier for the industry to live with the VOC tax, which is planned as part of the Swiss Environmental Law. This is a perfect example of how a statement of intent can be instrumental in implementing further-reaching measures. Industry is much more willing to accept such measures if the necessary changes are already under way.

## 7. Conclusions

Covenants will clearly play an important role in the market-oriented policies of the future. The cooperative principle will greatly help to overcome the shortcomings of present legislation and its implementation. The existence of clear-cut conditions -- i.e. a guarantee for industry that no more restrictive legislation will be adopted if it keeps its part of the bargain -- will increase mutual confidence. The fact that there are so few covenants at present has nothing to do with the weakness of the instrument *per se* but with the lack of confidence and dialogue. The tradition of command and control, compounded by reactive behaviour by industry, still prevails.

Research on present covenants must be pursued if a realistic assessment of their value is to be obtained. Since they are part of a general policy mix, more studies need to be carried out into their overall impact in combination with other measures: for instance, what are the implications of the covenant agreed by the paint and varnish industry for the implementation of the VOC tax? New combinations of instruments, such as the "dual instrument" (see Juen *et al* 1992) -- a covenant combined with a tax -- should also be considered.

What is needed now is creative, innovative solutions. And -- as the former Dutch Minister for the Environment told us during our investigations -- this seems to be only a matter of time: "As government and producers in the industrialized nations gather more experience and confidence in the process of self-regulation, the number of experimental schemes is bound to increase." Swiss environmental policy is about to embark on such experiments, and the prospects are exciting.

## References

**Brugger, E. A.; Steib, S.** (1992) Umweltpolitik aus unternehmerischer Sicht, Bundesamt für Konjunkturfragen, Bern.

**CEPS**, Centre for European Policy Studies (1993) Economic Instruments in EC Environmental Policy, Working Paper Report Nr. 8, Brussels.

**Frey, R. L.** *et al* (1991) Mit Ökonomie zur Ökologie - Analysen und Lösungen des Umweltproblems aus ökonomischer Sicht, Helbling & Lichtenhahn, Basel.

**Juen, C.** *et alii* (1992) Politique de l'environnement conforme à la logique de marché. SDES, Genève.

**Koeman, N.** (1993) Bilateral Agreements between Governments and Industry in Dutch Environmental Law, *in*: European Environmental Law Review, July.

**North, D. C.** (1992) Institutionen, institutioneller Wandel und Wirtschaftsleistungen, J.C.B. Mohr (Paul Siebeck), Tübingen.

**Tomer, J.** (1992) The Human Firm in the Natural Environment: a Socio-economic Analysis of its Behaviour, *in*: Ecological Economics 6, Amsterdam.

Concerning Swiss legislation:

**BUNDESGESETZ UEBER DEN UMWELTSCHUTZ** (1983) Stand 1. Januar 1993.

**BUNDESRAT** (1979) Botschaft zu einem Bundesgesetz über den Umweltschutz, 31. Oktober, Bern.

**BUNDESRAT** (1993) Botschaft zu einer Änderung des Bundesgesetzes über den Umweltschutz, 7. Juni, Bern.

**NATIONALRAT** (1994) Die Kohärenz staatlicher Aktivitäten: das Beispiel des Vollzugs der Luftreinhaltepolitik; Inspektionsbericht der Geschäftsprüfungskommission des Nationalrates zuhanden des Bundesrates, 5. Mai, Bern.

**STÄNDERAT** (1994) Umweltschutzgesetz-Änderung; Entwurf des Bundesrates/ Anträge der Kommission Umwelt, Raumplanung und Energie des Ständerates, Bern.

Concerning the case study:

**INFRACONSULT** (1992) Strukturelle Effekte von Umweltregelungen am Beispiel der Lack- und Farbenfabrikanten, Bundesamt für Konjunkturfragen, Bern.

**KEL-CH** (1986) Gemeinsame Erklärung zuhanden des Bundesamtes für Umweltschutz betreffend die Reduktion organischer Lösemittel bei der Herstellung und Verwendung von Anstrichstoffen und Hilfsmitteln dazu, Zürich.

**KEL-CH** (1993) Reglement für die Vergabe von KEL-CH-Zeichen für lösemittelfreie- und lösemittelarme Anstrichstoffe, Zürich.

# *6. Acceptability of Economic Instruments*

Environmental Policy Between Regulation and Market
ed. by C. Jeanrenaud


# 6.1 Economic Instruments and Social Acceptability: A Debate about Values

**Beat Bürgenmeier**
**University of Geneva, Geneva, Switzerland**

## 1. Introduction

The theoretical foundations of economic instruments for environmental protection have long been known and their efficiency gains widely recognized (Baumol and Oates, 1988). Yet the application of these instruments raises important conceptual problems.

For economists, these difficulties are due to factors outside their discipline. Accordingly, they either recommend going beyond the sole economic dimension (Bürgenmeier, 1994) or delegate the study of social acceptability problems to other disciplines, particularly the behavioural sciences.

This contribution examines another proposition: that the economic approach itself causes a social acceptability problem, because it has to face formidable conceptual issues. In fact, economics considers the natural environment as a free good, having no monetary cost until its overexploitation due to "free-riding" makes it scarce and converts it into an economic good. The theory that tries to model this transformation subjects this process to the market and considers the environment as an externality to the economic sphere. This surprises most researchers in the social sciences. Robert Hettlage expresses this astonishment as follows:

"Externalities are built, whose exploration can be delegated to other sciences lest they contaminate the model (this would at once clean the theory), if this explanation effort were necessary" (Hettlage, 1993, p. 84). This fundamental conceptual problem cannot be solved using a theoretical approach. Instead of accepting the intellectual limitations of every economic theory built on a "pure" model, most economists continue to recommend the efficient market -- albeit corrected by internalizing the externality -- as a means of better environmental protection. In our opinion, this procedure is an important barrier to exploring new approaches, particularly those which try to analyse interactions between the economy and the environment in global models. Thus, we also consider it to be at the origin of social acceptability problems.

The conceptual aspects arising in the economics environment relationship are treated in four parts. The first discusses the notion of externality through the concept of social cost. The second examines the problem of the determination of values that are not exclusively economical. The third analyses how new costs are taken into account by economic modelling. Finally, the fourth part stresses the relationship between these conceptual aspects and ethics.

## 2. The notion of externality

When it treats the environment as an externality, economics is subject to criticism, particularly with respect to the concept of social cost which originates in reasoning that attempts to give a scientific basis to income transfers by relying on internalization techniques rooted in welfare economics. The externality argument admits that market failures can arise, where private benefits from some economic activity impose a cost on someone else. Such a case directly pertains to pollution, but has a more general scope. Every time there is an immediate benefit from an economic activity whose marginal cost is less than the expected marginal revenue because a third party bears part of the total costs, we are confronted with market failure. No incentive to reduce or to stop such an activity exists. The total cost to society is greater than the cost borne by he who receives the private benefit. This situation calls for government intervention. The recommended policy is to ask the government to invoice the individual for the social cost created, for example through a tax on the polluting activity. This method of justifying government intervention consistent with market rules raises a daunting question: to what degree can the private marginal cost be compared to the social one? Such a comparison would only be conceptually acceptable if the costs could be measured objectively (Buchanan, 1969).

In the opinion of L. Robbins, the narrow interpretation of the market confronted with an externality has no scientific foundation. In 1933, in his famous essay about the nature and the meaning of economics, Robbins criticized A.C. Pigou for formulating concrete policy recommendations. The economic policy recommended by Pigou is based on welfare theory, that is, on the assumption of a decreasing marginal utility function. Indeed, for Pigou (1918), as each monetary unit of supplementary income is worth less for a rich person than for a poor person, transferring income between both is enough for global economic welfare to increase. In more popular terms, this means that a dollar does not have the same value for everybody and that fiscal transfers can correct this inequality.

This conclusion has been considered for many decades as the result of scientific reasoning. It has also been at the root of the theoretical concept of an ecological tax, compatible with the smooth operation of markets facing social costs because of pollution. Thus, it deals with the scientific foundation of values. Yet Robbins' challenge deals precisely with this point. According to him, all values are subjective, thus excluding any interpersonal utility comparison. Since the market faces external costs calling for environmental protection, no scientific response is possible and the debate remains primarily a normative one. We cannot objectively add numbers about individual valuations. According to this view, economists have no scientific criteria by which to evaluate the variation of a social value. If pollution reduces this value and requires a tax to be introduced, it is followed by income transfers which ultimately cannot be evaluated without value judgements. To our knowledge, this argument has never been refuted. No economist belonging to the marginalist school has ever proposed a theory that takes Robbins' objection into account. Accordingly, the whole dream of scientific economic policy recommendations based on theoretical modelling vanishes. Although it is widely acknowledged by now that welfare theory is based on value judgments (Feldmann, 1983), many economists feel comfortable with the microeconomic modelling of the kind proposed in widespread textbooks (e.g. Varian, 1992). According to this kind of modelling, any exchange on competing markets increases global economic welfare, whatever the initial income distribution is, notwithstanding Robbins' warning. They follow Roy Harrod, who, in turn, criticized Robbins, explaining in 1938 that if Robbins were right, this would destroy any applied economics, because any policy favours some people to the detriment of others. If no subjective interpersonal utility comparisons can be made, economists should remain forever silent on the valuation of a society's economic welfare following any decision, be it individual and private or collective and public. According to Harrod, such a conclusion is too radical and therefore cannot be justified (Harrod, 1938). This position led Robbins to recant his own criticism of Pigou's welfare theory (Robbins, 1938); however, like all economists after him, he never gave arguments justifying this retraction. It may be that Robbins followed Harrod with a view to perpetuating economists' influence on the formulation of political recommendations. Yet this withdrawal was motivated by strategic, not scientific arguments. It was a question of not harming a discipline that experienced great difficulty in dissociating itself from a past where it formed a part of moral and social philosophy. This scientific ambition carefully separates positive from normative aspects of economics. In the field of environmental protection, market logic still has to be defended at any price. Among more recent authors, we found one of the sole expressions of this epistemological concern in the writings of James Buchanan (1969). If the debate on the most appropriate means to protect the environment is normative, it cannot be skirted by a logic that purports to be scientific, neutral and free of values. Economists have to be more explicit about the norms and values underpinning their discipline.

## 3. Determination of values

The value of the environment as determined by a market refers to an economic use of it. Yet the value of the environment is also determined outside the market and independently of any economic use. The beauty of a landscape, amazement before scientific and artistic creation and inspiration are but a few examples of the value of the natural environment outside any exchange value. Because the natural environment has an intrinsic value, its worth cannot be determined exclusively by the market. Likewise, failure to subject the natural environment to economic use -- albeit possible on both a technological and a normative plane -- can give the environment a "non-use" value that economists readily associate with an opportunity cost.

This cost is part of every economic choice, and is defined by the fact that any economic decision related to managing scarce productive resources necessarily implies giving something up. Not turning a mountain pasture into a tourist resort is evaluated economically in terms of loss of an alternative use. Yet such a decision also expresses values defined outside the economic field, where the concept of opportunity cost does not apply, and therefore hinges on society's perception of nature. The fact that we have progressively replaced the word "nature" by the word "environment" may be a sign of the current change in this perception (Baudrillard, 1972). An environmental protection policy essentially finds its legitimacy in such a change, itself determined by the social perception of non-market values. This change takes place over a long time, and is codified in laws, regulations, morals and customs which reflect the fact that people are increasingly aware of problems related to the protection of the environment, despite the fact that the intensity of this awareness can vary. Social concern about an undamaged environment can be supplanted by other, temporarily more worrying problems such as unemployment. However, this growing awareness has released a process that influences all the symbols expressing the values to which our society adheres.

Finally, the change in these symbols also influences the behaviour of economic actors. It is therefore illusory to think that changing social perceptions of the natural environment will only change the institutions of a society and leave its perceptions of the economy untouched. Accordingly, we cannot observe constant and inalterable economic behaviour, but rather a permanent interaction between social values and human behaviour. The origin of changing motivations in individual as well as collective behaviour thus becomes crucial to the building of an operational environmental protection policy. Here, mainstream economic theory sees no more than an optimization problem subject to the constraint of satisfying new needs. This problem is addressed through an improved ability to solve environmental problems, which can be expressed, in terms of the standard economic rationale, by a change in the utility function. This time, however, this problem can no longer be treated identically from an individual and a collective point of view, due to the reasons set out above. There is no scientific criterion that enables us to move from an individual to a collective level. Society's economic welfare cannot be defined by simply adding up individual behaviours, then treated as a simple allocative problem. A change in the social utility function does not take into account the factors which induced it. This brings us back to an old question in the history of economic thought: the definition of the "right" price.

This question cannot be solved by simply referring to a price determined by supply and demand on a competitive market. It also implies the normative valuation of social values and costs. If economic theory does not recognize this dilemma, it lays itself open to the criticism of removing its normative reference, and thus of serving a market ideology. Instead of pursuing the difficult path to more scientific accuracy, it becomes the main obstacle to the analysis of current economic practices. After all, a discipline that does not itself admit questioning must remain suspicious, because scientific debate has always progressed through a critical dialectic and continuous questioning of fundamental hypotheses (Bürgenmeier, 1994). Under the pretence of having answered once and for all the question of the "right" price with a market model, mainstream economic theory resists a much-needed renewal. It thus continues to treat the environment as an externality that has to be internalized in monetary terms on the market. It would then become possible to apply the economic optimality calculus required by a cost-benefit analysis to the pollution problem, in which case the environment is considered as a public good revealed by market failure. Tasks not fulfilled by the market are, in practice, handed over to the government; this strengthens the widespread opposition between the market and the State, between individualism and collectivism and between market society and civil society.

The public choice school points out that government failures also exist. Correcting market failures through government is often illusory. This concern has led us to favour a normative approach once again. If market and government failures exist, we might as well trust the market. Before handing over a task to the government, it has to be clearly established that it cannot be achieved by market forces. If it is readily admitted that, from the standpoint of externality theory, the market alone is no safeguard for the protection of the environment and that public intervention is therefore justified, the latter has to be implemented in accordance with market rules.

Our discussion of the fundamental problems raised by interpersonal comparisons and their aggregation into social values and costs shows that the opposition between the market and the state does not make much sense, because the evaluation criteria cannot be the same for the market and the state. Therefore, market failures cannot be compared with government failures either. If we nevertheless try to make this comparison and conclude that markets are superior, we have to use the trick of public choice theory, and credit political actors with the same motivations as economic actors. If economic rationality is taken to be the unique behavioural norm, a comparison can only be made with economic criteria. In this case, nobody is surprised by the conclusion of the market's superiority, and the argument becomes perfectly circular.

From a scientific standpoint, the aforegoing cannot be justified. An operational environmental protection policy is ill served by the simplistic opposition between the market and the state, or between individual competition and collective planning, and would benefit from a process which, in the end, leads to institutional changes in the market as well as to an increased consumer awareness motivating a new behavioral pattern. All values cannot be economic in nature.

## 4. Cost-benefit analysis versus social learning process

The supposition that social costs related to increased awareness of environmental protection can be considered as economic values in terms of growth raises daunting problems of evaluation. As any new economic activity over a certain size is subject by law to an environmental impact assessment in many countries, the consequences on the natural environment are, as a matter of fact, very difficult to determine. Mainstream economic theory recommends addressing this problem with a valuation based on a cost-benefit analysis reflecting the optimization concept. A large-scale investment project, such as drilling a new tunnel through the Alps, may increase economic growth and provide a means of reducing pollution, but may also add to existing environmental costs. Accordingly, the valuation of a new investment project's environmental consequences must meet many criteria, where the question is to evaluate not only new costs, but also new expected benefits.

In practice, however, it can be observed that cost-benefit analysis, despite its theoretical foundations, is rarely applied. The research departments in charge of studying the environmental impact of some economic activity mainly rely on engineering expertise. Even if, due to the complexity of the criteria used, impact assessments are made by teams which also include social scientists, most analyses use a multi-criteria approach, where the weighting of the different explanatory factors becomes essential. Engineers apparently prefer the weighted sum approach, which does not differentiate between explanatory factors on a qualitative level. Thus, if one variable is considered less important, the low weighting assigned to it cannot do justice to cases where this variable -- for example above a certain emission level and sometimes only in relation to another variable -- can, during the valuation procedure, take on a totally different meaning. Thus, considering new costs for the protection of the environment therefore requires analytical methods more sophisticated than those of cost-benefit analysis or the weighted sum.

Although the relevant literature offers many other methods for weighting the variables entered in an evaluation procedure, it should be stressed that none of them attaches sufficient importance to the fact that an impact assessment is a process that also includes a learning aspect. This element not only makes it likely that the value function of the criteria will be changed, but also increases the information of the actors concerned by the project and their knowledge of possible interactions between different variables. Consequently, we encounter a problem that has become familiar by now, namely the estimation of the social values and costs submitted to normative judgments, which varies over time according to the changing social perception of environmental problems. In other words, taking new environmental protection costs into account is a normative process, dependent on a politico-legal context which is also supposed to reflect these changes. This conclusion thus rules out any allegedly objective valuation method.

In order to take these objections into account, Jean Simos (1990) proposes an original approach that combines a multi-criteria analysis with a negotiation procedure that includes a cognitive process for the actors concerned by the project evaluated. Therefore, not only does the actors' information improve during the negotiation, but the learning function is clearly taken into account. The normative dimension of any impact assessment then becomes part of the analysis. Important economic actors such as private firms and special interest groups have long stressed that any assessment must be combined with a negotiation process. They expect increased transparency in applying valuation criteria, as well as a learning effect for all the partners concerned by a project, in order to improve the instruments that are to be introduced with a view to enhancing environmental protection.

Thus, the future of impact assessments lies in going beyond approaches inspired by economic theory. Cost-benefit analysis is unable to dispel the ambiguity field to the economic treatment of social costs which could be considered beneficial in terms of growth. It cannot solve the conceptual problem of defining social costs as the aggregation of individual costs. Consequently, the definition of new costs will require increasingly complex analytical techniques that not only refer to socio-economic aspects such as the changing institutional frame provided by legislation but also give more importance to the actors concerned. Such a trend is clearly interdisciplinary (Brown, 1994). It explicitly takes into account the fact that, ultimately, the positive aspect of an analysis cannot be separated from its normative aspect. On the contrary, these two aspects interact directly with each other and cannot be separated, for analytical needs, as economic theory tries to do.

Taking a normative dimension into account in any valuation of new environmental costs is also justified from the point of view of the economic theory of welfare, which shows income distribution to have a normative content. Yet internalizing social costs into the market necessarily implies a redistribution which, when converted into monetary units, generates income transfers between different actors on the market. Hence, cost distribution is subject to normative judgments, which concern not only the distribution between actors of the same generation but also distribution between generations. Accordingly, consideration of new costs is subject to the collective perception of social justice. We saw that the growing awareness of environmental problems also depends on the evolution of this perception. The importance of social costs is therefore valued through a normative interactive process.

This conclusion applies to concrete markets. The weight of new environmental costs for the collectivity and their distribution among different actors depend on the structure of these markets. Among the elements that define these structures, we find not only price elasticities of supply and demand linked to the internalization of external costs but also imperfect competition situations. Therefore, the environmental costs expressed in monetary units are more or less borne by producers or by consumers according to these prevailing conditions.

The resulting prices thus reflect the market structure and value judgments with respect to distributive aspects.

If new environmental costs are taken into account on the markets, for example by means of a fuel tax, the expected result is usually an increase in fuel prices. Yet it is not certain whether this increase is fair. As any price increase has an exclusion effect dependent on income and wealth, this increase may reduce demand from low income earners while probably having very little impact on high income earners. Insofar as market structures are strengthened by coalitions, fixing the price influences the cost distribution between not only producers and consumers but also generations. A standard economic analysis requires separating the positive and the normative aspect and necessitates a strict separation between economic policy -- in our example, it would take the form of a fuel tax -- and social policy. An environmental protection policy should be accompanied by a distinct social policy with corrective income transfers, for example in the form of a subsidy or direct payments for low incomes. Such a strategy again follows the illusion that the internalization of social costs can be objectively subjected to markets.

## 5. Ethical aspects

When the market integrates social values and costs by converting them into monetary units, the problem of non-market environment valuation no longer arises. If this problem is also recalled in moral terms, we necessarily have a dilemma, because non-market environment valuation cannot be exclusively determined by economic criteria. The strategy of profit maximization by producers and utility maximization by consumers may lead to an economic optimum, but if the environment has to have a value per se, determined outside the market and following non economic criteria, a normative space has to be preserved for other strategies. Yet the more competitive the markets, the better the economic efficiency criterion is served and the more profits are constrained; at the same time, this approach reduces the margin for following non-economic motivations. Therefore, a policy inspired by ethical criteria which are intrinsic to environmental values is easier with imperfect competition, whereby economic efficiency is reduced but profit margins are higher. Imperfect competition situations are typical cases of market failures. The possibilities of protecting the environment on ethical grounds increase with the number of market failures. The dilemma is to locate the estimation of social values and costs associated with environmental protection between these two extreme limits, without being able to rely on a common criterion allowing a comparison between economic and non-economic valuation. Either markets are very competitive and high efficiency becomes the criterion guiding an exclusively economic valuation of the environment, or markets are less efficient, allowing higher profits and an ethical valuation of the environment.

It has to be proved in concrete situations that it may be optimal, from a certain perspective, not to respect the economic principles of optimal allocation. For example, if a business gives up the production of a good that, given market prices and resources available for production, would be efficient, the economic cost of this decision should be compared with its social utility, which means that we would be comparing apples and pears. As no means of comparing individual and social costs exists, a business that had acted on ethical grounds would incur individual costs that would have to be related to the social estimate of positive environmental consequences, without the help of a unique comparison criterion. Consequently, ethical action is not subject to economic valuation following objective criteria. In such a case, the economic sphere has to manage a fundamental conflict stemming from a reduction of its activity on ethical grounds.

However, it can be argued that the techniques which subject the environment to market forces are governed by an ethical conception that belongs to the economic sphere and is not distinct from it (Novak, 1987). As a decentralized collective decision mechanism which best preserves participants' freedom of choice, the market expresses ethic on behalf of the community if, through competition, markets also ensure an optimal allocation of resources. Only in case of failure would the market diverge from this objective. Again, market failure in this case justifies public intervention on ethical grounds.

Of course, it is difficult to understand in practice the link between market failures and ethics. This difficulty comes from the behavioural hypotheses underlying the economic model. The rationality credited to economic actors is, indeed, a simplifying assumption that makes modelling more forceful, but it also stems from the philosophical current of utilitarianism. The question here is not how relevant is this current; it is enough to note that, despite the fact that society has codified utilitarianism in different ways in order to justify its objectives, it has never been accepted as a unique reference. Government action to protect the environment cannot be exclusively evaluated from a utilitarian perspective: it is also motivated by other references, such as Rousseaus' social contract or the Aristotelian practice of happiness (Bien, 1985).

## 6. Conclusion

The debate about values, the internalization of social costs and the explicit recognition of ethical aspects confront economic theory with conceptual issues as soon as it tries to apply modelling to the physical and biological environment. New approaches are being brought to bear on this theoretical debate. Today, we can no longer refer to a single, exclusive economic approach. Many research strategies are being followed. We can basically distinguish four categories belonging to two groups of models.

The first group consists of conventional models which take the environment into account, either through public goods theory or through property rights theory. These two theories cover the main issues in environmental economics. The second group refers to more global models, which try to feature relationships between economics and the biosphere. In the same way, models which add a social dimension in order to come closer to the concept of sustainable development (where these three levels: economic, environmental, social are included) also belong to this group, and cover the field of ecological economics. The first difference between environmental economics and ecological economics is one of methodology. Whereas environmental economics mainly refers to a mechanistic vision of the working of our societies and treats social institutions as an exogenous factor in its models, ecological economics refers to a more evolutionary perception of our societies and treats their institutional changes as important endogenous variables in its theoretical representations.

This methodological difference has an important consequence for the behavioural hypothesis underlying theoretical representations. Environmental economics explicitly refers to economic rationality, crediting social actors with only one type of behaviour, that is, maximizing either profits or utility, under the constraint of production costs or income respectively. On the contrary, ecological economics reminds us that the market, already in its temporal dimension, is short-sighted with regard to the ethical responsibility of individuals of the same generation and especially of one generation to another. Therefore, the crucial point is the integration of the ethical dimension in the traditional theoretical representations of the economy that society has adopted (Bürgenmeier, 1993). The trend is to delegate ethical questions to the human sciences and thereby to "clean" the economic model, which then constitutes the main reference. Ethics, as well as the environment, are then treated as an externality to the market. This distinction makes it possible to work on a "scientific" level and root criteria in a positive basis. Theology and moral philosophy are assigned the task of maintaining the social representation of the workings of the economy as intact as possible, treating these aspects outside the economic sphere. The intellectual challenge lies not only in preventing such a development (which is already well under way if one is to judge by the many institutes specialized in economical ethics), but also in integrating ethical considerations in economic reasoning itself. The assumption of behaviour guided by a single motivation cannot be maintained, and has to be replaced by social representations which view economics with its entire ethical dimension. This extension inevitably exposes economics to a dilemma.

If, on competitive markets, actors act rationally, they contribute to the best possible resource allocation. Their activities follow an ethical logic that also seems to help preserve the environment. However, perfect competition rules any action on non-economic grounds. On the contrary, in situations of imperfect competition, some room is made for other objectives, but economic efficiency is less. Yet such situations are put in the same category as market failures that become a prerequisite for the pursuit of non-economic objectives. For example, the possibility of taking the ethical dimensions of intergenerational responsibility into account is dependent on market failures. This conclusion reveals the ambiguity inherent in an exclusively economic treatment of environmental problems. The following two modes of reasoning cannot be easily reconciled.

Economic reasoning tries to reduce pollution at the lowest possible cost. It proposes an optimization calculus that equalizes the marginal cost with the marginal benefit of any economic activity. If many actors on a market are responsible for a given polluting emission, it recommends a policy reducing these where the marginal costs of undertaking such a reduction are smallest. Following the efficiency criterion, economic reasoning therefore allows for a differentiated treatment of actors on a market according to their cost or utility functions.

On the contrary, moral reasoning, often legally codified, is concerned with equality of treatment, which originates in a normative judgment (for example, through the collective memory of a community). This normative judgment may also come out of a political process. If this process is democratic, it is supposed to express the opinion of a majority. Once economic theory recognizes that there may be as many normative judgments as actors on the markets (and as many individual actions), legal reasoning only refers to collective preferences. As has been showed, an economic theory cannot be derived from individual preferences (Arrow, 1977), and we have no common criteria to evaluate these two logics.

However, it is quite possible for society to choose economic behaviour as the norm. In this case, efficiency belongs to the moral dimension of collective decisions and the conflict between legal and economic reasoning disappears. However, it is open to question in this connection whether society has solved the problem of determining non-market values of the environment.

## References

**Arrow, K. J.** (1977) The Organization of Economic Activity Issues Pertinent to the Choice of Market Versus Non-market Allocation. *In:* Public Expenditure and Policy Analysis, R.H. Haveman and J. Mangolis (eds), Rand MacNally College Publishing Company, Chicago.

**Baudrillard, J.** (1972) Pour une critique de l'économie politique du signe. Editions Gallimard, Paris.

**Baumol, W. J. and Oates, W. E.** (1988) The Theory of Environmental Policy. Cambridge University Press, Cambridge.

**Bien, G.** (1985) Das Glück, die "erste Freude" und die menschliche Natur. *In*: Glück und geglücktes Leben, Engelhardt P. (ed.), Matthias-Grünewald-Verlag, Mainz.

**Brown, G.** (1994) Estimating Non-use Values Requires Interdisciplinary Research. *In:* Economy, Environment and Technology, A Socio-Economic Approach, B. Bürgenmeier (ed.), M.E. Sharpe Inc., Armonk, New York.

**Buchanan, J.** (1969) Cost and Choice, An Inquiry in Economic Theory. Chicago University Press, Chicago.

**Bürgenmeier, B.** (1993) Ethical Aspects of Environmental Protection. *In:* Environmental Contamination. J.P. Vernet (ed.), Studies in Environmental Sciences, 1 : 55, Elsevier, Amsterdam, London, New York, Tokyo.

**Bürgenmeier, B.** (1994) Economy, Environment and Technology : A Socio-Economic Approach. M.E. Sharpe Inc., Armonk, New York.

**Feldmann, A. M.** (1983) Welfare Economics and Social Choice Theory. Kluwer Academic Publisher, Boston, London, Amsterdam.

**Harrod, R. F.** (1938) Scope and Method of Economics. *Economic Journal, XLVIII.*

**Hettlage, R.** (1993) Ökonomie auf dem Weg zur Sozialwissenschaft ? *Revue Suisse de Sociologie,* 19 (1).

**Novak, M.** (1987) Une éthique économique, les valeurs de l'économie de marché. Institut La Boétie, Cerf, Paris.

**Pigou, A. C.** (1918) Economics of Welfare. London.

**Robbins, L.** (1933) An Essay on the Nature and Significance of Economic Sciences (2nd edition). St-Martin, New York.

**Robbins, L.** (1938) Interpersonal Comparaisons of Utility: A Comment. *Economic Journal,* XLVIII.

**Simos, J.** (1990) Evaluer l'impact sur l'environnement. Presses polytechniques et universitaires romandes, Lausanne.

**Varian, H. R.** (1992) Microeconomic Analysis (3rd edition). W.W. Norton & Co., New York, London.

# *7. Direct Regulation and Economic Incentives: Opposition or Complementarity?*

Environmental Policy Between Regulation and Market
ed. by C. Jeanrenaud


# 7.1 Direct Regulation and Economic Instruments: Antagonists or Allies?

**Hans Vos**
**DHV Environment and Infrastructure, Amersfoort, Holland**

## 1. Application of environmental policy instruments in practice

*Instruments of environmental policy*

Environmental policy programmes frequently consist of mixes of policy instruments. Such mixes contain individual measures in varying compositions. Before dealing with such instrument mixes, individual instruments are briefly described.

Environmental policy-makers have at their disposal three types of action aimed at modifying behaviour by target groups (Opschoor *et al*, 1994):

1) actions aimed at alteration of available options;
2) actions aimed at alteration of costs and benefits relevant to target groups;
3) actions aimed at alteration of the priorities and significance target groups attach to environmental quality.

Examples of the first apporach are environmental permits, standards, product bans, and other restrictions. This "command-and-control" policy is the traditional policy type, most often applied in practice. Environmental permits regulate the environmental behaviour of firms, either by prescribing technology or by setting limits to pollution. Standards regulate environmental characteristics of production equipment or of products. Bans rule out the use of certain products or substances.

Examples of the second category of environmental policy are typically economic instruments. Economic instruments include effluent charges, product charges (among which tax differentiation), financial assistance, market creation, deposit-refund systems, and enforcement incentives.

Effluent charges are charges to be paid on discharges into the environment and are based in principle on (a proxy of) the quantity and/or quality of discharged pollutants. Emission charges may take the form of user charges; these are payments for the costs of collective or public treatment of effluents.

Product charges/taxes are charges or taxes on products that are polluting in the manufacturing or consumption phase or for which a disposal system has been organized. Product charges can be based on some product characteristic (e.g. charge on sulphur content in mineral oil) or on the product itself (mineral oil charge).

A special form of product charges is that of tax differentiation leading to more favourable prices for "environmentally friendly" products and vice versa.

The next type of economic instruments concerns various forms of financial assistance. This type includes grants -- non-repayable forms of financial assistance -- soft loans (loans with interest rates below market prices -- and tax allowances, including tax exemptions, rebates and accelerated depreciation.

In deposit-refund systems, a surcharge is added to the price of potentially polluting products. When pollution is avoided by returning these products or its residuals to a collection system, this surcharge is refund.

Markets can be created where actors might buy "rights" for actual or potential pollution or where they can sell their "pollution rights" or their process residuals. The well-known type is emission trading, where dischargers operate under a single or multi-source emission limit and trade is allowed in permits adding up to that limit. A second type is market intervention, price intervention in case market prices fall below certain levels, or *ex ante* price guarantees that might create or facilitate the continued existence of a market. Finally, liability of polluters for environmental damage or clean-up costs could also be counted as an economic instrument, in so far as it leads to the creation of a market in which risks for damage penalties are transferred to insurance companies.

Enforcement incentives are aimed at 'punishing' noncompliance, either *ex ante* (by requiring a payment returnable upon compliance) or *ex post* (by levying a fine when noncompliance is occurring).

The third approach to modify behaviour by polluters concerns so-called "suasive" instruments and includes communicative tools like information (also aimed at social pressure), education, training, etc. The purpose is to change the attitude of target groups towards environmental quality, which is a necessary, though not always sufficient, condition for changing behaviour. This category of instruments is included for sake of completeness. It is not an element of the analysis of sections 1 and 2 that is restricted to direct regulation and economic instruments, but it briefly returns in section 3 in the form of covenants, which increasingly are an important part of Dutch environmental policy, including acid deposition policy.

*Conceivable mixes of policy instruments*

Much of the economic literature on environmental policy tools is devoted to analysing the values of economic incentives as alternatives to direct regulation. Many authors have dealt with practical aspects of choice and design of environmental policy instruments (e.g. Stavins, 1990, Tietenberg, 1992, Government of Canada, 1992). A general approach to examine such aspects has been proposed in Opschoor *et al* (1994). This study focused on economic instruments and took a comprehensive approach, covering, among other things, circumstances in which specific economic instruments could be of use.

Few contributions however deal with combinations of instruments of different categories, that is combinations of e.g. charges and permits. Skou Andersen (1994) assessed the role of charges as a component of water quality policy in four countries. De Savornin Lohman (1994) analysed the difficulties of applying economic instruments in a policy environment dominated by direct regulation. In an OECD report (1994), mixed approaches for tackling global environmental problems are discussed.

An in-depth analysis of mixed approaches is outside the scope of this paper. As to what one could expect in terms of mixes of environmental policy instruments, the following provides some general notions. If limited to broad categories of policy instruments, the following combinations could be distinguished (Table 1):

Table 1. Conceivable mixes of environmental policy instruments.

| | Permits | Standards | Bans |
|---|---|---|---|
| Charges | + | + | + |
| Subsidies | + | + | + |
| Tradeable Emission Permits | + | + | - |
| Deposit-refund Systems | - | + | + |
| Enforcement Incentives | + | + | + |

+ = combination conceivable
- = combination not conceivable

Economic theory suggests that correctly measured pollution charges have a better allocative and dynamic efficiency than pollution permits, which are uniformly issued across polluters. Accordingly, a mix of permits and charges is less obvious, although practical circumstances will differ from the situation assumed in theory, rendering combinations of such instruments quite feasible (Barde, 1992). Issuing uniform permits is hardly, if at all, practised, nor are environmental damage functions generally taken into account in establishing allowable pollution levels. The more permits are made-to-measure (taking into account general directives), the smaller the allocative efficiency advantage of charges over permits will be. Consequently, there is no reason not to contemplate certain combinations of permits and charges. An example are permit-cum-charge systems in water quality policy in France and the Netherlands (Skou Andersen, 1994), in which charges are levied on permitted residual pollution levels, with the purpose of raising revenues for financing collective wastewater purification.

Imposing standards for applied technology can be seen as an alternative for issuing permits with an equal outcome in terms of allowable pollution levels. A similar reasoning is valid, pointing at adequate combinations of types of instruments in practice. The combination of charges and standards can be found in the German water quality policy (Sprenger *et al*, 1994).

Charges and bans, e.g. for the use of certain toxic substances in production processes, cannot be combined simultaneously: once the product is banned, nothing is left to charge. However, successive application of charges and bans is quite conceivable, and actually found in practice in the case of banning CFCs and halons. Denmark and the US imposed a charge on these chemicals in anticipation of their disappearance.

Provision of subsidies as an addition to direct regulation is common practice in environmental policy. Subsidies may be considered as in conflict with the Polluter Pays Principle, but some exceptions to this rule are widely accepted (OECD, 1986). One main condition for application is that temporary subsidies may be provided in order to avoid economic hardship, if it is guaranteed that trade and competitiveness are not distorted.

The system of tradeable emission permits (TEP) was developed as an additional instrument that could provide some room for polluters in the rather rigid environment of permits and standards in US air quality policy (Liroff, 1986). Combinations of TEPs and permits or standards thus are normal cases. TEPs and bans do not combine; moreover, successive application is not probable, since establishing TEP is a complicated matter.

Combinations of deposit-refund systems (DRS) and pollution permits are not likely. DRS typically constitute an instrument for products, whereas permits are associated with processes and environmental output. Combinations of DRSs on the one hand and of standards and bans on the other hand do exist. The Danish regulation with regard to drink containers is an example. The size and shape of returnable beer bottles is to some extent standardized, and non-returnable bottles are forbidden.

As enforcement incentives are meant to stimulate compliance with various forms of direct regulation (permits, standards and bans), such instruments are logical counterparts of direct regulation, by definition.

## 2. Factors in the choice of instruments

*Direct regulation vs economic instruments*

The combinations of different types of policy instruments, dealt with in section 1, raise the question of the "why" of application of these instruments. What are the necessary conditions for the successful implementation of these instruments? Which factors influence the choice of instruments? And how do economic instruments differ from direct regulation? The answers to these questions can be found by examining the characteristics of both instruments and the context of their application. In general, our aspiration is to establish instruments that are effective, efficient, practicable and enforceable and, consequently, characteristics of instruments and application contexts should match.

Which characteristics determine the difference between direct regulation and economic instruments? Five factors can be discerned:

Table 2. Differences between direct regulations and economic instruments.

| Direct regulation | | Economic Instruments | |
|---|---|---|---|
| * | Direct impact | * | Market-oriented |
| * | Coerced action | * | Voluntary action |
| * | Industry | * | All market parties |
| * | No financial transfer | * | Financial transfer |
| * | Unique impact | * | Continuous incentive |

Instruments of the direct regulative type have a direct bearing on those who are regulated. Compliance is mandatory, and no other courses of action are open. Economic instruments work through the market and leave room for alternatives. Direct regulation is aimed at the production sector, whereas economic instruments have a wider field of application. Contrary to economic instruments, direct regulation implies no financial transfer and provides no continuous incentive for modifying behaviour.

*Context-specific and instrument characteristics*

Which characteristics determine differences between specific application contexts? This question must be answered in two different ways: first, with regard to situations where environmental effectiveness of the chosen instrument is of overriding importance, and second, in situations where effectiveness is one criterion among others.

In the first situation, overriding conditions will restrict policy options. Such conditions may include:

1) harmful environmental (e.g. toxic) impacts;
2) short-term problems;
3) great risks and uncertainties.

In such circumstances, the use of direct regulation (bans, strict standards) is strongly advocated, because we are in a situation where effectiveness of the instrument is by far the most important decision criterion. Enforcement is of decisive importance.

In the other situation, environmental effectiveness is one criterion among others. With regard to the context of application to be examined, two sets of concrete factors then are basic to environmental policy situations in which one must decide on instruments. They relate to 1) sources of pollution, and to 2) polluting substances and their environmental impact. These context-specific characteristics should have a relationship with the decision criteria such as effectiveness, efficiency, practicability, and enforceability. The following list of characteristics may be considered (adapted from WRR, 1992; see also Opschoor *et al*, 1994).

*Characteristics related to sources of pollution:*

- number and variety of sources;
- costs of compliance and variety therein;
- scale of the market and (international) competitiveness;
- availability of alternative processes and products as substitutes;
- potential for technological innovation.

*Characteristics related to polluting substances and environmental impacts:*

- recognizability of pollution/opportunity for monitoring;
- structure of the linkage between emissions and environmental impact (carrying capacity/thresholds).

Measuring these characteristics indicates how instruments would score in terms of evaluation criteria. For example, in a situation with a large number of sources which vary in size and in relative costs of compliance, economic instruments might be more efficient than direct regulation. However, if these sources operate on an international market, substantial, unilaterally implemented, charges would not be acceptable. If sources compete with foreign firms in the domestic market, then emission charges would negatively influence competitiveness, unlike product charges. Generic, rigid charges may not be effective, if alternative, less-polluting processes or products are not available. Gradual strengthening of standards may be advisable. If a strong case for technological innovation exists, charges might create a dynamic impulse for innovation.

If recognizability of pollution is low, e.g. with respect to diffuse sources, charges or TEP may create large administration costs, negatively affecting efficiency. In cases where no linear relationship exists between emissions and environmental impact, e.g. when a threshold exists below which emissions do not create environmental damage, charges and TEP could be a non-efficient solution.

To further illustrate the relevance of different contexts of application, we refer to an example of an implementation-oriented, indicative approach, developed by the Dutch Scientific Council for Government Policy in its report on environmental policy instruments (WRR, 1992). It attempts to translate specific characteristics of application contexts into indications of types of instruments to be applied.

This approach is based on the assumption that instruments should be tuned to general characteristics such as those described above, which are inherent in any context of application, and which can be subdivided into three groups:

1) characteristics related to the recognizability of emissions (possibility of monitoring) and of environmental impacts;

2) characteristics related to the structure of the target group; market structure and competitiveness are relevant issues in this respect;
3) characteristics determining the resistance in the target group against the instrument investigated; among these, costs of compliance and characteristics of demand (e.g. elasticities) play a major role.

In order to illustrate its approach, WRR distinguishes and characterizes a number of distinct contexts of application by varying the values of the main characteristics as discerned above. As indicators for the groups of characteristics they use: (1) monitoring of emissions, (2) number of sources, and (3) relative costs and cost differences of compliance. These indicators are all source-related. Table 3 presents an overview of these application contexts, which have been given names for characterizing.

Table 3. Seven distinct application contexts and the possible choice of instruments.

| Context of application | Monitoring of emissions | Number of sources | Costs of compliance | Suggested policy instruments |
|---|---|---|---|---|
| "Clear" | Good | Low | Low | DR |
| "Distribution" | Good | Low | High | EI |
| "Manageable" | Good | High | Low | EI/SI |
| "Heterogeneous" | Good | High | High | EI |
| "Invisible" | Poor | Low | High | DR |
| "Diffuse" | Poor | High | Low | SI |
| "Complicated" | Poor | High | High | ? |

DR=Direct regulation; EI=Economic instruments; SI=Suasive instruments

Source: adapted from WRR (1992).

Illustrations of some of these application contexts are given and suggestions for possible policy instruments are made below. These examples are derived from WRR (1992).

*Emissions of acidifying substances by power generation plants* can be characterized as a "distribution" problem application context, in which a small number of sources emit pollutants which can be relatively easily monitored. Costs of abatement are high, and may vary considerably among (old and new) sources. Here, economic instruments, notably emission charges, could be an appropriate instrument. Common standards are less efficient; implementation of emission charges does not create large administrative burdens.

A reverse situation occurs with regard to the problem of *litter in natural areas* ("diffuse" application context), i.e. large number of sources, difficult to monitor, for which costs of compliance are low. Here, suasive instruments could be advisable, because other types of instruments would not work properly due to monitoring problems and inaccessibility of sources. Moreover, if costs of compliance are low, "soft" instruments do have a chance.

The case of *reducing CFC emissions* could be described as a "manageable" context of application, in which a high number of sources would probably face relatively low costs of compliance. However, if it is assumed that production of CFCs could act as a proxy for emissions, monitoring pollution at a low number of sources (producers) would be relatively easy. Then, both a product charge and suasive instruments, or a combination thereof, could be envisaged. Other factors like urgency of compliance, however, may render non-coercive instruments unreliable.

*$CO_2$ emissions by residential heating* offer a similar case, but for the fact that costs of compliance with reduction targets are relatively high. This is indicated as a "heterogeneous" context of application. A product charge on energy could be envisaged, possibly in combination with suasive instruments. The latter would probably be ineffective, if applied alone.

*Direct regulation* and *economic instruments*

The subject of this chapter relates to the question whether or not different types of instruments, notably direct regulation and economic incentives, could be combined. The approach described in section 2 (context-specific and instrument characteristics) attempts to shed light on circumstances in which single instruments can be effective, efficient and practicable tools of environmental policy. Such a contextual analysis, however practical, remains sketchy, and ignores the incremental approach often found in day-to-day life of policy design.

In practice, the choice of instruments is influenced not only by criteria of effectiveness and efficiency but also by other considerations. Instrument choice is subject to the turmoil in the political arena, where many interests struggle for recognition. Instrument choice may imply a process lasting several years and involving different institutions (government levels, administrative bodies). Instrument choice is also influenced by instruments that are already in application. Combinations of various types of instruments ("mixed" applications) can therefore be advantageous for a number of reasons. First, adding an economic instrument can provide an additional dynamic incentive for compliance over and above compulsory, but often static direct regulations. Second, economic instruments may precede adjustments of existing or new direct regulations. It takes time to define and implement direct regulation, in particular where it concerns legislation on the international level. Third, combinations of different types of economic instruments might create a forceful instrument, e.g. a combination of a deposit-refund system and a product charge on non-returnable products. Finally, keeping direct regulation or economic instruments as alternative options may help stimulate negotiations aimed at achieving voluntary agreements with industry for changing environmental behaviour.

Thus, the question of the title of this chapter is not simply answered in one unique way. On the contrary, the best answer could be that direct regulation and economic instruments are either antagonists or allies, dependent on the specific contexts of their application. More research of complex application contexts could further clarify the linkages between (combinations of) policy instruments and of specific policy situations.

In order to illustrate the question of combination of different policy instruments, section 3 describes the case of tradeable emission permits (TEPs) as discussed, as an additional instrument in Dutch acid deposition policy.

## 3. A new instrument for acid rain policy?

*Dutch acid deposition problem: contributors and policy targets*

A major target of Dutch acid deposition policy is to reduce to acceptable levels the emissions of acidifying substances from all relevant domestic sources. Acceptable levels were defined on the basis of acceptable deposition, taking transboundary pollution into account. Table 4 provides an overview of the main sources of acid emissions, emission levels in 1990, and of policy targets in 2000 and 2010. Ammonia, which is an important acidifying pollutant on the local level, is not covered.

Table 4. Emission levels and policy targets of $SO_2$ and $NO_x$, by source.

| (1,000 tons) | Emission 1990 | | Target 2000 | | Target 2010 | |
|---|---|---|---|---|---|---|
| | $SO_2$ | $NO_x$ | $SO_2$ | $NO_x$ | $SO_2$ | $NO_x$ |
| Industry | 50 | 91 | 15 | 28 | 7 | 7 |
| Oil refineries | 66 | 20 | 36 | 9 | 18 | 3 |
| Power plants | 45 | 72 | 18 | 30 | 18 | 16 |
| | 161 | 183 | 69 | 67 | 43 (-38%) | 26 (-62%) |
| Transport | 33 | 315 | | | | |
| Other | 13 | 77 | | | | |
| Total | 207 | 575 | | | | |

Source: RIVM, Nationale Milieuverkenningen III, 1993.

The table shows that as far as stationary sources are concerned, impressive reductions of $NO_x$ and $SO_2$ are expected in the period of 1990 to 2000, although much remains to be done beyond 2000. Emissions of $SO_2$ need to be reduced a further 38%, and emissions of $NO_x$ as much as 62%. Compared with the situation at the time of the start of acid deposition policy, reductions of 80-90% have been achieved. For the final ten or so per cent, marginal abatement costs are likely to be fairly high. This calls for a cost-effective set of policy instruments.

*Current policy instruments*

Present acid deposition policy is dominated by instruments of the direct-regulation type. Any new industrial firms must apply for an environmental licence and any modification of existing activities also requires a licence. Relevant regulations were recently reformulated in the Environmental Management Act, which came into force in 1992. Key elements in this law include the following:

- Requirements regarding emission characteristics of industrial activities must be based on the ALARA-principle: environmental impact should be As Low As Reasonably Achievable. Since the adoption of this basic principle, much discussion has been devoted to the question of how to interpret "reasonably".
- Issuing licences is the competence of regional (provinces) or local (municipalities) government. Provinces provide for licences for about the 450 largest firms in the country.
- There is a noticeable tendency to replace technical requirements in licences by emission target requirements, in order to provide more freedom in choosing the optimal method of pollution abatement for applying firms.
- Two major pieces of legislation serve as a background for licensing authorities. First, the Dutch Emission Directive which lays down general rules for larger plants. Second, the Decree on Emission Requirements for Combustion Plants that is directly applicable. This decree is the Dutch implementation of the EU Large Combustion Plant Directive, but is stricter.
- Dutch environmental policy is characterized by a sector-by-sector approach, which is also reflected in the recent policy approach of voluntary agreements (covenants), concluded between the Dutch government and a large number of industrial branches.

Such covenants are becoming an increasingly important pillar of environmental policy, and of acid deposition policy too. Covenants exist for major sectors, such as power plants, the metal industry and the chemical sector. The characteristics of these covenants include the following:

- Target groups express commitment of achieving the emission reduction targets set in the covenants;
- Authorities expect that equipment be installed with reasonable cost-effectiveness;

- Participating firms have the obligation to install an Environmental Management and Audit System (EMAS);
- Less stringent application of the Decree on Emission Requirements for Combustion Plants.

*Direct regulation, convenants and tradeable emission permits*

The question to be answered then is what changes in policy are necessary if TEPs are introduced as an instrument in acid deposition policy. A preliminary remark that must be made is that removing all existing policy instruments is virtually out of the question. Against the background of over 25 years of environmental policy evolution, current instruments such as the integrated environmental licence, as well as the practice of seeking consensus on individual environmental measures, laid down in covenants, are not easily abandoned. TEPs will be an additional instrument, if indeed they are ever introduced.

However, introduction of TEPs should require a number of changes in current policy and policy instruments. Major changes include (Versteege and Vos, 1995):

- TEPs will have to be included in the Environmental Management Act as an instrument of environmental policy.
- The position of the ALARA-principle as a basic principle of environmental policy in this field needs to be reformulated. In the framework of TEPs, it is to a certain extent left to the discretion of parties whether they "abate and save", or "pollute and pay". Strict execution of ALARA considerably limits the firms' manoeuvring room, so that no additional efficiency might be expected from introduction of the TEPs.
- The obligation for renewing the environmental licence in case of extended activities or increased emissions must be eliminated. Firms should have an opportunity to cover additional pollution by buying permits in the market, and should not be obliged to apply for a modified licence.
- The Dutch Emission Directive and the Decree on Emission Requirements for Combustion Plants should remain unchanged during the operation of the TEP system.
- Finally, a general legal regulation is needed for settling the practical details of TEPs.

This should include, among other things:

- central registration of base-line emissions and trading;
- initial allocation of base-line emissions and adjustment of provincial permits;
- a communication plan.

The policy changes needed are not insubstantial. Apart from the lengthy and complicated process of changing a major law like the Environmental Management Act, which, all in all, may take up to five years, adopting tradeable emission permits implies an essential change in policy concept. Current policy is characterized by two major elements. The first element is the continuous integration of environmental policies on the instrument level, culminating in the issuing of one license for all of the firm's environmentally relevant activities. The second element is the sector-by-sector approach in environmental policy, reflected by the successes of the many covenants, concluded on a sectoral basis.

Tradeable emission permits constitute an instrument that is diametrically opposed to the current policy approach. As to the first element, namely integration, TEPs are specifically an instrument directed at one polluting substance, or at best a few, highly comparable pollutants (for example, in the case of acid deposition policy, so-called acid equivalents). Licensing with regard to the pollutant(s) under the TEP system should be frozen. This implies splitting integrated licensing up into different procedures. The second difference concerns the sectoral approach. The cost-effectiveness of TEPs increases with the differences in marginal abatement costs. This presupposes many different actors in the market, preferably from different sectors. This should require a reassessment of covenants concluded on a sectoral basis.

Furthermore, there is the question of competence. Larger industrial firms have to apply for an environmental licence at the provincial level. In case of a national system of tradeable permits, the role of the provinces would inevitably decrease. The goal of an efficient operation of a TEP system is not served by overly close observation, by for example the means of a renewal of licences. Consequently, provinces would lose their grip on the environmental output of economic activities within their jurisdiction, and their role would likely be reduced to "hot spot" warners.

In terms of the central question of this chapter, the main conclusion could be that, in theory, environmental licensing (direct regulation) and tradeable emission permits (economic instrument) could very well be allies. Licences issued on the basis of the Dutch Emission Directive and the Decree on Emission Requirements for Combustion Plants could provide the base-line for a TEP system that further promotes a solution of the acid deposition problem with increased cost-effectiveness. This is a case of a partly simultaneous, partly successive application of direct regulation and economic instruments. However, many practical considerations, mainly of an institutional nature, considerably complicate the situation, and the conclusion could very well be that, in this case, given the context of application, these types of instruments may turn out to oppose each other.

## 4. Conclusions

Observed from a theoretical as well as from a practical point of view, various types of instruments of environmental policy -- both of a market-oriented and of a direct regulatory nature -- could form acceptable combinations. Indeed, practice shows that many such combinations exist. Charges, subsidies and tradeable permits are hardly, if at all, applied in isolation from some form of direct regulation, be it licensing, standards or bans.

Factors determining the choice of specific instruments and combinations are related to the characteristics of such instruments, and to the characteristics of the policy contexts in which they are applied. When the analysis is restricted to direct regulation and economic instruments, five typical differences in character can be discerned. They concern the way these instruments work (directly or via the market, coerced or voluntary), their target groups (industry or all market parties), the involvement of financial transfers, and the dynamics of their impact.

Factors related to the application contexts of instruments can unconditionally point to the use of direct regulatory instruments, for example if toxic impact and considerable short-term risks are involved. Otherwise, specifying a number of pollution and source-related factors could point to specific instruments. Major characteristics include the number and variety and sources at which policy measures should be directed, the costs of compliance with such measures, and the possibility of monitoring emissions.

It further appears that the choice of instruments is very much a matter of practical consideration. Institutional and practical circumstances often result in the existence of mixed approaches, in which direct regulation and economic instruments are actually to be considered as allies. This leads to the conclusion that direct regulation and economic instruments are either antagonists or allies, dependent on the specific contexts of their application.

The question of combining different policy instruments was illustrated with the description of the case discussed, i.e. tradeable emission permits as an additional instrument in Dutch acid deposition policy. Present policy is dominated by direct regulation and covenants. Introduction of tradeable emission permits would no doubt require quite fundamental changes in acid deposition policy. Present, policy is based on integration over polluting substances, whereas TEPs are a single-pollutant instrument. Currently, policy is also organized on a sector-by-sector basis, whereas TEPs should flourish in a much broader context. Although a combination of direct regulation and economic instruments is quite conceivable, many practical considerations complicate a mixed approach.

## References

**Barde, J. Ph.** (1992) Economie et Politique de l'Environnement. Press Universitaire de France, Paris.

**de Savornin Lohman, A.** (1994) Economic Incentives in Environmental Policy: Why are they White Ravens? *In*: Economic Incentives and Environmental Policies: Principles and Practice, Opschoor H. and Turner K., Kluwer Academic Publishers, Dordrecht.

**Government of Canada** (1992) Economic Instruments for Environmental Protection: Discussion Paper, Ottawa.

**Liroff, R. A.** (1986) Reforming Air Pollution Regulation. The Conservation Foundation, Washinghton.

**OECD** (1986) OECD and the Environment. OECD, Paris.

**OECD** (1991) Environmental Policy : How to Apply Economic Instruments? Paris.

**Opschoor, J. B.; de Savornin Lohman, A. and Vos, H. B.** (1994) Managing the Environment : The Role of Economic Instruments. OECD, Paris.

**RIVM** (1993) Nationale Milieuverkenningen III (Third National Environmental Assessment).

**Skou Andersen, M.** (1994) Governance by Green Taxes. Manchester University Press, Manchester.

**Sprenger, R. U. et al** (1994) Das Deutsche Steuer- und Abgabensystem aus Umweltpolitiker Sicht. Ifo-Institut, München.

**Stavins, R. N.** (1990) Innovative Policies for Sustainable Development in the 1990's: Economic Incentives for Environmental Protection. Paper prepared for the UNECE, Washington, D.C.

**Tietenberg T. H.** (1992) Relevant Experiences with Tradeable Entitlements. *In*: Combating Global Warming, UNCTAD, New York

**Versteege, H. J. M. and Vos J. B.** (1995) Verhandelbare Emissierechten in het Nederlandse Verzuringsbeleid (Tradeable Emission Permits in Dutch Acid Deposition Policy). Ministry of the Environment, The Hague.

**Wetenschappelijke Raad voor het Regeringsbeleid (Scientific Council for Government Policy)** (1992) Milieubeleid : Stategie, Instrumenten en Handhaafbaarheid (Environmen-tal Policy: Strategy, Instruments and Enforceability). The Hague.

Environmental Policy Between Regulation and Market
ed. by C. Jeanrenaud


# 7.2 Regulations and Market-based Instruments in Swiss Environmental Policy

**Christian Albrecht**
**Federal Office of Environment, Forests, and Landscape, Bern, Switzerland**

## 1. Introduction

Swiss environmental policy is at a turning point. Up until now, it has depended heavily on direct regulation and to some extent on subsidies. This policy has been relatively successful. However, for some time now the limits of traditional environmental policy have been apparent. The cost of enforcing new regulations is increasing, while the level of public acceptance has been declining because of the sheer number of rules.

A modern environmental policy must be founded mainly on market-based instruments. Accordingly, incentives will play an increasingly important role in Switzerland's environmental policy over the next few years and should make it more efficient. A market-oriented environmental policy will make it possible to bring economic and ecological interests into greater harmony. This new policy dovetails with the concept of sustainable development, the principle adopted by the United Nations Conference on Environment and Development held in Rio de Janeiro in 1992.

Incentive-based instruments rely on the market mechanism to achieve environmental targets and to give polluters a price signal to make them aware of the true cost of their activities for the population at large, thereby enhancing market transparency. Products and substances harmful to the environment are made more expensive by means of incentive taxes. Revenue-raising taxes also make it possible to ensure that pollution control costs are borned by those responsible. Environmental management and the eco-audit are important ways of evaluating environmental protection progress made by firms. With "soft law" instruments such as the eco-label, information is provided to enable environmentally conscious consumers to follow their environmental inclinations.

This paper outlines the traditional environmental policy of direct regulation and the successes which it has achieved. It goes on to explain how we are now moving away from an environmental protection policy based on direct controls to one which uses market incentives. It then describes the economic instruments at present being considered for use in Switzerland.

## 2. The successes of traditional environmental policy

Environmental policy regulations already existed in Switzerland at the beginning of the twentieth century (see Table 1):

- A milestone in the history of Swiss environmental policy was the *Law on Forest* of 1902. Thanks to this law, the surface area of our forests (28.7% of the country's total surface) has been maintained.
- The threat to water resources was recognized at a very early stage. The *Water Protection Laws* of 1955, 1971 and 1991 established a basis for protecting the country's lakes and rivers effectively. As a result, the quality of water has improved steadily.
- The *Law on Hunting and Bird Protection* of 1925, the *Law on the Protection of Nature and the National Heritage* of 1966 and the *Land Planning Law* of 1979 were further steps in Swiss environmental policy.

Table 1. Federal environmental laws in Switzerland (a selection).

| | |
|---|---|
| Factory Laws | 1877 / 1914 |
| The Law on Forests | 1902 |
| Water Protection Laws | 1955 / 1971 / 1991 |
| The Law on the Protection of Nature and the National Heritage | 1966 |
| The Land Planning Law | 1979 |
| The Environmental Protection Law | 1985 |
| Revision of the Environmental Protection Law | 1993-1995 |

In 1971, almost 93% of voters approved a new article on environmental protection in the Federal Constitution (Article 24 SEPTIES). It provided the basis for a comprehensive *Environmental Protection Law*, although this was not enacted by Parliament until 1983. The law, together with its ordinances, regulates the fields of air pollution, noise levels, environmentally hazardous substances, wastes, soil pollutants and protection against major accidents. This wide-ranging environmental protection legislation has left its mark on the environment:

- *Air pollution* levels have dropped substantially for some substances. Measurements made until 1993 showed emissions falling from their highest recorded levels by 57% for $SO_2$, 30% for $NO_x$ and 23% for VOCs.
- *Wastes* are increasingly sorted and disposed of in an environmental friendly manner. From 1989 to 1993, the amount of waste from residential areas dropped from 2.95 million to 2.8 million tons. Return of glass and aluminium cans for recycling increased by an excellent 78% and 80%, respectively, while corresponding figure for paper and cardboard recycling was 55%.
- Various *environmentally hazardous substances* have been restricted or totally banned (total ban on PCPs, prohibition of CFCs in aerosols, upper limits for mercury (Hg) in alkali manganese batteries).

## 3. The limits of traditional environmental policy: from direct controls to incentive-based regulations

Yet much remains to be done to protect the environment:

- The condition of forest resources remains a cause for concern;
- The objectives contained in the Clean Air Ordinance were not reached as required by 1 March, 1994, for some pollutants (particularly $NO_X$ and VOCs);
- The Ordinance on Noise Pollution must be tightened in the next few years, particularly regarding traffic;
- In the field of biotechnology, Switzerland is now one of the few industrialized countries without an environmentally relevant legislative basis;
- Concentrations of carbon dioxide ($CO_2$) and other gases responsible for the greenhouse effect have continued to increase ever since the beginnings of industrialization.

Generally speaking, we are still far from the objectives of an environmentally compatible economy in the fields of production, consumption and waste disposal. For some time now, the limits of traditional environmental policy have been apparent.

- *Effectiveness*: It has not always been possible to enforce regulations and prohibitions everywhere (e.g. VOC emissions, energy consumption).
- *Efficiency*: The cost of enforcing regulations is increasing.
- *Acceptance:* In view of the sheer number of rules now in existence, public acceptance of new regulations is declining as against a strong demand for deregulation.

Consequently, direct controls must be supplemented where appropriate by market-based instruments in an optimal policy mix. Increased use of these tools is central to any modern, environmental policy. In other words, more incentives instead of a steady increase in rules are required. Market-based instruments strengthen the polluter-pays principle, reduce the cost of achieving environmental targets, and make it possible to do without ever more regulations.

The first draft of an environmental protection law produced in 1973 (the Schürmann draft) contained provisions for environmental taxes, but did not make it through the consultation process. In view of the fact that the private sector spoke out strongly against such taxes, they were not included in the Environmental Protection Law of 1983 for fear of overloading it, thus triggering a referendum request and once more postponing the entry into force of environmental protection legislation. Nevertheless, in the last few years Switzerland has been able to gain some initial experience in the use of economic instruments. We have introduced:

- excise differentiation between leaded and unleaded petrol;
- airport landing taxes depending on noise levels;
- emission trading in the half cantons of Basle City and Basle Rural (see Staehelin-Witt in this volume);
- municipal tax on rubbish bags.

## 4. Market-based instruments in Swiss environmental policy

In the last few years, a great deal of work has been done at the federal level on designing economic instruments (see Table 2).

Table. 2. Economic instruments in Swiss environmental policy.

| **Incentive taxes** | **Soft law** |
|---|---|
| * The tax on VOCs | * The eco-audit |
| * The tax on "extra-light" heating oil with a sulphur content of more than 0.1% | * The eco-label |
| * The $CO_2$ tax | * Voluntary agreements |
| **Revenue-raising taxes** | **Other instruments** |
| * Disposal charges payable in advance | * Tradeable permits |
| * The waste deposit tax | * Ecological tax reform |
| * The tax on heavy traffic depending on capacity | |

*Environmental taxes*

Environmental taxes make up the most important economic instrument in Swiss environmental policy. In the general matter of environmental taxes, a distinction must be made on the basis of their ultimate purpose, between incentive taxes and revenue-raising taxes.

*Incentive taxes*

Changing producer and consumer behaviour is the main role of incentive taxes. If a substance represents a threat for the environment, the price increase due to the tax must be high enough to reduce production and consumption to the extent required. The resulting revenue may be seen as an unavoidable side-effect of taxation: compensation must be provided, i.e. the revenue raised must be refunded to the private sector.

With the 1993-1995 revision of the Swiss *Environmental Protection Law,* incentive taxes are to be levied on volatile organic compounds (VOCs) and "extra-light" heating oil with a sulphur content of more than 0.1%.

In mid-June 1995, the National Council followed the Council of States in approving these two specifically targeted incentive taxes. It may be assumed that they will come into force some time during 1997. They will be incorporated in Swiss federal law as the first environmental incentive taxes and will make a decisive contribution to reaching the objectives of the Clean Air Ordinance in the field of VOCs and of further reducing $SO_2$ emissions according to the prevention principle.

For some time now, a $CO_2$ incentive tax has been in preparation within the framework of another law. This should help to reach the Federal Council's objectives in the fields of climate and energy (stabilizing and then reducing $CO_2$ emissions). On 31 May 1995, following the consultation process, the Federal Council authorized the Federal Department of the Interior to work out a law to reduce $CO_2$ emissions by mid-1996 and to provide a legal basis for the $CO_2$ incentive tax as part of this law as an instrument to reduce $CO_2$ emissions. Its step-by-step introduction and its rate would depend on the effectiveness of other measures such as voluntary agreements. A $CO_2$ incentive tax is not likely to be introduced before the year 2000.

Revenue raising-taxes will be used to pay for environmental protection measures such as the treatment of waste water and the disposal of solid wastes. Raising revenue, is its main purpose. Consequently, the rate of these taxes for each specific purpose will depend on the financing requirement in question.

The introduction of revenue-raising taxes is now being prepared within the framework of the changes in environmental protection law. Two measures are in the planning stage: a retail disposal charge that gives a price signal at the point of product purchase to cover the expenses of an environmentally friendly disposal of products (batteries, fluorescent lamps, refrigerators, drinks packages, synthetic substances, electronic equipment, vehicle tyres) and a tax on waste deposits to finance the cleaning-up of disposal sites. Both Houses of Parliament have already approved these revenue-raising taxes.

Another important revenue-raising tax is the tax on heavy vehicles depending on vehicle weight and number of kilometers travelled. On 19 June 1995, the Federal Council released a bill for consultation. The planned tax is intended to ensure that heavy vehicles cover the costs for which they are responsible. It represents a first step in the direction of internalization of external costs in the transport sector. This means that external costs such as those of accidents and noise as well as of damage to buildings resulting from air pollution will be taken into account and internalized (Dienst für Gesamtverkehrsfragen, EVED, GVF-Bericht 3, 1993). The tax serves as a basic measure intended to ensure that future goods traffic through the Alps will be shifted from road to rail as provided for by the new article on protection of the Alps in the Federal Constitution (art. 36 $^{\text{SEXIES}}$ BV).

*The eco-audit, the eco-label, voluntary agreements*

In addition to environmental taxes, market economics and environmental protection must involve ever greater effort and self-responsibility on the part of the private sector. This includes the environmental management and eco-audit system, the eco-label and voluntary agreements. The Swiss Parliament has expressed the opinion that these should play a role as instruments of environmental protection legislation.

*The eco-audit*

The environmental management and eco-audit system is intended to evaluate and improve environmental protection efforts by firms and to provide appropriate information to the public. A number of important companies have already gained experience in environmental management systems in Switzerland and carry out internal audits. In 1993 the EU regulation on voluntary participation by firms in a community environmental management and eco-audit system came into force (European Commission, 1993). Since the middle of 1995, ISO draft texts on the environmental management system have also been available. These should be given final approval in the course of 1996.

*The eco-label*

The *eco-label* is intended to promote methods of development, manufacture, marketing and use of products which reduce negative environmental impact throughout their entire lifespan. Adopting this environmental symbol further enhances market transparency.

*Voluntary agreements*

As with the eco-audit and the eco-label, it is intended that *voluntary agreements* should be included in the field covered by environmental protection legislation and thereby institutionalized. The law to reduce $CO_2$ emissions being drafted by the Federal Department of the Interior for mid-1996 should also rely on voluntary agreements to achieve the $CO_2$ reduction target.

There is no doubt that voluntary agreements will play an important role in the future. They strengthen the cooperation between emitters and environmental authorities and the self-responsibility of households and firms. But for such agreements to be effective the two general conditions and prerequisites must be fulfilled:

- Agreements must be binding and verifiable.
- There is a need to establish working arrangements between the negotiating partners and to ensure that no major party is excluded from the negotiations.

*Other instruments*

Other possible instruments of Swiss environmental policy include tradeable permits and ecological fiscal reform. The process of elaborating these instruments at the federal level is not as far along as that for environmental taxes.

*Tradeable permits*

Tradeable permits limit the total amount of pollutants which may be emitted. In turn, this amount must be compatible with the environmental target which the government has set. Price is determined by the market. Anyone whose activities pollute the environment must hold the necessary number of permits (see Stritt and Nichols in this volume).

At the federal level tradeable permits have not yet been introduced into the portfolio of market-based instruments. They are viewed as less effective than environmental taxes, the other classical instrument. The real problem is implementation. In the case of tradeable permits, emission monitoring at every single source is necessary. The large number of small sources makes this virtually impossible in Switzerland. Notwithstanding, the possibility of using tradeable permits in Switzerland cannot be definitely excluded. Tradeable permits are considered at present by the Federal Office of Energy as a possible strategy to reduce electricity consumption by installations, vehicles, and equipment.

*Ecological tax reform*

Making the tax system more environmental friendly is a question of taxing the pollution of natural resources (energy stands in the forefront), while existing taxes may be removed or reduced so that the total tax level remains unchanged (see Vallender and Jacobs in this volume). The question of such a transformation of the tax system is not likely to be ready for serious political discussion for some considerable time, probably well after the turn of the century. However, the proposed $CO_2$ tax may be seen as a first step in this direction.

## 5. Final remarks

Increased use of market-based instruments form the lynch-pin of any future-oriented environmental policy or policy for sustainable development. It is important to determine the optimal mix of market incentive and direct controls. If this is done, environmental policy will become more in tune with market economics and hence more efficient.

Market-based instruments should be introduced in a targeted manner where environmental policy action is most needed and where their advantages can be best exploited. A large number of incentive taxes at the federal level is undesirable. Accordingly, the Swiss environmental authorities intend to concentrate on just a few areas.

The effective use of economic instruments should make it possible to avoid too many new regulations in future. But new regulations can probably never be foregone altogether, particularly in fields such as risk prevention.

In the final analysis, environmental policy is tied to sustainable development. The increased use of market-based instruments represents an important step in transforming this concept into reality. The process is at present in the making. Provided that the necessary political determination exists, it should be possible to reconcile economic and social factors with the needs of environmental protection.

## References

**European Commission** (1993) Council regulation No. 93/1836 of allowing voluntary participation by companies in the industries sector in a Community ECO-Management and Audit regulation.

**Dienst für Gesamtverkehrsfragen, EVED, GVF-Bericht 3** (1993) Grundlagen zur Kostenarbeit im Verkehr, Bern.

# *8. Conclusion*

Environmental Policy Between Regulation and Market
ed. by C. Jeanrenaud


# What Have We Learned About Market-based Instruments?

**A. Markandya,**
**Harvard Institute for International Development, Harvard University, Cambridge, USA**

## 1. Introduction

*Background*

The use of market-based instruments in environmental policy regulation has a long history on the theoretical side and a short but rapidly growing one on the empirical side. Over 60 years ago, the British economist Pigou (1932) observed that environmental problems could be effectively addressed through a special tax based on the level of pollution generated. Ever since, economists have, to varying degrees, extolled the virtues of instruments such as taxes, subsidies, tradeable permits and others where the polluters have an economic incentive to reduce the level of pollution they are generating. In this paper, I will consider all such policies as market-based instruments (MBIs). In the early 1970s, Baumol and Oates (1972), Beckerman and Markandya (1974), Anderson *et al* (1977) and a number of other environmental economists stressed the advantages of MBIs. At the time, policy-makers did not seem to take notice; they believed that the problems could be addressed adequately using the forms of direct technology-based controls with which they were familiar. It was not until the regulatory authorities became aware of the considerable costs of direct controls, and of the little flexibility that they offered industry in responding to stricter environmental standards, that the attraction of MBIs became more apparent.

In the static model that is presented in all the textbooks on environmental economics, a tax equal to the marginal damage created by the pollutant, at the level of pollution where the marginal damage and the marginal cost of abatement are equal, will guide the industry to the 'optimum' level of pollution at least cost. The position is optimum because there is no other level at which net benefits to society can be increased (referred to as the Pareto optimum). The same position can, in theory, also be reached by giving polluters a subsidy equal to the marginal damage for every unit that they reduce pollution, or by issuing permits to pollute, equal to the optimum level of pollution, and letting the polluters trade these permits among themselves. Thus, one can use the 'carrot' (the subsidy) or the 'stick' (the tax), or the auction house to achieve the optimum position. Moreover, with the MBI in place, there will be an ongoing, dynamic, incentive to reduce pollution, as such an improvement will benefit the 'bottom line' -- i.e. the profitability of the company.

Of course, there are all kinds of problems with using this simplified analysis as the basis of policy. First and foremost, the marginal damages and marginal costs of abatement are not known. If all the information were available there would be no need for an MBI -- one could simply set the pollution levels for the enterprises at these optimum levels. Thus, the simple model has an internal contradiction -- it can only provide a justification for MBIs when they are not necessary. This problem of the omniscient authority recurs in the simulation analysis used to estimate the benefits of MBIs, and I will revert to it later.

Given uncertainty about the marginal damages and marginal costs, economists have analysed the options of charges versus taxes in terms of minimizing the expected losses of welfare, and concluded that charges are preferable if the marginal damages are relatively constant as pollution increases, while permits are preferable if marginal costs are relatively constant (Weitzmann, 1974). Although this analysis should be useful in selecting between tradeable permits and charges, it does not appear to have been so used. By and large, charges have been employed as a mild additional instrument of regulation, while permits have been introduced on an *ad hoc* basis, to address the needs for flexibility in responding to the clean air and water legislation in the US.

Second, there is a fundamental distinction between the tax and the subsidy -- the carrot and the stick. Whereas the tax generates revenue, the subsidy requires revenue from some other source. This, combined with the fact that subsidies raise problems of equity (what levels of pollution should be taken as the benchmark from which the subsidy is calculated?), as well as those of corruption and size of the polluting sector, means that economists have favoured the tax approach (Tietenberg, 1991). An issue that has become fashionable in recent years is that of the double dividend'. The idea is that an environmental charge is beneficial, not only because it addresses the environmental problem, but also because it provides revenue to the government from a (paradoxically) 'clean' source; clean in the sense that there is no loss of welfare associated with its collection. This issue was analysed recently by Bovenberg and de Mooij (1994), Goulder (1994) and others, including the paper by van Regemorter (1995) in this volume. Certainly, there is no demonstration that, because of the double dividend argument, environmental charges should be raised above their Pareto optimum point, starting from a zero base. What has been established is that the costs of a scheme where the revenues from the environmental tax are returned on a lump-sum basis are less than those where the revenues are used to reduce distortionary taxes. This is called the 'weak double dividend argument'. The strong argument contends that there are net benefits from substituting an environmental tax with a distortionary tax, leaving total revenues unchanged. This claim is not generally correct. More information on precisely what taxes are being considered is required before a statement can be made of the net gains/losses (Goulder, 1994). Since it is generally the stronger version that people have in mind, care should be taken when using the double dividend argument.

Third, the static economic model which to this day provides the main economic rationale for such taxes says nothing about the inevitable compromises required in any regulatory framework. 'Pollution' is a rather general term, and the question of what to tax has to be defined more precisely. If it is the level of depositions of pollution, then there is a serious problem of measurement. If it is the level of emissions from the polluter, account has to be taken of the fact that emissions from different sources are not all equally damaging, and therefore should not be taxed at the same rate (the spatial dimension). In some situations, the tax may have to be levied on the product -- input or output -- rather than the pollutant. In this case, what are the losses of welfare from the use of the 'indirect' point of taxation, and how do they compare with the benefits of the tax? Where tradeable permits are employed, how does one deal with the fact that emissions from different sources have different environmental impacts? All these issues take us into the realm of the second best. While one can derive rules for taxes and other instruments in this framework, and such rules have been derived within the theory of optimal taxation (Atkinson and Stiglitz, 1980), much less has been done for MBIs (see, however, Sandmo (1975)). Furthermore, these rules for optimal taxation have not provided much direct practical advice on the setting of tax rates, etc. Consequently, it is necessary to rely on empirical data and on the experience of authorities that have implemented such taxes.

*Structure of this paper*

For the reasons given above, the debate on MBIs is mainly an empirical one, although the theoretical structure has some useful directions to offer. Consequently, I will look at a range of MBIs that have been introduced and ask the following questions for each kind of instrument:

a) How does the actual experience fit into the theoretical framework?
b) What does the 'simulation' evidence tell us about the effectiveness of the MBIs?
c) What conclusions can be drawn about the achievements of the MBI from the 'empirical' evidence?

A distinction is made between the simulation 'evidence' and the empirical evidence because a large part of the evaluation of different MBIs involves comparing the costs of the instrument against an alternative direct control option. This makes a number of assumptions about the costs of abatement and, possibly, the damages, that are not always verified from the actual data. Moreover, some assumption has to be made about what the regulators know when they set the direct controls or fix the MBIs. If they are taken to be omniscient, the result may be very different from that obtained if they are assumed to have only partial knowledge. This issue has been noted and addressed by Tietenberg (1985) and others, but there is no basis for deciding what the policy-makers actually do know.

The range of MBIs is very large; in fact, any policy measure that has a bearing on the choice of technology or on the level of environmental quality could be referred to as an MBI. It would be a hopeless task to try and address them all in this short paper. This being so I have focussed my attention on emissions charges and tradeable permits (and their variants). This excludes product charges, the use of differential taxes, incentives for environmental investments, deposit refund schemes, betterment charges, congestion fees and a myriad of others (for a complete description of such taxes, see Opschoor *et al* (1994) and Kågeson (1993)). Although there is a great deal of interest in the use of product charges for environmental purposes (particularly on energy sources), the justification for not addressing them here is that this paper has as a major focus the links between the theoretical justification for MBIs and the actual instruments. For product charges, these links are even more tenuous than for the MBIs considered.

Finally, it should be noted that the discussion is mainly focussed on the experience of the OECD countries (and to some extent the economies in transition). I do not comment, except in passing, on the experience of developing countries, where MBIs are increasingly being employed (for an evaluation of MBIs in developing countries see Markandya and Shibli (1995), Panayotou (1994a)).

## 2. Emissions charges and subsidies

### *Air pollutants*

The use of emissions charges is still limited in most countries. For air emissions, taxes have been introduced for $CO_2$[1] in the Scandinavian countries (Sweden, Denmark, Finland and Norway) and in the Netherlands. Sweden has charges for $NO_x$ (for large power stations only), HC and sulphur, Norway has charges on sulphur and lead and France has very low charges on

---

1 Some commentators refer to the carbon tax as a product tax because it is levied at the fuel stage. Given the lack of realistic technologies for collecting carbon dioxide after combustion, this is equivalent to an emissions tax. Where the tax is partly an energy tax, however, the situation is different.

HCL, $NO_x$, $SO_2$ and $H_2S$ (about one per cent of those in Sweden). Switzerland is considering the introduction of a charge on VOCs and $CO_2$ (Jeanrenaud and Stritt, 1994). Rates even within this small group of countries vary widely (OECD, 1995).

Several countries in Eastern Europe have pollution emissions charges, including the Baltic States, Czech Republic, Slovak Republic, Poland and Russia. The levels range from the extremely low (Baltic States) to those higher than, but of the same order of magnitude as, France (Poland, Czech and Slovak Republics ). The countries of the former Soviet Union are in the process of reforming their taxes, but the present structure is extremely complex and the number of pollutants taxed is very large (Markandya and Lehoczki, 1994; Klarer, 1994).

*Relation to the theory*

These taxes can claim only a distant relationship to the theoretical structure, although the Swedish taxes are perhaps the closest to the Pigovian framework. First, in no case do the taxes claim to be related to marginal damages. In fact, no mention is made of damages in relation to the taxes, and therefore the question of equating marginal costs to a given marginal damage is redundant.[2] Second, in most cases the taxes do not have a stated goal in terms of environmental quality. Only in the case of Sweden is there a statement that the tax is set at the marginal cost of achieving a 30 per cent reduction in $NO_x$ by 1995. Rather, the objective seems to be to collect some revenue, part or all of which is then used to assist the polluters to reduce their emissions. Third, in all cases there are other regulations on the enterprises that have a direct bearing on the emissions of the polluters (choice of technology, choice of fuel, maximum permitted emissions). This renders the system a hybrid that is difficult to evaluate except on a detailed case-by-case basis.

In the theoretical discussion about the choice between environmental taxes and tradeable permits, it was noted that if the marginal cost function is relatively steep, and the marginal damage function relatively flat, taxes are preferred to permits on uncertainty grounds. In fact, evidence such as there is suggests that this is the case - marginal damages are relatively

---

2 In the Eastern European countries, emissions taxes vary by location and are increased if emissions go above the maximum permitted concentrations. Thus, they attempt to emulate the marginal damage, without, of course, knowing what that damage is.

constant and marginal costs vary widely across sectors (EC, 1995; Jeanrenaud and Stritt, 1994). At the same time, the trend in many cases is to go for tradeable permits. There are other, non-economic, reasons for this, although none have been well articulated. It is worth noting, however, that the theoretical framework has not been used in this regard.

*Simulation evidence*

The simulation evidence on these taxes is very limited. Tietenberg and others have shown, in a strictly simulation context, what the potential benefits of a Pigovian tax could be over a direct regulation regime (Tietenberg, 1985), but the actual situation is not like this. The taxes in place have not been compared to a realistic direct control option, except for the carbon taxes. For $CO_2$, studies have been carried out to show the benefits of a tax solution at the national and international level (for a summary see Pearce, 1991). The required tax levels will depend on the target reductions and, among the countries that have imposed $CO_2$ taxes, Sweden has a rate of ECU 36/ton of $CO_2$, Denmark ECU 13/ton, Finland ECU 1/ton, Norway ECU 1-4/ton and Netherlands ECU 2/ton.[3] For the EU, Barrett (1992) has estimated that a tax of ECU 20/ton is needed to achieve the goal of freezing emissions at 1990 levels by 2000. Except for the case of Sweden, therefore, the imposed taxes would not by themselves lead to the desired solution. Even in the case of Sweden, it is estimated that this will not be sufficient to achieve the desired goal (Naturvårdsverket, 1992), mainly because Sweden had already reduced emissions by 40 per cent prior to 1990, leaving the country with a much higher marginal cost.

The carbon tax has been subject to extensive analysis of its macroeconomic impacts as well as its costs (EC, 1991; Shackleton, 1992) and income distribution (Poterba, 1991). The analysis is generally favourable for this tax. The same cannot be said categorically for the other taxes, because the analysis has not been carried out. However, it is unlikely that these small taxes have had any major macroeconomic impacts.

---

3 These taxes are on top of existing taxes on fuels that on average are estimated to equal ECU 84/ton in the OECD. Sweden has reduced other taxes on fuels, so that the total tax on carbon dioxide is the figure given in the text.

The other issue that has been studied in a simulation model is that of spatial differentiation. It is clear that the optimal situation cannot tax all emissions equally, because damages from emissions will depend on where they originate, on stack heigh, etc. For sulphur emissions Mäler (1990) has compared the solution with a uniform sulphur tax with the full optimum. The total additional cost is only one per cent, which is well within the bounds of error. For individual countries not to lose out from the imposition of this tax, they would, of course, have to be compensated from the revenues collected. While this may hold for sulphur (and there are many simplifying assumptions in his model), it is not necessarily true for all air pollution problems, especially where a greater percentage of the total impact is at the local level.

*Empirical evidence*

To date, there is very little empirical evidence on actual emissions taxes. It can be said with near certainty that most of them have had little in the way of an incentive effect, (the Swedish taxes are possibly an exception). The revenues collected have been important in some cases, particularly in Eastern Europe, where they have bankrolled environmental investments that would not otherwise have been funded.[4] As a signal that the government is taking environmental issues seriously they may have been more useful, although I am unaware of any formal study of this impact.

The experience in Eastern Europe, where a complex system was in place in some countries prior to the transition, suggests that comprehensive emissions taxes, levied at the point of emission, are complex to administer. A system whereby the tax is computed on the basis of the fuel used and a standard technology, with rebates for technologies that reduce emissions, seems to be the easiest administrative procedure. Differences based on region (urban, rural) and/or stack height have also been used although, as was noted earlier, the basis for the differences remains unclear.

---

4 Being extra budgetary, most of the revenue does not go to the Ministry of Finance or its equivalent, but to the environmental ministries and agencies which can then allocate them for environmental purposes.

*Water pollutants*

All countries levy some charge for the collection of waste water and regulate in some form the discharge of water effluent directly into rivers, lakes, and coastal waters. The use of emissions charges can be said to be present when either the sewerage fees or the fees for direct discharge depend directly on the composition of the waste water. The less the charges distinguish between different types of waste and between the waste from different polluters, the less the system can be said to have an 'emissions charge' structure.

The countries of the OECD that have some water emissions fees are Austria, Belgium, Ireland, Finland, France, Germany and the Netherlands (OECD, 1993; OECD, 1995). Of these, detailed information was available for France, Germany and the Netherlands.

*France*

The discharge of waste water and withdrawal of water from the rivers and lakes by industry and by water supply and sewerage companies is controlled by the basin agencies which impose charges based on the quantity and quality of the water. The water pollution charges are considered as too low to have an incentive impact. What funds are collected are made available for use within the agency to assist industry in making investments in financing its own activities. Opschoor and Vos, 1989 comment that there is general reluctance to accept higher charges even though most of the funds are recycled. Thus, the effectiveness of the system is curtailed by a low charge-low abatement expenditure syndrome.

*Germany*

German standards are more stringent than the EC directives and among the most stringent in the world. Whereas the EC technology standards are based on the 'best available technology not entailing excessive cost' or BATNEEC, the German ones require the use of 'state-of-the-art' technology. Control of discharge is by permit but there is also a waste water charge based on the residual pollution load. This charge is in addition to charges for municipal sewerage services and is used solely for water pollution control projects. The rate was last revised in 1990 and is now fixed at DM 50 per unit of 'toxicity'. Furthermore, the rate is set to rise *at a known and publicly announced rate* until the year 2000, when it will be around DM 90 per unit of toxicity. Table 1 below gives the waste water charges. Charges are structured so that there is no payment for emissions below threshold levels. Furthermore, there are several exemptions. Dischargers using state-of-the-art treatment systems are eligible for a 75 per cent discount, reduced to 40 per cent after 4 years and 20 per cent after 8 years. Any investment that reduces effluent emissions by 20 per cent or more can be deducted from the charge. Thus, the system is highly regulated, with several incentives to invest in water treatment facilities, and with little prospect of raising a large amount in revenue. It is important to note, however, that these charges operate in addition to a system which itself has very strict standards.

The German system has been criticized partly for the fact that the charges are themselves preceded by direct controls so that the system is overregulated. Furthermore, there are weak incentives for reducing discharges into the sewerage system. It appears that low emissions limits and strict enforcement of standards have done more to encourage reductions in pollution than the charges described above.

*The Netherlands*

The Netherlands instituted a system of water pollution charges in the 1960s and 1970s, largely to address a major problem of organic pollution but also including charges for heavy metals and other pollutants. Charges are based on estimates of the waste content of water users, with large users having their waste monitored and smaller users having their waste estimated according to industry averages. Thus, the system of waste water charges is most closely based on the actual content of the water than in most other countries. The rates are determined by the cost of the measures to clean polluted surface waters and apply only to waste water released directly into public water bodies such as rivers, etc.

The revenues collected are used to subsidize the construction of sewage treatment plants. Most industrial enterprises have treatment plants, with about 85 per cent of the waste water being treated before discharge. According to Opschoor and Vos, these charges have had an impact in reducing water effluent from large enterprises but not from small enterprises or households which pay the average charge for all users in that category. The point has further been made that basing the charges on clean-up costs and earmarking the revenues for this purpose have resulted in over- investment in water pollution abatement relative to other forms of pollution.

Table 1. German waste water charges.

| Pollutants | Definition of Toxicity | Thresholds and Concentrations and annual loads |
|---|---|---|
| Chemical Oxygen Demand (COD) | 50 kg oxygen | 20mg/l and 250 kg |
| Phosphorous | 3 kg | 20mg/l and 15 kg |
| Nitrogen | 25 kg | 5mg/l and 125 kg |
| Halogenated Organic Compounds (AOX) | 2 kg halogens as organic chlorine | 100µg/3 and 10 kg |
| Metals and Metal Compounds | | |
| Mercury | 20 g metal | 1µg/l and 100 g |
| Cadmium | 100 g metal | 5µg/l and 500 g |
| Chromium | 500 g metal | 50µg/l and 2.5 kg |
| Nickel | 500 g metal | 50µg/l and 2.5 kg |
| Lead | 500 g metal | 50µg/l and 2.5 kg |
| Copper | 1000 g metal | 100µg/l and 5 kg |
| Fish Toxicity | 3000 $M^3$ effluent/$G_F$* | *$G_F = 2$ |

* $G_F$ is the factor of dilution at which the effluents are no longer toxic.

Source: Johnston (1992).

*Eastern Europe*

In many economies in transition, there is a system of charges for effluents released into water. Substances are classified in categories of toxicity and a charge is levied per ton of waste released. In most countries, the charge is set at one figure for waste emissions within the permitted norms, and at a higher rate, which can be three times or more the standard rate, for emissions beyond the norm. These charges are levied on the final emitter, so that an enterprise pays the charge directly only if it releases wastes into a water body. If it is collected and disposed of by a sewage company the latter is responsible for the charge. The extent to which the charge is reflected in its tariff structure is a matter for the company. Generally, the sewage charges do not reflect fully the final pollution charges.

The charge levels are generally very low, and in many countries have been eroded considerably by inflation. The only Eastern European country with significant water charges, lower than some of the German ones but of the same order of magnitude, is Poland.

The main rationale behind instituting pollution and resource taxes has been to raise revenues, while the second has been to provide incentives for the protection of the environment. While the first objective has been met to a limited extent, the impact on water use and on effluent releases has not been noticeable.

*Relation to theory*

As with air pollution taxes, water effluent charges have little relation to marginal damages, insofar as the latter have been estimated and the charges determined accordingly. In the case of the Netherlands, it is explicitly stated that the basis is cost of abatement and not marginal damage. Unlike the air pollution charges, however, the water charges are tied to water quality objectives, as measured by the concentrations of water pollutants in a water body. In order to achieve these goals, however, the charges play only a small part in the system of regulations. Much more important are limits on total discharges and/or maximum allowable concentrations.

In addition to the above, there are a number of specific issues that have a theoretical perspective. These are:

a) the collective versus the private treatment of waste and disposal charges for water; and
b) allowing for inter-media substitution.

*The collective versus the private treatment of waste and disposal charges for water.* Since waste water can be treated private and publicly, the relative costs of the two methods should be reflected in the prices that enterprises face. If private abatement is more costly than public abatement at all levels of treatment, it is appropriate that all abatement should be done by the public sector, and vice versa. But both of these will be unusual cases. In general, both activities will be required. If the social cost of removing a ton of waste from the water is

ECUx in the public facility, the charge levied on the private polluter for each ton *emitted* should also be ECUx. In this way, the polluter will undertake private abatement and demand public abatement services in the least cost fashion (i.e. equating the marginal costs of abatement in both cases).

At the same time, the charges levied for waste water treatment and the charges levied for effluent released (i.e. the pollution charges) should be linked. Unfortunately, in most cases they are not. If the charges imposed by the water authorities are too low compared to the effluent fees, enterprises will have an incentive to take advantage of the lower fees and will release too much into the sewerage system. Conversely, if the effluent fees are too low, and the water charges take full account of the costs of treatment, enterprises will have an incentive to go for direct releases. The correct pricing of water discharges should equate the costs of the two methods for a given composition of waste. As with pollution charges, this does not appear to be the case.

*Inter-media substitution.*[5] Waste can be generated in solid, liquid or gaseous form. The choice is, in part, up to the polluter. Hence, the amounts of each will depend on what the relative prices are and on the costs of converting waste from one form to another. For example, the recovery of waste from water as sludge makes for cleaner water. But the extent to which this is done will depend on the costs of sludge disposal as well as the costs of recovery.

If all disposal methods bear their full social costs, the choice can be left to the polluter. As this is not the case, charges for waste water and sewage sludge have to be set with some recognition of their linkages. If more encouragement is required for converting from liquid to solid waste, charges for liquid waste should be raised, and vice versa. In all of the countries surveyed, the two appear to be set independently.

---

5 This point also applies, of course, to the other charges considered here, for air emissions and solid waste emissions.

*Simulation evidence*

There are simulation studies, carried out mainly in the US, that show the benefits of charges versus direct controls in regulating water effluent (Freeman, 1990). However, they do not analyse the actual water charge schemes that are in place, with their multi-tiered structure of regulations. Hence, it is not possible to say precisely what the contribution of the emissions charges has been over and above that of the complex direct controls.

As with the air emissions taxes, it is unlikely that there have been major socioeconomic or macroeconomic impacts from these taxes *per se*, although the increasingly strict system of water quality controls in public water bodies is raising water charges significantly in Europe, a factor which may well have some impact on income distribution and the macroeconomy.

*Empirical evidence*

As has been noted above, water effluent charges are primarily a means of collecting revenue, although there have been some incentive impacts, for example in the case of the Netherlands. It is interesting to note that, whereas the low French water effluent charges have been cited as resulting in too little investment in water clean-up, the higher Dutch charges have been cited as resulting in too high a level of investment in that sector. This provides one example of the possible dangers of earmarking the tax revenues. As with air pollution charges, however, the presence of earmarking has been of some benefit to the environmental sector in Eastern Europe, where the amount of investment in pollution abatement would probably have been even less if the revenues had gone into the general budget.

*Solid waste charges*

Solid waste disposal has a similar structure to water in that the waste can either be collected by a municipal authority and then disposed of, or can be disposed of directly at special sites. Normal disposal fees should cover the direct costs of operating the dump, although in most countries these 'fees' take the form of a tax on waste generators (usually a property tax). Charges based on the content or volume of the waste only apply in special cases (e.g. hazardous wastes). The waste disposal sites are sometimes managed by the same authority that is responsible for collection, so the 'costing' of the disposal sites services is an internal matter that rarely involves full 'cost-centered accounting' (i.e. the department delivering the waste has to pay the department receiving the waste).

Recently, however, some countries have imposed charges on solid waste delivered to sites, payable by the authority operating the site to the regional or national authority responsible for waste management.[6] It is then up to the operator to pass on the charge to the persons or entities responsible for generating the waste. In addition, many countries have instituted a number of recycling schemes. One is the deposit refund scheme whereby manufacturers are obliged to charge a deposit for containers, the deposit being returned when the container is brought to a specified place. Another is a system in which the manufacturer of goods sold with packing contributes to a service which collects that packaging and recycles it.

General waste disposal charges exist in all OECD countries, but volume-based ones are to be found in Australia, Belgium, Canada, Denmark, France, Germany, Italy, Netherlands, Switzerland, and the United States (OECD, 1993; OECD,1995). The rates vary widely, from ECU 1/ton (Australia) to ECU 26/ton (Denmark). Rates may also depend on whether the waste goes to landfill or whether it is incinerated (Denmark, Germany and Switzerland). The state or regional authority charges for hazardous waste disposal in Austria, Finland, Germany and the US. The known rates are ECU 15/ton (Austria), ECU 40/120 ton (Germany) ECU 90/ton (USA, maximum rate, for Minnesota) (Kågeson, 1993; OECD, 1993).

Deposit refund schemes operate in 12 OECD countries, namely, Austria, Belgium, Canada, Denmark, Finland, France, Germany, Netherlands, Norway, Sweden, Turkey and the United States. (OECD, 1993). Recycling schemes involving manufacturers have been set up in France and Germany. In the United Kingdom, disposal authorities have to compensate collection authorities when the latter retain materials for recycling, the payment depending on the net saving in disposal costs.

In Eastern Europe, waste charges for solid waste exist in several countries; with the rates being determined by the location of the site, conformity with sanitary standards, etc. In most cases, these charges are imposed on operators of waste dumps and are not passed on to the

---

6 In addition, a number of countries impose charges for products that clearly have a disposal cost associated with them, such as batteries (Canada, Norway and Sweden), plastic bags (Italy), tyres (Canada), and oil products (several countries).

waste generators. That is certainly true for household waste and also for much of industrial waste. The main purpose is given as revenue raising, but there is considerable evidence that the authorities are unable to collect the full amounts imposed, and the amounts collected are generally small (again Poland is an exception). There is no evidence that the charges are acting as an incentive to reduce waste.

*Relation to theory*

As regards solid waste, users should be charged the social cost of its disposal, and any other means of reducing waste should be provided at an amount such that the marginal social cost of the reduction in waste is equal to the social cost of disposal. If recycling results in a reduction of one ton of waste per annum at a cost of ECU 100 and the marginal social cost of one ton disposed of is ECU 50, then there is too much recycling; it would be better to dispose of the last ton of waste and save ECU 50. Conversely, if the marginal cost of the recycling option is ECU 50 and that of the disposal is ECU 100, it is better to increase the recycling.

The marginal social costs of the different options that are available, i.e. deposit refund schemes, waste landfill, waste incineration and manufacturer recycling schemes, should therefore be equalized.[7] In order to know whether this is the case, the correct marginal costs of each of the options have to be ascertained. As far as I am aware, such a calculation has not been carried out, although some estimate must have been made of the relative costs of landfill versus incineration. It is clear, however, that a full social cost comparison has not been carried out before deciding on the selection and relative amounts of the different instruments. At the same time, the rationale for subsidizing recycling has not been fully analysed. As a second-best policy, it may be justified to subsidizing one form of waste reduction when other forms are underpriced. The analogy is with subsidizing public transport when private transport is priced below its social cost. But it is generally better to price private transport at its social cost, as it is to price waste disposal properly. The economics of what the subsidy should be and how it should be phased out as waste disposal prices increase have not been studied in a policy context.

---

7 This assumes that each method is used for each type of waste. If some method is not used, then its marginal social cost may exceed that of the method that is used. Formally, there may be 'corner solutions'.

Knowing the social costs should help us guide the selection of the MBIs, but by itself it is not enough. The generators of the waste must face the appropriate prices based on the social costs. Thus, charging the waste managing authority a fee equal to the external cost of waste disposed of there (assuming that the private costs of the disposal are internalized) will only work if that authority then passes on that external cost to the households and industry that produce the waste. In many countries, waste collection is not priced in this way. Perhaps the practice of charging according to the amount of waste will come about once the charges on the disposal are high enough. This is beginning to happen in some countries (several cities in Switzerland, 14 cities in the US and some *Länder* in Germany).

*Simulation evidence*

Repetto *et al* (1992) have shown that a system of household waste collection where payment was based on weight and reflected social costs would result in economic savings of around $ 650mn per annum for the United States. Waste levels would decrease by between 6 and 12 per cent depending on the amount of the charges (which reflect social costs). The same paper looks at introducing curbside recycling at the same time as the higher charges and concludes that recycling is justified if the area has a high cost of waste disposal but not if the costs are 'low or moderate'. This analysis defines low -- and high-cost regions depending on land values, tipping fees and external disposal costs. With recycling, however, waste loads drop substantially -- by up to 37 per cent in the high-cost regions.

This study is only partly a simulation, as it is based in part on the experience of 14 US communities that instituted volume-based fees between 1980 and 1989.

*Empirical evidence*

For the European waste programmes, both in Eastern and Western Europe, there is little evidence that the charges have reduced waste. Kågeson notes that the 44 per cent increase in waste charges in Denmark is expected to decrease waste generation further but is not expected to reduce the amount of waste from packaging significantly. From US data, Jenkins (1994) estimates the price elasticities of demand for solid waste as -0.12 for households and -0.29 for commercial entities.

There is more evidence that the recycling schemes have reduced waste (OECD, 1993). Bottle bills in the US have reduced litter by 10 to 39 per cent and solid waste by 1-6 per cent. In the same country, kerbside collection programmes obtain recycling percentages of around 35 per cent for glass containers and 25 to 56 per cent for aluminium cans. As far as the impact of the producer-based recycling schemes is concerned, it is considered too early to make an evaluation (that was the case in 1993). Nevertheless, it is encouraging to note that the German Green Dot scheme has obtained the participation of over 50 per cent of households by 1993, and that a figure of over 80 per cent is expected by this year.

*Other emissions charges*

Emissions charges are applied in the case of aircraft noise, as a supplement to landing fees in Belgium, France, Germany, Netherlands, Norway, Portugal, Sweden and Switzerland. Uniformly, these are very low and, according to the OECD evaluation (OECD, 1990), have not had an incentive effect. It is worth noting, however, that there is some evidence of a reduction in aircraft noise levels. Walter *et al*, 1993 report that noise charges in Germany have resulted in an increased level of conformity with local standards -- from 26 per cent in 1987 to 43 per cent in 1989. For Geneva Airport, the annual report of the authorities for 1991 discusses the effectiveness of noise charges, introduced in 1985. Noisy aircraft pay higher landing fees, with night landings having a higher weighting because of the greater impact of the noise. The result has been a steady decline in the number of aircraft landings classified as 'noisy' -- from 30% in 1982 to 14% in 1991. It is unclear, however, to what extent this is due to the changes in the design of aircraft (which are becoming quieter) and to what extent it is due to the charge schemes.

Charges on chlorofluorocarbons (chemicals that damage the stratospheric ozone layer) have been adopted in Denmark and the US. These taxes are in addition to import controls and production controls (where appropriate). The US tax was a means of collecting windfall profits arising from restrictions in production, rather than as an incentive to reduce consumption (see Section 3). The incentive impacts of the Danish tax have not been analysed.

## 3. Tradeable Permits

*Background*

The basic idea behind tradeable permits is very simple. If you want to reduce the use of something by X units, from a level of 100, in a population of 10 persons, you can ask each of the 10 to reduce use by X/10, or you can give the users as a group the 'right' to consume 90 units in total. This allows those who do not want to cut back, perhaps because it is more expensive to do so or because they have large benefits from consuming the good, to purchase more consumption rights. By the same token, it allows those who want to 'cash in' their rights to do so.

Trade of this kind has a tendency to develop spontaneously in many situations; with food rationing for example, or with cigarettes or liquor when these products are allocated on a *per capita* basis. How successful and efficient the market is depends on whether trading is officially sanctioned and encouraged or whether it is prohibited.

For environmental problems, the notion that a market in the scarce rights was a way of solving the problem was noted by Mishan (1967) and Dales (1968). The application in a real world context, however, arose gradually. Earlier attempts all took place in the US (Hahn, 1987; Sorrell, 1994) and started with the air pollution control laws in that country in the late 1970s. At first 'trades' were very limited, and a new set of 'instruments' was developed: netting, offsets, bubbles and banking. The point of departure was the notion that a firm had to meet a given reduction for each of its sources of pollution. This was clearly inflexible, and netting allowed the firm that wanted to generate pollution from a new source to count greater-than-required reductions elsewhere. Initially, netting was only an internal trade. Offsets were introduced to allow new firms to set up in areas where the quality standards were not being met and where the normal regulations would not permit another enterprise to establish itself. If, however, the new firm could acquire a reduction in emissions from another firm by more than the amount the latter would have emitted, it could set up operations (as the degree of non-attainment would in fact be reduced). Bubbles were an instrument similar to netting and offsets, except that trading was only permitted within a firm and for existing sources of pollution. Finally, there was banking, which allowed a firm to save emissions reductions over and above those required for future trading. All these developments took place between 1974 and 1979.

To an outside observer, with the hindsight of even 15 years, these seem unnecessarily complex notions. Why not simply allow trade in emissions, one asks? The reasons of course lie in the fact that policy implementors are naturally cautious, and will not try something new unless it is forced upon them by circumstances. The need for flexibility was forced upon them and the instruments evolved gradually, as a wide range of administrative issues were addressed (Rico, 1995).

More recent programmes of emissions trading such as that applied to the reductions in $SO_2$ and $NO_x$ under the US Clean Air Act take a more liberalized approach to emissions trading. Each trading unit is allocated an allowance based on historical emissions which are "generally issued in perpetuity" (Rico, 1995). Total allowances equal the amount of emissions permitted in that year across the region. New sources have to purchase allowances from existing users, and either party may bank them. Since this programme is not concerned with addressing local pollution problems, the issue of 'hot spots' -- places where emissions become concentrated, affecting the health of the local population -- is not of direct relevance. All utilities are, however, subject to meeting local pollution standards as well, so any spatial problem arising from excessive local emissions is addressed in that way. Rico reports that an allowance market has been evolving slowly, with only about 9 per cent of utility boilers relying on allowance trading to meet compliance requirements. Allowance prices have dropped in the last two years of trading and there is no sign of market dominance by a single buyer (which could result in smaller firms being driven out with all that this implies for competition).[8] Surplus allowances are sold in 'first price' auctions (all bidders pay the price they bid rather than the clearing price).

For water pollution, tradeable permits have had more limited success. Hahn (1990) reports on the Wisconsin Fox River scheme where tradeable permits were introduced to reduce the costs of limiting biochemical oxygen demand on sections of the river. The scheme did not work, mainly because trades were very restricted; both by location (firms could only purchase

---

8 This confirms the earlier analysis of Tietenberg, 1985, who concluded that market power is a relatively rare problem that can be dealt with on a case-by-case basis.

from other firms in their location group, which made the market very thin) and by a number of administrative requirement for each trade to be sanctioned. Finally, the participants did not meet the profile of competitive market agents. Either they were oligopolistic firms or they were municipal waste treatment plants that did not necessarily behave as cost minimizers. As a result, the simulated model, which had predicted benefits of around $ 7 mn a year, proved to be incorrect, and the actual benefits were negligible.

A water problem that has been addressed using marketable permits is that of point versus non-point sources. The Lake Dillon watershed case is reported by Tietenberg (1991). Phosphorous emissions were making the lake eutrophic and, although point sources such as waste water treatment plants were employing advanced treatment methods releases from non-point sources were not being controlled and would have been enough on their own to ensure eutrophication. To address this problem, the authorities introduced a trading scheme combined with incentives for the introduction of state-of-the-art non-point control systems. Anyone introducing such a system received a credit of one unit for every two units of waste reduced from non-point sources. These credits could then be sold to point source users. To date, there are no reports on the success of the scheme.

Marketable permits have also been used in other areas. For the phase-out of lead in gasoline in the US, it was recognized that the costs of meeting the short deadlines could be very high for some petroleum refiners. Accordingly, the Environmental Protection Agency instituted a trading system whereby refiners who achieved more than the required reduction could sell the surplus to those who did not meet the reductions by the appropriate dates. Refiners could also bank reductions against future targets. The consensus is that the scheme resulted in a more cost-effective phase-out of lead (although there were several problems). By the end of December 1987, complete phase-out had been achieved.

A similar application was the phase-out of CFCs in the US. The international Protocol for the phase-out of these ozone-depleting substances called for reductions of specified percentages by specified dates. Each producer of CFCs was allocated an allowance equal to that percentage of his/her original production. The allocation was tradeable, so that those who could not make the cuts were able to buy from those who overachieved their targets. The trades were, however, restricted; the amounts by which any producer could increase production were very limited, even if he/she acquired the permits. The process of acquiring the permits on the basis of past pollution resulted in windfall profits for the producers who were then taxed on the sale of CFCs.

More generally, the idea of allowing some trading in an attempt to achieve environmental goals has caught on. One form that has had some applications is that of transferable development rights (TDRs) (Panayotou, 1994). The idea is that if a conservation area is declared, the owner of the land or buildings does not completely lose the right to develop the property. Rather, he/she can exercise that right elsewhere, in a place where further development is restricted and the right has some value. The terms usually allow an increase in development beyond existing legal limits by about 10-15 per cent. Such rights have been used in urban areas in the US and for coastal development in Puerto Rico. TDRs are somewhat different from marketable permits, in that they consist of giving groups which would lose out from the development compensation that is acceptable and that does not involve raising a large amount of resources through taxation. Marketable permits exploit trade between different agents, all of whom have to meet a given environmental objective.

In Western Europe, there is no real experience with tradeable permits, although they are being evaluated for the control of nitrogen oxides and VOCs in the canton of Basle in Switzerland (Stritt, 1994). The scheme was set up in 1992 but no trade has occurred as yet. In Eastern Europe, Poland has experimented with this instrument in the city of Chorzów, with some success (Dudek *et al*, 1992). However, its wider application both there and elsewhere in Eastern Europe will require some amendments in the law (Anderson and Ylich, 1993).

*Relation to theory*

The main questions that must be resolved with tradeable permits are:

a) How should permits be allocated ?
b) How should the spatial or locational dimension be dealt with ?
c) What are the impacts of such schemes on the development of technology for environmental pollution control?
d) How should the problems of market power and imperfections be addressed ?

The allocation of permits is simply a means of distributing the costs of achieving a reduction in pollution. A system where the allocation is in proportion to existing emissions has the advantage that existing polluters are no worse off than they would be in a command and control system. As a result, their objection to the system will be limited. On the other hand, allocating permits in this manner (known as grandfathering) creates an advantage for existing producers vis-a-vis new entrants and makes it necessary to consider the impact on industrial structure and power. One way of solving this would be to reserve some permits for new firms, but this has rarely been done (Tietenberg, 1985).

Auctions are used to facilitate market transactions, although in most cases the supply of permits comes from existing producers who have had them allocated under a grandfathering arrangement. The spatial question is clearly one of the most important and most difficult. The ultimate objective is ambient environmental quality in a region, or set of regions. Because an emission from one location does not have the same impact on quality as that from another location, emissions cannot be traded on a one-for-one basis and cannot be expected to have an effect on environmental quality.

There are a number of ways of addressing this problem. One is to define transfer coefficients, so that emissions from different locations are traded in ratios reflecting their relative damages. If $a_{ij}$ is the amount of an emission from location i that ends up as a deposition in location j, and

$V_j$ is the relative damage of a deposition in location j (any location can be set as the numeraire), then the cost of an emission from location i is $E_i$, where

$$E_i = a_{ij}.V_j$$

Trade between emissions in locations i and k would then be on the basis of the ratio $E_i/E_k$. The problem is knowing what the values of the V's are.[9]

[9] The analysis can easily be extended to emitters with different stack heights -- it adds another index to the emission vector.

Another approach is for the authority to define permits for each receptor point j, and issue enough such permits to ensure that the quality goal is met. This, too, would need trade between different emitters to be defined on a basis different from one for one. In this case, the trading ratio between emitters in locations i and k for receptor j would be $a_{ij}/a_{kj}$. With many receptors j, this could result in a complex set of trading rules.

A third approach is to restrict trades between emitters, so that only those who had roughly the same impact on all receptors would be allowed to trade (on a one-for-one basis). The problem here is that the markets could be quite thin. Enlarging the region would entail a cost in that quality goals could be compromised, or the costs of meeting the goals could be increased. The trade-off between the benefits of a larger region and the environmental costs of less 'correct' trading ratios has to be resolved at the empirical level (see below).

The impacts of marketable permit schemes on technological development should be positive. The incentive for continued reductions in emissions remains as long as the gains in terms of needing less permits are there. If, however, the permits are grandfathered and the number remains constant, the pressure to undertake technological developments will be less, as firms will always be able to count on receiving permits to cover existing procedures.

Finally, the design of the system has to address the problems of market power. If a few actors in the market had enough power, they could drive up the price of permits temporarily, thereby driving out their competitors and resulting in increased market share for themselves. There are suggestions that such behaviour was practiced in New Zealand, when the government attempted to issue permits for fish catches. Outside of that example, there is the possibility that those with market power could manipulate any auction price (bidding it down by not demanding permits or by holding down the number of permits issued). There is some theoretical work on the issue of strategic behaviour in these markets. Most of it suggests that the impacts of such behaviour on the costs of achieving a given level of control on environmental quality are small (Hahn, 1984; Tietenberg, 1985). This does not, however, address the question of its impact on industrial structure through market entry and exit. On this issue, Misiolek and Elder (1989) and Tietenberg (1985) suggest that the problem is rare but may need to be addressed on a case-by-case basis.

*Simulation evidence*

Much of the early simulation work on tradeable permits concluded that the benefits of trading would be large compared to conventional command and control regimes. For water pollution the Wisconsin Fox River, which predicted large savings that did not materialize, has been cited. For air pollution, Tietenberg (1995) reviews the evidence from several studies comparing the costs from command and control (CAC) systems with those from emissions based standards (EBS). The CAC costs are generally higher, but there is a wide range, going from 11 times the EBS to 0.4 times. Thus, in three of the eight studies looked at, the EBS standards had higher costs and in one the EBS costs could be higher, depending on the air quality requirements at the worst receptor. The reason why the simulation studies come up with higher costs in some cases is simply that CAC measures allow more targeting of the reductions than do EBS methods. Consequently, if the EBS is to achieve a given reduction in depositions at a given location, it may be necessary to make a larger reduction in emissions than would be required under a well-planned CAC system; and the benefits of tradeable permits are ambiguous and have to be looked at on a case-by-case basis. If there are a few large polluters near a receptor and many that are far away, all of which form part of a single trading area, the EBS system will result in quite strict limitations on permits, which will in turn raise the control costs. On the other hand, if the polluters are more evenly spread out and affect the receptor more or less equally, the cost advantage of the EBS system will be larger.

As Tietenberg notes, the simulations are based on the EBS system being designed to meet the quality standards at least cost, given whatever constraints are imposed on the trading system. In practice, the designers of the system will not have this knowledge and will be acting under uncertainty. To avoid the risk of undershooting their environmental quality target, they will tend to restrict permits even more, which will raise the costs of the EBS solution.

*Empirical evidence*

Empirical evidence on the use of tradeable permits has been gathered in relation to the US Clean Air Acts as well as for water pollution in river systems and lead in gasoline. In retrospect, the schemes have had mixed success. The early air emissions schemes applied, in varying ways, to VOCs, carbon monoxide, $SO_2$, particulates and $N0_X$ were studied by Hahn and Hester, 1989, who estimated a wide possible range of benefits. Clearly, there was much uncertainty about the possible gains. For water quality, most schemes have not demonstrated any significant gains. The main findings of interest can be summarized as follows:

- There are substantial gains to be made in simply allowing 'trade' within an enterprise. A number of studies found that this was where the biggest advantages lay (Hahn, 1987, ICF, 1989).
- The spatial dimension is clearly an important one but, following from the above, significant gains can be made even if the region is defined rather narrowly. Although with small zones trading possibilities are limited, the advantages of defining the emissions permits more accurately in terms of their impact on the environment seem to outweigh the losses arising from limited extra-firm trading. Smaller zones will, however, only result in lower control costs if it is assumed that full information on which to base the allocation of permits between zones is available. If that allocation is based on some rules of thumb, reducing the size of the zone does not necessarily help (Spofford, 1984).
- With small zones, there can still be a problem of 'hot spots' -- areas where emissions create an environmental problem because of their concentration. The issue appears to be "stack height" for air pollution, and the fact that this is generally ignored. It may be necessary to take variations in stack height into account in designing permit schemes.
- Most schemes restrict trades in some way -- across zones, or to address hot spots, etc. There is little theoretical work to indicate what restrictions are the 'best' ones to achieve the desired goals of the scheme. Three schemes have been studied in some detail: (a) the pollution offset approach where trades are allowed as long as no ambient air quality standard is violated; (b) the non-degradation offset approach which requires the same as (a) plus the condition that total emissions do not increase, and (c) the modified pollution offset, in which trades must not result in a violation of either the pre-trade air quality nor the concentration target, whichever is the stricter (Tietenberg, 1995). In general, one would expect the pollution offset to be the least costly, but it is also administratively complex, as to some extent are all three rules. This being so, it is probably preferable to

operate a system with basic emissions trading and then to evaluate a number of rules to address specific problems. The selection of the rules can be *ad hoc* for the present.

Transactions costs can be significant and can eat into the benefits of tradeable permit schemes. The extent to which they do will depend on what schemes are being considered. However, one study that has analysed transactions costs in some detail (Kerr and Maré, 1994) found that, for the lead phase-down programme described earlier, these costs amounted to about 10-15 per cent of the gains from trading. Of course, that program was simpler than some now being considered. As the complexity increases, so will the transactions costs.

## 4. Conclusions

To most economists, apart from a group who can be described as extremely liberal and who are against the government imposing any kind of corrective taxes, the idea of using MBIs in preference to CAC instruments is obvious. They are more flexible, achieve a given standard at least cost, provide a dynamic incentive and can even be used to attain the Pareto Optimum. The underlying theory is somewhat disingenuous, however, in that it finesses the issues of uncertainty and the costs of information, without which the benefits of MBIs over CACs are much less obvious. Furthermore, there are several issues of importance in the design of an MBIs, in relation to which basic theory has only some rather general advice to offer -- where to levy the instrument, how to define the 'market', etc. On these questions, the policy-maker has to strike a balance between efficiency and administrative factors. Experience is the main guide to the right practice.

One conclusion from the theory is that in the presence of uncertainty about the costs of abatement, it is better to use taxes rather than marketable permits if the marginal damages are relatively constant and the marginal costs of abatement increase sharply with the level of emissions. The evidence on these curves seems to suggest that this is the case. However, it has not been a major factor in the choice of instruments.

In this paper, I have looked in some detail at the experience of using emissions charges and subsidies; and tradeable permits. They are the MBIs that come closest to the textbook cases and provide a great deal of evidence on what works and what does not. I have also focussed, for lack of space, on OECD and Eastern European experience and ignored the growing use of MBIs in developing countries.

Emissions charges and subsidies are used in relation to air, water, solid waste and noise. A few countries in Europe have charges for air emissions. They are relatively low and, in most cases have little incentive effects. The main purpose is to collect some revenue and use it to finance pollution abatement activities. This is a general pattern for the other emissions charges as well. They are not based on marginal damage and, in many cases, not on marginal cost either.

From a theoretical viewpoint, charges should be differentiated according to the relative damage done by the emissions. On the other hand, the administrative costs increase as the number of differentiated charges rises. Some evidence seems to suggest that uniform charges may not result in too large a loss compared with a full optimum, but this finding may be dependent on the fact that the pollutant considered had a small local impact compared to the total impact.

The use of air emissions charges is set in a framework with many other constraints on the polluter -- technology-based standards, maximum allowable emissions, location, etc. These result in a complex system that is difficult to analyse. The tentative conclusion appears to be that these other instruments have been more influential in determining the level of pollution than the charges.

In Eastern Europe, the system of charges is more comprehensive and more complex but of even less significance from an incentive viewpoint, because most polluters have faced soft budget constraints and because the charges have been very low. The main advantage at present is that the system provides some revenue for environmental protection. The earmarking of this tax is useful in this context at least, but it can only remain an earmarked tax as long as it is not too large.

Very similar considerations apply for water effluent charges. Here, I wish to point out that the system of fees is *ad hoc* in a number of respects. In addition to those noted for air emissions, charges for direct releases into the body of water and releases through waste water treatment plants are not related (which they should be). No attention is paid to the possibility of inter-media substitution in setting the fees (i.e. waste can be converted from liquid to solid or gaseous form).

For solid waste, like water, pricing has played only a small role, i.e. in the recovery of private costs, using average cost prices. This is the least efficient pricing system, economically speaking. It has been accompanied by subsidies on recycling that could be justified on second-best grounds, although there are no thorough studies on what the optimum subsidy should be. In this context, it is interesting to note that a number of authorities are now moving towards charges schemes that are based on marginal cost and attempt to take some account of external damages.

Unlike the water and air charges, some of the MBIs used in connection with solid waste have had a major incentive impact -- the recycling and deposit refund schemes.

The use of tradeable permits has been mainly confined to the US, although a number of applications are being considered in Canada and Western Europe. The system grew out of the need for a degree of flexibility in meeting pollution targets and many of the benefits derive from such simple changes as allowing an enterprise to meet emissions standards as a whole, rather than for each of its activities. Several complex systems have been tried, but the most recent -- in connection with acid rain -- has a relatively simple structure.

Permits are judged mainly by their impact on cost, relative to a system of direct controls. Most simulation studies find considerable benefits for the permit schemes, but in practice the benefits are often less. The schemes have been applied to air pollution emissions, lead in gasoline, CFCs, and water effluent. Recently, there have been some applications of tradeable development rights to conservation areas, but these are not conceptually similar to the other permit schemes considered.

Permits are mainly allocated on some kind of grandfathering basis. This makes it easier to convince existing polluters to join the scheme. They have been found to be positive in their impact on continued technological development in pollution abatement and, in general, worries about market power have not materialized. In the latter context, however, it is important to look at each application with some care and to take appropriate measures when necessary.

The main concern with permits has been over the spatial dimension. How large should the trading area be, and should trades between emitters with different impacts on the receptors have a trading ratio other than one-for-one? Emission-based permit schemes can be more or less costly in terms of achieving certain environmental goals depending on how these issues are handled, and some studies seem to indicate that the actual schemes implemented have been more costly than a direct command and control scheme (presumably because they had to restrict emissions by more than the CAC to meet concentrations targets).

The main benefits of tradeable permits can be obtained with relatively small trading zones. To take full advantage of these zones, the allocation of permits between zones should be based on maximum information about the impacts of emissions on ambient concentrations, and should allow for differences in stack height. More complex schemes that allow trade across zones can increase cost efficiency but will entail much higher administrative costs. It is not clear that benefits exceed costs.

To conclude, MBIs have a role to play in environmental regulation, but the difficulties should not be underestimated. Some of the issues raised here need further research -- the second-best trade-offs. If product charges are adopted, such as additional charges on fuels, it becomes even more important to evaluate these second-best trade-offs. For other issues, the directions are more evident. Some increases in the levels of charges (accompanied by reductions in other carefully selected taxes) would probably be beneficial, despite the doubts cast by the recent double dividend arguments. Where more than one instrument is available for a pollutant, the design of the system as a whole needs to be considered. In the case of tradeable permits, simple increases in flexibility in meeting standards can achieve a great deal. Applications outside the areas that have been attempted so far should also be considered.

*Acknowledgements*
*I would like to thank C. Jeanrenaud, M.-A. Stritt and N. Wallart for comments on an earlier draft. I would also like to thank the participants of the Conference on Environmental Policy: Between Regulation and Market, held in Neuchâtel, 19-20 June 1995. All errors that remain are my own.*

## References

**Anderson, G. and Zylicz, T.** (1993) Environmental Policy in Poland : Prospects for Economic Instruments. Paper presented at a seminar on Industrial Pollution Permitting in Central and Eastern Europe, Vilnius, Lithuania.

**Anderson, F. R. et al** (1977) Environmental Improvement Through Economic Incentives. Johns Hopkins University Press for Resources for the Future, Baltimore.

**Atkinson, A. B. and Stiglitz, J. E.** (1980) Public Economics. McGraw Hill, New York.

**Atkinson, S. E. and Tietenberg. T. H.** (1982) The Empirical Properties of Two Classes of Designs for Transferable Discharge Market. *Journal of Environmental Economics and Management*, 9 : 101-121.

**Baumol, W. J.** (1972) On Taxation and the Control of Externalities. *American Economic Review*, 62 : 307-322.

**Barrett, S.** (1992) Global Warming : Economics of a Carbon Tax. *In*: Blueprint 2: Greening the World Economy, D.W. Pearce (ed.), Earthscan Publications, London.

**Beckerman, W. and Markandya, A.** (1974) Pollution and Optimal Taxation - a Static Analysis. *Journal of Environmental Economics and Management*, 1 : 43-52.

**Bovenberg, A. and de Mooij, B.** (1994) Environmental Levies and Distortionary Taxation. *American Economic Review*, 84 : 1084-1089.

**Dales, J. H.** (1968) Pollution, Property and Prices. University of Toronto Press, Toronto.

**Dudek, D. J. et al** (1992) Implementing Tradeable Pollution Rights in Poland. Paper presented at the Third Annual Conference of the European Association of Environmental and Resource Economists, Cracow, Poland.

**EC** (1991) Impact of Energy and Carbon Taxes on $CO_2$ Emissions. European Commission, DGXII, Brussels.

**EC** (1995) Externalities of Fuel Cycles: ExternE Project Report No. 9: Economic Valuation. European Commission, DGXII, Brussels.

**Freeman, A. M.** (1990) Water Pollution Policy. *In*: Public Policies for Environmental Protection, Resources for the Future, Washington, D.C.

**Gilder, L. P.** (1994) Environmental Taxation and the Double Dividend: A Reader's Guide. Paper presented to the 50th Congress of the International Institute of Public Finance, Harvard University, Cambridge, USA.

**Hahn, R. W.** (1984) Market Power and Transferable Property Rights. *Quarterly Journal of Economics*, 99 : 753-765.

**Hahn, R. W.** (1987) Economic Prescriptions for Environmental Problems: Not Exactly What the Doctor Ordered. School of Urban and Public Affairs, Working Paper No. 88-4, Carnegie-Mellon University, USA.

**Hahn, R. W.** (1990) Meeting the Growing Demand for Environmental Protection: A Practical Guide to the Economist's Toolchest. American Enterprise Institute, Washington, D.C.

**Hahn, R. W. and Hester, G. L.** (1989) Where Did All the Markets Go? An Analysis of EPA's Emissions Trading Program. *Yale Journal of Regulation*, 6 : 109-153.

**ICF** (1989) Economic Analysis of Title V (Acid Rain Provisions) of the Administration's Proposed Clean Air Act Amendments (HR 3030/ S 1490). Washington, D.C.

**Jeanrenaud, C. and Stritt, M.-A.** (1994) Market-Based Instruments : A Way to Reduce the Cost of Air Pollution Control: Some Empirical Results for Switzerland. IRER, Working Paper No. 9301, Neuchâtel University, Switzerland.

**Jenkins, G. P.** (1994) International Modernization of Tax Administrations: Revenue Boards and Privatization as Instruments for Change. *Bulletin for International Fiscal Documentation*, 48 : 75-81.

**Johnston, N.** (1992) Water Resource and Opportunity. Centre for Exploitation of Science and Technology, London.

**Kågeson, P.** (1993) Economic Instruments in European Environmental Policy. European Environmental Bureau, Stockholm.

**Kerr, S. and Maré, D.** (1994) Trading Behavior in Tradeable Permits with Transactions Costs: Empirical Evidence from the US Lead Program. Mimeo, Harvard University, Cambridge, USA.

**Klarer, J. (ed.)** (1994) Use of Economic Instruments in Environmental Policy in Central and Eastern Europe: Case Studies of Bulgaria, The Czech Republic, Hungary, Poland, Romania, The Slovak Republic and Slovenia. The Regional Environmental Center, Budapest, Hungary.

**Mäler, K.-G.** (1990) International Environmental Problems. Oxford Review of Economic Policy, 6 : 80-108.

**Markandya, A. and Lehoczki, Z.** (1994) Environmental Taxation: A Review of OECD Countries and Prospects for Economies in Transition. The Regional Environmental Center, REC Paper Series n° 1, Budapest, Hungary.

**Markandya, A. and Shibli, A.** (1995) Industrial Pollution Control Policies in Asia: How Successful are the Strategies? *Asian Journal of Environmental Management*, 3 : 87-118.

**Mishan, E. J.** (1967) The Costs of Economic Growth. Frederick A. Praeger, New York.

**Misiolek, W. S. and Elder, H. W.** (1989) Exclusionary Manipulation of Markets for Pollution Rights. *Journal of Environmental Economics and Management*, 15 : 395-418.

**Naturvårdsverket** (1992) Åtgarder mot Klimatförändringar. SNV Rapport 4120, Solna, Sweden.

**OECD** (1990) Recent Developments in the Use of Economic Instruments for Environmental Protection in OECD Member Countries. OECD, Paris.

**OECD** (1993) Environmental Taxes in OECD Countries : A Survey. OECD Monographs No. 71, OECD, Paris.

**OECD** (1993a) Applying Economic Instruments to Packaging Waste: Practical Issues for Product Charges and Deposit-Refund Systems. OECD Environment Monographs No. 82, OECD, Paris.

**OECD.** (1995) Environmental Taxes in OECD Countries. OECD, Paris.

**Opschoor, J. B. and Vos, H. B.** (1989) Economic Instruments for Environmental Protection. OECD, Paris.

**Opschoor, J. B.** (1994) Developments in the Use of Economic Instruments for Environmental Policies in OECD Countries. *In*: Economic Instruments for Air Pollution Control, G. Klaassen and F. Førsund (eds.), Kluwer Academic Publishers, Dordecht.

**Panayotou, T.** (1994) Conservation of Biodiversity and Economic Development : The Concept of Transferable Development Rights. *Environmental and Resource Economics*, 4 : 95-110.

**Panayotou, T.** (1994a) The Use of Economic Instruments in Developing Countries. UNEP, Nairobi.

**Pearce, D. W.** (1991) The Role of Carbon Taxes in Adjusting to Global Warming. *Economic Journal*, 101 : 938-948.

**Pigou, A. C.** (1932) The Economics of Welfare (4th Edition). Macmillan, London.

**Poterba, J. M.** (1991) Tax Policy to Combat Global Warming: On Designing a Carbon Tax. *In*: Global Warming: Economic Policy Responses, R. Dornbush and J.M. Poterba (eds.), MIT Press, Cambridge, USA.

**Repetto. R.** (1992) Promoting Environmentally Sound Economic Progress: What the North Can Do. *In*: Change: Threat or opportunity for human progress ? Kirdar-Uner (ed.), Volume 5. Ecological change: Environment, development, and poverty linkages. 46-87. UN Sales No. E 91 III B 9, New York : United Nations.

**Rico, R.** (1995) The US Allowance Trading System for Sulfur Dioxide: An Update of Market Experience. *Environmental and Resource Economics*, 5 : 115-129.

**Sandmo, A.** (1975) Optimal Taxation in the Presence of Externalities. *Swedish Journal of Economics*, 77: 86-98.

**Shackleton, R. et al** (1992) The Efficiency Value of Carbon Tax Revenues. USEPA, Washington, D.C.

**Sorrell, S.** (1994) Pollution on the Market: The US Experience With Emissions Trading for the Control of Air Pollution. STEEP Report n° 1, Science Policy Research Unit, University of Sussex.

**Spofford, W. O.** (1984) Efficiency Properties of Alternative Source Control Policies for Meeting Ambient Air Quality Standards: An Empirical Application to the Lower Delaware Valley. Discussion Paper D-118, Resources for the Future, Washington, D.C.

**Stritt, M.-A.** (1994) Certificats échangeables: applications possible dans le domaine de l'environnment. IRER, Working Paper No. 9401, Neuchâtel University, Switzerland.

**Tietenberg, T. H.** (1985) Emissions Trading: An Exercise in Reforming Pollution Policy. Resources for the Future, Washington, D.C.

**Tietenberg, T. H.** (1988) Environmental and Natural Resource Economics (2nd Edition). Scott Foresman and Co., Glenview Ill.

**Tietenberg, T. H.** (1991) Reductions in Emissions: Command and Control of Market Based Mechanisms? Paper presented at the Conference on Economy and Environment in the 90s, Neuchâtel University, Switzerland.

**Tietenberg, T. H.** (1995) Tradeable Permits for Pollution Control When Emission Location Matters: What Have We Learned? *Environmental and Resource Economics*, 5 : 95-113.

**Walter, F. and Suter, S.** (1993) Umweltabgaben in Europa. *Schriftenreihe Umwelt*, 198, Bundesamt für Umwelt, Wald and Landschaft (ed.), Ökonomie, Berne.

**Weitzmann, M.** (1974) Prices vs Quantities. *Review of Economic Studies*, 4 : 477-491.

GPSR Compliance

*The European Union's (EU) General Product Safety Regulation (GPSR) is a set of rules that requires consumer products to be safe and our obligations to ensure this.*

*If you have any concerns about our products, you can contact us on ProductSafety@springernature.com*

In case Publisher is established outside the EU, the EU authorized representative is:

Springer Nature Customer Service Center GmbH
Europaplatz 3
69115 Heidelberg, Germany

**Batch number: 10726298**

Printed by Printforce, the Netherlands